Birds of Prey of the West

A Field Guide

Illustrations and Text
by Brian K. Wheeler

Range maps researched by
John Economidy and Brian K. Wheeler
and produced by John Economidy

PRINCETON UNIVERSITY PRESS
PRINCETON AND OXFORD

Copyright © 2018 by Princeton University Press

Published by Princeton University Press,
41 William Street, Princeton, New Jersey 08540

In the United Kingdom: Princeton University Press,
6 Oxford Street, Woodstock, Oxfordshire OX20 1TR

press.princeton.edu

COVER ARTWORK

Golden Eagle: A regal member of the order Accipitriformes, which, since 2012, includes all diurnal raptors, except falcons and vultures. This impressive bird of prey is uncommon but widespread west of the Great Plains and throughout w. Canada and Alaska. It is found east of the Great Plains during winter. Golden Eagles suffer from ingestion of lead bullet fragments when scavenging, leg hold traps, shooting, and wind turbines.

All photographs were taken by the author unless individually credited

All Rights Reserved

ISBN (pbk.) 978-0-691-11718-8

Library of Congress Control Number: 2017949016

British Library Cataloging-in-Publication Data is available

This book has been composed in Minion Pro and Myriad Pro

Printed on acid-free paper. ∞

Printed in China

10 9 8 7 6 5 4 3 2 1

To my wife, Lisa,
and my son, Garrett,
with all my love

A special thank you to
Ned and Linda Harris

In memory of Joe Harrison and
Wayne Johnston

Contents

List of Plates

Grand Canyon National Park (South Rim), Coconino Co., Ariz. (Jun. 2008)
California Condor, adult (#33). As with all Condors, this 12-year-old bird must survive the perils of lead poisoning.

Preface

This book is a culmination of data and knowledge from over 50 years of painting birds and studying raptors.

The journey began when I was seven years old, when I earnestly started drawing birds and mammals, first on old canvas pieces, then on cardboard, and then on white watercolor paper with transparent watercolor paint. Later, I used the thicker opaque watercolor medium called gouache. This paint allowed me to paint on surfaces other than white paper. Gouache also allows more detail than transparent watercolor. Since the 1980s, all paintings have been done on variously colored 100 percent rag, acid-free mat boards, which brought more atmosphere to my developing style of painting.

An innate desire to replicate realism motivated me even at that early age. My first watercolor paintings of wildlife date back to age 12, and I sold my first painting when I was 14. Painting wildlife and especially birds was my passion, and I drew or painted every day. By my early 20s, I concentrated mainly on birds and only rarely painted mammals (a life-size Red Fox in the snow was one of my favorites among the last mammals I painted—and the largest mammal I painted life-size).

My late teens and 20s were spent learning bird anatomy. I spent much time with waterfowl hunters and preparing road- and window-killed specimens for museums, first for Central Michigan University, Mount Pleasant, Mich., and later for Yale University's Peabody Museum, New Haven, Conn. I have always "constructed" my bird figures based on my anatomy studies rather than by copying photographs, though I have always used photos as references, especially for accurate color of fleshy areas and proportions—especially later for flight images. The only photography I did at the time was of close-ups of the heads and feet of freshly collected waterfowl to capture fresh coloration. Most of my time was spent doing a massive amount of field sketching of birds using a spotting scope.

I was naturally inspired by the impressive life-size works of John James Audubon. Even though his figures were anatomically distorted and stylized, he was a highly skilled technician. The exquisite paintings of Louis Agassiz Furetes (in *Birds of America*, Doubleday & Co., 1936 [especially plates 43–52]) inspired me for many years. J. F. Landsdowne, the awesome bird painter from British Columbia, Canada, impressed me the most, with his super-realistic, close-focus vignette bird illustrations. Though much different from what I do, the full-background, mood-laden shorebird paintings by Robert Verity Clem are incredible.

At 20, I moved from Michigan to Connecticut to attend the Paier School of Art (now Paier College of Art). I selected this school because it had a strong culture of realism and superbly talented instructors. I was self-taught in my preferred medium of gouache, which I started using crudely in my early painting career. When I got to art school, I was told—in so many words—I was not using gouache in the "proper method." Other paint mediums and aspects of art were of course explored, but gouache suited my style, and I perfected my "improper" painting method even more. After two and a half years of formal training—and out of funds to attend the balance of the four years—I embarked on a freelance career as a gallery-type life-size-bird painter.

Virtually all of my bird painting—except in my early years—consisted of life-size renditions, mainly in vignette style, that captured the intimate character of the bird with limited, close-focus, detailed background. (I love painting wood grain in branches and logs.) The use of the colored mat boards gave more breadth and character to my limited-background works.

I did full background paintings on occasion, also in close-focus format: a life-size Barn Owl perched in a barn window, with broken glass pieces on the edge of the window pane and red paint peeling off the wood, is my all-time favorite.

My paintings were sold through galleries to private and corporate clients (mainly in Michigan, Connecticut, and New York City). I succeeded in making a living creating life-size bird paintings for 12 years (1978–90). The recession of the late 1980s set in, which closed galleries and, by 1991, forced me to find other means to put food on the table and pay bills.

Most of my life-size works were of commercially appealing, decorative types of birds, particularly waterfowl, upland game birds, and of course pretty songbirds (I painted lots and lots of Northern Cardinals). All paintings, including book illustrations, were done with a museum specimen in my left hand—or on a nearby table if a large bird—and a paintbrush in my right hand. (I was always fortunate to have ready access to major natural history museums, and I would check bird specimens out like library books; specimens were safely housed and transported in a metal container.)

One of my greatest accomplishments—and far different from any other works—was a commissioned four-by-five-foot mural to commemorate the extinct Passenger Pigeon. This impressive painting is on permanent display at the Chippewa Nature Center, Midland, Mich. The painting was done in oil on Masonite board and depicts a full-background market-hunting scene from the mid-1800s. It has life-size pigeons in the foreground and market hunters with wagons in the middle ground. It is also the only painting I have done since my art-school days that had human figures in it. (A friend modeled various poses for me with a shotgun; figures are about 10 inches tall.)

Birds of prey have been my "thing" since my childhood days, even though I did not regularly paint raptors for gallery sales, because—except for falcons and owls—most do not possess the commercial qualities of songbirds and game birds. I did, however, spend much of my time studying diurnal raptors: finding nests, banding birds, and watching and photographing them across the United States, Canada, and Mexico.

After the collapse of gallery-type bird painting, my first commercial art stint was *Hawks of North America* with W. S. Clark, for Houghton Mifflin (Clark and Wheeler 1987). Bill Clark and I met by happenstance in 1981 at Cape May, N.J.; he was looking for an artist to illustrate a potential hawk field guide. Bill garnered a contract with Houghton Mifflin—for the first illustrated raptor guide ever published—and I began painting the 26 illustrations (plates) on stark white watercolor paper for the book in 1983 and finished in 1986.

Painting miniature field-guide-size figures, I found out, was much different from painting life-size figures: even a single fine-point brushstroke can make or break the outcome on such a small figure! Feather detail: none.

My second commercial publication gig was the second edition of the *Hawks* guide, which I worked on during the latter half of the 1990s (Clark and Wheeler 2001). With the success of the original guide, the ante was upped, and 40 new color plates were painted, which depicted more needed variations. The same simple format of comparative poses was followed, and I again painted the plates on stark white watercolor paper.

During the 1980s and 1990s, I delved into raptor photography. It was initially for reference for book illustrations—especially for accuracy of flight figures; however, it soon evolved into its own art form. Raptor photography produced *A Photographic Guide to North American Raptors*, also with Bill Clark (Wheeler and Clark 1995, 2003), and the highly acclaimed and unique (and out of print) *Raptors of Western North America* and *Raptors of Eastern North America* (Wheeler 2003b, Wheeler 2003a), both published by Princeton University Press. My raptor images were also published in an untold number of bird books, bird guides, calendars, and birding magazines during the 1990s and 2000s. A few of my images are used herein to separate geographic secctions and for added decoration

Bird photography became the rage with the advent of the digital revolution, beginning in the late 1990s. It seemed everyone began photographing birds, including raptors, and capturing images of impressive quality.

Photographic bird guides also mushroomed during this period. Photo-type guides are very attractive but, in my opinion, not the best learning tool. Lighting variables and millisecond nuances captured by the camera do not lend themselves to consistent comparison of figures. Field marks are easily masked by irregular lighting or distorted angles of view. Backgrounds may also be distracting. Digital manipulation corrects some inconsistencies but cannot consistently exhibit absolute comparison of lighting and angles of all figures. That is not saying photographs do not work—they can, but not always consistently for the broad application of a guide.

Illustrated guides make for a *far* better educational tool. Direct comparison of *all* figures with *consistent* lighting and same angle of view in a neat arrangement—whether perched or flying—is the optimal formula. This may be "boring" but is a highly effective learning format. Field marks are depicted better and more consistently with illustrations.

Field guides present visual comparison data but minimal interpretive data. Handbooks present a large amount of interpretive data but minimal visual comparison. General bird field guides—whether photographic or illustrative—are great for general birding knowledge but lack adequate space to depict the multitude of age, sex, and color morph variations found in birds of prey.

The desire to produce a *very* comprehensive North American guide to hawks, eagles, and falcons using the same-angle-of-view illustrated format prompted me to create this book. An "enhanced" field guide seemed the perfect solution: larger, with more data, than a standard field guide but smaller and more user-friendly than a handbook.

This book is packed full of illustrations in an easy-to-compare and easy-to-read field guide format. Except for Red-tailed Hawk and Aplomado Falcon, for which a considerable amount of new data needed to be presented, text for each species is short and concise but highly informative. To keep the book easily readable, I used layperson's terminology for anatomical features (e.g., *flight feathers* rather than *remiges*, *tail feathers* rather than *rectrices*).

Large range maps, completely updated, are carried over from *Raptors of Eastern North America/Western North America*. They give more detailed range plotting than found in *any* bird or raptor book.

Habitat images adjacent to the maps aesthetically enhance the book and assist visual comprehension of habitat zones within a species' mapped areas.

"Repetitive" field-guide-style perching and flying figures are used for optimal educational simplicity and comparison, but small tweaks were incorporated to add a bit more flair than in my previous illustrated guides. This included a small amount of habitat when space permitted—providing it enhanced layout—and use of some "character" postures of birds. Typical of book illustrations, the original plates were painted double the printed book size (at 10½ by 16 inches).

Doing the head portraits was fun. The large images add impact and show field markings better than typically small field guide images. They also permitted me to put more detail and character into the figures. I got to do all but the Bald and Golden Eagles and the California Condor life-size—just as I do on my life-size gallery works. Also, even birds of the most nondescript species and age classes can be identified by head pattern alone. (The exceptions are some juvenile Broad-winged Hawks and Red-shouldered Hawks.)

I have disdained white backgrounds for years—just as I dislike watching or especially photographing hawks against a white sky/background—because of harsh contrast, and Princeton University Press permitted me to paint the plates on variously colored neutral background stock. All plates were painted on 100 percent rag, acid-free, mat boards in blues, browns, grays, and greens, using my age-old favorite medium: gouache (Winsor & Newton brand). Painting on a color surface also allowed me to control light-dark contrast, especially pale trailing edges of wings and tails and pale translucent wing windows of buteos. (Having an art department "drop" figures painted on a white surface onto a colored surface prevents artistic control of needed contrast situations.)

I also used two colors of transparent watercolor paint to assist in creating three-dimensional form: Vandyke Brown and Lamp Black (Winsor & Newton). These two colors are used as "glazes" that deepen a shadow or create darker 3-D form but still retain underlying detail.

I used only high-quality paintbrushes (namely Winsor & Newton Series 7) for many years but found Jack Richeson brushes (7000 and 9000 Series) do a superb job at a lower cost. I prefer medium-size brushes (#4 and #5 are my favorites) that hold a sharp point for detail. Older, less-pointed brushes are used for texture or painting broad areas. Large brushes (#6–#8) are used too, but less frequently.

This longer-than-anticipated venture was contracted soon after I submitted *Raptors of Eastern North America* and *Raptors of Western North America* in late 2002. It has been a 13-year project of much learning, painting, and writing.

Brian K. Wheeler
Firestone, Colorado
November 2017

Acknowledgments

My wife, Lisa, who endured the whole process of the above-mentioned Princeton University Press books, lived through the very long process of this book. She was always there to give support and encouragement—and, as before, technical help with most computer issues. She has been an awesome partner. I always made time for my son, Garrett, too. I started the book when he was four years old, and he had grown to be a fine young man of 17 by the time I submitted it to Princeton. I always made time for him: from playground tag games after elementary school to the long hours watching him at skate parks to looking for his first car. It has been a memorable journey.

Thank you to Lisa's mom, Virginia Greeno, for helping with Garrett in his before-driving-age years when I was bogged down at the drawing table or computer. She also proofread parts of the book.

Though I have always been a "bird freak" to my family, they have always supported my endeavors: my mother, Marian, and siblings Wade, Nancy, Loren, and Marilee.

John Economidy helped research the awesome range maps and then created them on computer during the entire 13-year stretch. An attorney in San Antonio, Tex., he was incredibly diligent in researching the most recent distribution and status of most species, querying and corresponding with hundreds of people. He spent untold hours—amid his busy professional schedule—researching and creating and updating range maps on computer. John's love of raptors shows in the time and effort it took to create these maps. His compassion for birds of prey shows in the quality of these maps.

Robert Kirk, publisher of field guides and executive editor at Princeton University Press, has been an incredible editor He has been easily accessible and has always provided guidance throughout the 13-year endeavor. He allowed me *utmost* control over this project from start to finish.

This book came to life under the professional and talented expertise of the following people: Ellen Foos, senior production editor at Princeton University Press; freelance copy editor Amy K. Hughes; Princeton's illustration manager Dimitri Karetnikov; and Robert Still, publishing director of Princeton WildGuides. It was a *very* rewarding experience to work with each of them.

The following museums and their personnel allowed me access to their raptor collections to study plumages or sent me close-up digital images of specimens: American Museum of Natural History, New York, N.Y. (Paul Sweet, Peg Hart: full access to specimen collection); Bell Museum of Natural History, Minneapolis, Minn. (Robert Zinc and John Klicka: granted full access to the collection); Brigham Young University, Provo, Utah (Clayton White: sent specimens of and supplied data on non-native Peregrine Falcons and "Peale's" Peregrine Falcons); Canadian Museum of Nature, Ottawa, Ont. (Michel Gosselin: sent images of Canadian Red-tailed Hawks); Carnegie Museum of Natural History, Pittsburgh, Pa. (Stephen Rogers: sent images of Canadian Red-tailed Hawks taken by Mindy McNaugher); Conner Museum, Washington State University, Pullman, Wash. (Kelly M. Cassidy: sent images of "Western" and "Harlan's" Red-tailed Hawks); Denver Museum of Nature and Science, Denver, Colo. (Jeff Stephenson, Garth Spellman: full access to specimen collection); Louisiana State Museum, Patterson, La. (Donna Ditman: sent specimens of Hook-billed Kite, South

American Peregrine Falcon (*F. p. cassini*), and "Sutton's" Sharp-shinned Hawk); Royal British Columbia Museum, Victoria, Vancouver Island, B.C. (R. Wayne Campbell: sent numerous museum specimens of "Queen Charlotte" subspecies of Goshawk, Red-tailed Hawk, and Sharp-shinned Hawk); Royal Ontario Museum, Toronto, Ont. (Mark Peck: sent images of Canadian Red-tailed Hawks); U.S. National Museum of Natural History/ Smithsonian Institution (James Dean: sent images of Canadian Red-tailed Hawks) ; University of Alaska Museum, Fairbanks, Alaska (Jack Withrow: sent images of Merlin and Red-tailed Hawks from Alaska).

Special thanks to Bill Clark, who has always shared his love and knowledge of raptors with me. Bill also let me have access to his large number of unpublished Red-tailed Hawk images taken on his expeditions in w. Canada. Bill also shared his images, knowledge, and unpublished data on White-tailed Hawk, and on raptor molt.

Thank you to Jerry Liguori and Brian Sullivan for their expert opinions and discussions on Red-tailed Hawks.

I would like to thank Tom Carrolan and Paul Fritz, who, independently, constantly pointed out data pertinent for the book. Tom also provided Red-tailed Hawk and Rough-legged Hawk data from the East. Paul also provided data on "Harlan's" Red-tailed Hawks from Alaska.

Gary Vizniowski lives in the Northwest Territories and regularly provided images of Red-tailed Hawks from this boreal forest region. His data helped verify that the "Eastern" subspecies does indeed breed much farther north and west than previously known.

In the midst of her busy schedule as a romance author, Dianna Cosby shared many a Red-tailed Hawk image of resident pairs taken on her property in southern New Jersey, including some that match plumages once thought to be of more westerly types.

Chuck Susie has lived in many areas of the coastal mainland and islands of Alaska for the last few years, and he kept me posted as to what he saw on the little-studied "Black" subspecies of Merlin (*Falco columbarius suckleyi*). He had a blast watching and photographing various pairs of this Merlin in southeastern Alaska and regularly sent me images from the breeding season.

Steve Kirkland of the U.S. Fish and Wildlife Service graciously provided hand-drawn plotting of the California Condor's most up-to-date range as of early 2016.

Peter Pyle generously shared his extensive knowledge of molt on birds of prey and reviewed several plates that depicted molt sequences for accuracy. His help was most appreciated.

The following people assisted John Economidy and me, mainly in distribution and biological data, but also some plumage data; citing them in text would have been impossible in this book's format. They are listed alphabetically, with respective species and state/province to which they contributed information: Bob Anderson ("Richardson's" Merlin nesting, N.Dak.), Tom D. Anderson (Ferruginous Hawk winter range in sw. N.Dak. and nw. S.Dak.), John Arvin (Cooper's Hawk, Tex.), Randall Baker ("Peale's" Peregrine Falcon images), Carrie Battistone (Bald Eagle, Calif.), Alice Beauchmun ("Richardson's" Merlin nesting, N.Dak.), Steve Bentsen (Aplomado Falcon capturing Cattle Egret, Tex.), Gavin Bieber (Common Black Hawk spring migration, s. Ariz.), Gretchen Blatz (Bald Eagle, Wash.), Pete Bloom (Red-tailed Hawk, Calif.), Jim Boone (Zone-tailed Hawk, Nev.), Devin Bosler (Gray Hawk, se. New Mexico), Jack Bowling (Red-tailed Hawks, winter, cen. B.C.), Walter Boyce (Swainson's Hawk, Calif.), David

Bradford (Swainson's Hawk, Tex.), Ryan Brady (Harlan's Red-tailed Hawk, Idaho), Toby Burke ("Taiga" Merlin on Kodiak Island, Alaska), John Campbell (Peregrine Falcon, Red-tailed Hawk, Mackenzie River, N.W.T.), Leah Carlson (Swallow-tailed Kite, Minn.), Oscar Carmona (White-tailed Hawk, White-tailed Kite, Zone-tailed Hawk, ne. Tex.), Suzanne Carriere (Golden Eagle, Gyrfalcon, Peregrine Falcon, Nunavut, N.W.T.), Ed Clark ("Eastern" Red-tailed Hawk, Alaska), Hal Cohen (Swainson's Hawk, Calif.), Mich Coke (Common Black Hawk spring migration, s. Ariz.), David Coursey ("Black Merlin," Calif.), Gordon Court (Peregrine Falcon, Alta.), Warren Current ("Black" Merlin nesting, Wash.), David DeSante (Common Black Hawk, Calif.), Ken De Smet (Bald Eagle, Man.), Kara Donohue ("Harlan's" Red-tailed Hawk light morph, cen. Alberta), Norm Dougan (Red-tailed Hawk, B.C.), Nat Drumheller ("Harlan's" Red-tailed Hawk light morph, se. Alaska), Chris Earley (Rough-legged Hawk, Devon Island, Nunavut), Cameron Eckert ("Harlan's" Red-tailed Hawk light morph, juvenile plumage, Y.T.), Richard Erickson (Swainson's Hawk, Baja Calif. Sur, Mexico), Patrick Eriksson (Red-tailed Hawk, Alta.), Craig Flatten ("Queen Charlotte" Goshawk, in-hand images from Juneau; also from Douglas, Kupreanof, Kuiu, and Prince of Whales Islands, Alaska), Mark Flippo (Common Black Hawk, Gray Hawk, Mississippi Kite in Big Bend region, Tex.), Allistair Fraser ("Harlan's" Red-tailed Hawk in winter, B.C.), Cecily Fritz ("Harlan's" Red-tailed Hawk, n. B.C., Alaska), Jay Gilliam ("Krider's" Red-tailed Hawk, Iowa), David Gill (Osprey, Okla.), Dr. Branimir Gjetvaj ("Eastern" Red-tailed Hawk, Sask.), Peter Hamel and Margo Hearne (photos and data on "Queen Charlotte" Goshawk, "Queen Charlotte" Red-tailed Hawk, "Queen Charlotte" Sharp-shinned Hawk, and Merlin, Graham Island, Haida Gwaii [Queen Charlotte Islands], B.C.), Chuck Hayes (Bald Eagle, N.Mex.), Tom and Jo Heindel (Red-shouldered Hawk, Calif.), Steve Heinl ("Alaskan" Red-tailed Hawk, "Harlan's" Red-tailed Hawk light morph, se. Alaska), Reid Hildebrandt (Red-tailed Hawk, N.W.T.), Keith Hodson (Red-tailed Hawk, Mackenzie River, N.W.T.), Rick Howie (Red-tailed Hawks, c. B.C.), Stuart Houston (Turkey Vulture, migration data, Sask.), Rich Hoyer (Common Black Hawk spring migration, s. Ariz.), Marshall Iliff (Swainson's Hawk, Baja Calif. Sur, Mexico), Pete Janzen (Broad-winged Hawk nesting, Wilson Co., Kans.), Alan Jenkins (Bald Eagle, Okla.), Joel Jorgensen (Bald Eagle, Nebr.), Clive Keen (Red-tailed Hawks, winter, cen. B.C.), Richard Kinney (Crested Caracara, n. Tex.), Sandra Kinsey (Broad-winged Hawk, B.C., N.W.T.), Nathan Kuhnert (Sharp-shinned Hawk nesting, Okla.), Laird Law (Broad-winged Hawk, B.C., N.W.T.), Doug Leighton ("Eastern" Red-tailed Hawk, summer/winter records, cen., se. B.C.), Bill Lisowsky ("Black" Merlin, Ariz.), Mark Lockwood (White-tailed Hawk, Jeff Davis Co., Tex.), Chris Loggers ("Black" Merlin nesting in Wash.), Carl Lundblad (Red-shouldered Hawk, Nev.), Rich MacIntosh ("Taiga" Merlin, Kodiak Island, Alaska), Jeffrey Marks (Mississippi Kite, Mont.), Mark Martell (Golden Eagle movements, cen. and e. Canada, U.S.), Ron Martin (Bald Eagle, Merlin, N.Dak.), Terry and Ann Maxwell (Zone-tailed Hawk, Tex.), Scott McConnell (Ferruginous Hawk, Swainson's Hawk, Okla.), Kim McCormic ("Black" Merlin nesting, Wash.), Jeri McMahon (Osprey nesting, Okla.), Matthew Mega ("Black" Merlin nesting, Wash.), Matt Mendenhall (Crested Caracara records), Robert Mesta (Bald Eagle in Baja Calif., Sonora, Mexico), Steve Millard (Ferruginous Hawk winter range in sw. N.Dak. and nw. s.Dak.): Steve Mlodinow (Gray Hawk, "Harlan's" Red-tailed Hawk, Merlin, Short-tailed Hawk, Swainson's Hawk, Zone-tailed Hawk, Baja Calif. Sur, Mexico), Wayne Mollhoff (Mississippi Kite, Nebr.), Stan Moore (Common

Black Hawk, "California" Red-shouldered Hawk male head color, Calif.), Rick Morse ("Harlan's" Red-tailed Hawk in summer, Alta.), Dave Mossup ("Harlan's" Red-tailed Hawk, Y.T.), Frank Nicoletti (Northern Goshawk fall movement at Duluth, Minn.), Amanda Nicolson ("Scarlette" Red-tailed Hawk, Va.) Ryan O'Donnell (Mississippi Kite, Utah), Chuck Otte (Mississippi Kite, Kans.), Greg Page ("Black" Merlin in winter, Tex.), Bruce Paige ("Harlan's" Red-tailed Hawk light morph, se. Alaska), Jeff Palmer (Swallow-tailed Kite, S.Dak.), Chris Parish (California Condor ageing), Dave Parker (Osprey at Pt. Barrow, Alaska), Mark Phinney (Broad-winged Hawk, Ft. St. John, B.C.), Jim Pike ("Harlan's" Red-tailed Hawk, Baja Calif., Mexico), Andy Piston ("Alaskan" Red-tailed Hawk, "Harlan's" Red-tailed Hawk light morph, se. Alaska), Kim Poole (Golden Eagle, Nunavut, N.W.T.), Scott Price ("Black" Merlin nesting, Wash.), Lawrence Resseguie, Ph.D. (Swainson's Hawk, Calif.), Joe Roberts (White-tailed Kite, Okla.), Dale Robinson (Swainson's Hawk, Alta.), Mike Rogers (Zone-tailed Hawk, Calif.), Phillip Rogers (Ferruginous Hawk winter range in sw. N.Dak. and nw. S.Dak.), Gerald Romanchuk ("Harlan's" Red-tailed Hawk in summer, Alta.), Karen Rowe (Bald Eagle, Swallow-tailed Kite, Ark.), Rex Sallabanks (Bald Eagle, Idaho), Nick Saunders (Red-tailed Hawk, Sask.), Ross Silcock (Zone-tailed Hawk, Nebr.), Dave Silverman (Broad-winged Hawk nesting, Colo.), Pam Sinclair (Harlan's Red-tailed Hawk, Y.T.), Kent Skaggs (Mississippi Kite, Nebr.), Jeff Smith (Golden Eagle, Mont.), Noel and Helen Snyder (Peregrine Falcon, Ariz., N.Mex.), Eileen Dowd Stukel (Bald Eagle, S.Dak.), Paul Suchanek ("Harlan's" Red-tailed Hawk, Haines, Alaska), Brian Sullivan (Common Black Hawk spring migration, s. Ariz.), Chuck Susie ("Black" Merlin nesting, dark Red-tailed Hawk, se. Alaska), Chris Tessaglia-Hymes (Swainson's Hawk, Pt. Barrow, Alaska), Chris Tremblay (Swainson's Hawk, Pt. Barrow, Alaska), Dan Varland (Peregrine Falcon, Wash.), Gus Van Vliet (Harlan's Red-tailed Hawk light morph, se. Alaska), Thomas Walker (Broad-winged Hawk nesting North Platte, Nebr.), Bob Walters (Bald Eagle, Utah), Clayton White ("Peale's" Peregrine, non-native Peregrine Falcons), Jim Winner (Osprey, Okla.), Chris Wood (Mississippi Kite, Wyo.).

Introduction

Taxonomy

The American Ornithologists' Union (AOU) has been engaged in *major* reclassification of avian species during the last several years based on technological advancement of DNA studies. Some taxonomic changes are quite drastic from previous, historic treatment, which was based on physical aspects. Major taxonomic placement of our New World Vultures and especially of falcons has been *logically* altered based on recent DNA studies.

Falcons

The most drastic AOU reclassification occurred in 2012, when falcons were separated from other diurnal birds of prey: the eagles, harriers, hawks, kites, and Osprey (and, at the time, vultures). Falcons were kept in their own order, the Falconiformes, and the other species were put into a new order, Accipitriformes.

The AOU based its decision on two conclusive DNA studies. (1) An evolution divergent-time study by Ericson et al. (2006) said: "Our data recover a clade that includes the Secretarybird and accipitrid diurnal birds of prey (osprey, hawks, and allies) to the exclusion of falcons." (2) The clincher, a phylogenomic study done by Hackett et al. (2008), stated: "One of the most *unexpected* findings was the sister relationship between Passeriformes [songbirds] and Psittaciformes [parrots], with Falconidae (falcons) sister to this clade."

The wheels started grinding after these publications as to where falcons (and parrots) should be placed in the avian taxonomic format. DNA confirmed that songbirds and especially parrots shared a similar but distant ancestry with falcons.

Based on these two DNA studies, the AOU moved falcons (and parrots) to taxonomic locations in the checklist before Passeriformes and after non-passerines, following the woodpeckers (order Piciformes). This new taxonomic location is of course far away from the falcons' previous standing, which was based on physical foraging similarities with the diurnal raptors of Accipitriformes.

This taxonomic split goes back to just after the Cretaceous, the period of the dinosaurs, to 66 million years ago (mya). In a massive whole-genome DNA study, Jarvis et al. (2014) determined that diurnal raptors of Accipitriformes, falcons, and parrots (and many other orders) were dividing into their respective designations in the early part of the Paleogene period, which followed the Cretaceous (bird types, of course, evolved with dinosaurs in the Cretaceous). Accipitriformes and Falconiformes emerged early on, and parrots a tad later, in the 50 mya range. (Passeriformes evolved much later, in the 30 mya range.)

The revised taxonomic realignment of falcons, parrots, and passerines by the AOU in 2012 was verified in the Jarvis et al. (2014) study. Falcons are "sister" to parrots and passerines.

The recent taxonomic revisions by the AOU will undoubtedly stand. It is difficult to dispute multiple DNA studies that have arrived at the same conclusion.

Falcons are, unquestionably, birds of prey; they just evolved from a different lineage from the bird of prey species—and former brethren—of Accipitriformes.

Many raptor enthusiasts were undoubtedly perplexed at first by the actions of the AOU. "How dare the noble falcons be placed next to gaudy parrots!" was an understandable reaction. Although falcons did not derive from parrots, many parrot species still retain falcon-like attributes from a shared ancestor: fleshy area around the eyes, post in the nostrils, a notch near the tip of the upper mandible, and few build nests. However, parrots evolved zygodactyl feet (as did most woodpeckers), with two toes in front and two in back, which are far different from the feet of falcons.

DNA is extremely complex, and it is certainly difficult for most of us to discern its technical terminology.

A more tangible visual connection also exists between falcons and parrots (and passerines): wing molt between falcons and parrots is virtually identical—and tail molt of falcons, parrots, and passerines is identical (P. Pyle pers. comm.).

Based on 4,500 specimens of falcons and parrots, Pyle (2013) found they are the *only* two bird orders in the world to share this particular flight feather molt sequence.

Falcon Wing Molt: *Primaries.*—Molt starts on the 4th or sometimes 5th feather, counting from the inner part of the 10 primaries, and molts in a bidirectional sequential fashion. *Secondaries.*—Molt always begins on the 5th feather from the outer part of the secondaries and also proceeds in a bidirectional sequential fashion. From an innermost-point on the farthest inward feather of this tract, molt proceeds outward and meets the inward bidirectional molt of the inner wing. *Tail.*—Molt starts on central set, as on most birds, but extends in a sequential fashion outwardly to the outermost set.

Parrot Wing Molt: *Primaries.*—Molt starts on the 5th or sometimes 6th feather, counting from the inner part of the 10 primaries, and molts in a bidirectional sequential fashion. *Secondaries.*—Molt is identical to that of falcons. *Tail.*—Molt is identical to falcons (and Passeriformes).

Hawk (and Allies) Wing Molt: *Primaries.*—Molt starts on the innermost feather of the 10 primaries and molts sequentially outwardly to the outermost feather. *Secondaries.*—There are 3 different points where molt initiates: an outer point (1st feather), a midpoint (5th feather), with both molting sequentially inwardly, and an inner point, on the innermost feather (variable with size of bird), which molts sequentially outwardly, meeting where the 5th feather's initiating point ends its inward sequential extension (variable with size of bird). *Tail.*—From the central set, molt jumps to the outermost set, then to variable locations between center and outer sets (usually 3rd set).

Gray Hawk

A less dramatic taxonomic alteration also occurred in 2012, when the AOU changed Gray Hawk's scientific name from *Asturina nitida* to *Buteo plagiatus* (AOU 2012). This realignment illustrates that this hawk is more closely related to species in genus *Buteo* than previously determined. Its position in the AOU Checklist, however, remains the same.

White-tailed Hawk

A notable change occurred with this species in 2015 (AOU 2015). White-tailed Hawk's scientific name was changed from *Buteo albicaudatus* to *Geranoaetus albicaudatus*, and its taxonomic position in the AOU checklist sequence was realigned. This species was historically placed in the genus *Buteo* between Swainson's Hawk (*B. swainsoni*) and Zone-tailed Hawk (*B. albonotatus*). In the current U.S. checklist, White-tailed Hawk has been moved to a slot between Harris's Hawk (*Parabuteo unicinctus*) and the 1st *Buteo*, Gray Hawk (*B. plagiatus*). (In the full North American checklist, two Neotropical species sit between White-tailed Hawk and Gray Hawk.) Though in a different genus, White-tailed Hawk may be more closely related to Common Black Hawk (*Buteogallus anthracinus*) and Harris's Hawk, though the three are in different genera, than to its previous kin in the genus *Buteo*. (Physically, these three hawks tend to have defined, visible nasal posts, similar to those of falcons, though they are less obvious on some White-tailed Hawks.)

Vultures

After considerable taxonomic juggling in recent years, the AOU reclassified New World Vultures (yet again) in 2016. They are no longer considered part of the diurnal birds of prey order Accipitriformes. However, all recent DNA studies still consider them birds of prey. But, as with falcons, they originated from a different ancestral source, evolving near the time of Accipitriformes and Falconiformes in the early Paleogene period (per Jarvis et al. 2014). Historically, vultures were considered diurnal birds of prey in the original Falconiformes. In 1997, the AOU took vultures, which make up the family Cathartidae, out of Falconiformes and placed them with storks into the order Ciconiiformes (AOU 1997). However, neither Ericson et al. (2006) nor Hackett et al. (2008) found any DNA correlation between vultures and storks; this taxonomic alignment was based on ill-founded physical data. Cathartidae was "provisionally" returned to its original sequence in front of birds of prey within the original Falconiformes in 2007 (AOU 2007).

Since vultures did not fit the mold of diurnal birds of prey—or other avian forms for that matter—they were *logically* assigned their own order, Cathartiformes, in 2016 (AOU 2016). The AOU based this realignment on the whole-genome DNA study by Jarvis et al. (2014), which placed them in Accipitriformes but as sister clade to other species of that order. The AOU stated: "Cathartidae are sister to the rest of the Accipitriformes and … are as old or older than other lineages recognized as orders." Vultures are still taxonomically placed directly in front of the diurnal birds of prey of Accipitriformes (formerly Falconiformes), just in their own rightful order.

This taxonomic realignment may likely (finally) remain intact.

Summary

Falcons: These spirited birds are included in this book because they are predator and bird of prey. They just evolved from another ancestral lineage and fill different ecological niches than their former brethren of Accipitriformes. Actually, falcons are much more humane predators: They quickly kill their prey by immediately severing the spinal cord with their beak. Birds of the Accipitriformes kill prey slowly with their feet, by suffocation or by bleeding out—eating them until they die. Falcons and parrots (and passerines) are offshoots of a similar ancestor; thus, they are *logically* placed near each other in taxonomic status. Jarvis et al. (2014) considered falcons as birds of prey in their DNA sequencing (and also placed them adjacent to parrots and in front of passerines).

Vultures: Bare-headed Cathartiformes are included herein because of their historic and recent taxonomic relationship with the diurnal birds of prey—and very recent disassociation from them. All recent DNA studies consider them birds of prey, very closely aligned with Accipitriformes but different enough to warrant their own order.

We have been at the mercy of the taxonomy juggling performed by the AOU when it could not find an appropriate taxonomic location for these unique birds. It seems that, finally, they have been allocated their own taxonomic spot. Though they are primarily scavengers, all three North American species are carnivorous and will occasionally kill prey; Black Vulture is most notorious for such predatory behavior.

Wing molt sequence in the vultures is also similar to that of Accipitriformes (molt starts at the same points on the innermost primaries, and on the 1st, 5th, and innermost secondaries).

Note: Species format and layout in this book reflect the new AOU changes of the 7th edition of the 57th Supplement, as of Jul. 2016 (AOU 2016).

Great Slave Lake (south shore), N.W.T, May 2016
"Eastern" Red-tailed Hawk, adult. This subspecies inhabits the boreal forest–albeit in very low density. It shares its western Canadian range with far lesser numbers of "Harlan's" and "Western" Red-tailed Hawks. Interbreeding is prevalent. (Ventral view of this bird is in Fig. 3, p. 191.) Photo by Gary Vizniowski

Book Format

Scope

North America encompasses the continental United States and Canada, as well as the Caribbean, Central America, Mexico, and Greenland. This book's scope is *only* the w. continental United States north of the Mexican border and west of the Mississippi River, and w. Canada west of Manitoba and the w. shore of Hudson Bay, then due north into Nunavut and westward across the Arctic islands of Nunavut, the Northwest Territories, and Alaska west to the Aleutian Islands.

The Mississippi River is used as a demarcation line in the U.S., as it was for *Raptors of Western North America* (Wheeler 2003b), because it is a distinct geographic landmark, and western raptors mainly stay west of the river. Only a few western species regularly winter, in small numbers, east of the river, and a few stray east of the river in winter.

The w. shore of Hudson Bay and mainland e. Nunavut in Canada also form a diagnostic geographic border between East and West. Baffin Island, Nunavut, is not considered within this book's scope.

Format

Taxonomic order presides over most of the book; 1 short section contains birds grouped regionally. *Taxonomic layout.*—The taxonomic schematic of the 7th edition and 57th supplement of the American Ornithologists' Union Checklist (AOU 2016) is followed, except for placement of Bald Eagle. The eagle is placed directly in front of Golden Eagle, out of its normal place in the taxonomic order (between the kites and harrier), because of the 2 eagles' similarity in non-adult plumages (and size). As described in the "Taxonomy" section, falcons and vultures are included in the book as birds of prey. *Regional section.*—Behind the bulk of the species of the book is the regional section "Southwestern Specialty Species." These are species that are of Neotropical origin but make the sw. U.S. the n. extreme of their normal range. (The "Fuertes" Red-tailed Hawk is an exception; it is included with rest of Red-tailed Hawks rather than in this section.) All species within this section are still placed within respective taxonomic order.

Plates

Illustrations.—Figures are drawn and painted in repetitive same-position poses for *optimal* comparison between similar species. (The occasional exceptions include behavioral and/or feeding poses; e.g., Bald Eagle, Crested Caracara, Swallow-tailed Kite.) **Perching:** Simple profile angle with dorsal side of tail is shown. This best illustrates markings and shows wingtip-to-tail-tip ratio. Direct dorsal views are sometimes used. I like 3-dimensional figures, so some shadowing is used to convey 3-D form, including shadows when figures overlap. **Flying:** Four positions are used: (a) *Direct overhead soaring.*—Shows optimal wing size and shape and tail markings. (b) *Direct overhead gliding.*—Shows average wing position in moderate-speed glide with a partially closed wing. The position is used when space is limited and when a soaring figure is already shown. In both these flight positions, the birds are painted in shadowed backlit angles of view. "Shadow-side" rendering replicates real-life viewing with the sun's angle above the bird. This method also allows for optimal depiction of pale, translucent windows on species and/or ages that display this marking. Tail color and markings show best on several species when seen underneath in translucent lighting, especially when tail is fanned during soaring flight. (c) *Ventral flapping flight.*—Depicts figures lit by direct light, with 1 wing raised high in an upstroke to show shape and markings. Figures are off the perpendicular and flying slightly at the viewer. This figure saves space or unneeded redundancy in showing soaring or gliding positions. It also presents comparisons between backlit shadowed markings and directly lit markings—as in such species as Red-shouldered Hawk and Red-tailed Hawk. (d) *Dorsal flapping flight.*—Illustrates dorsal flight identification markings, such as wing panels and uppertail covert and tail color and/or markings. This angle also is used to show molt patterns. This figure is flying slightly away from the viewer, and position of wing and tail is at the beginning of an upstroke. *Note:* Mississippi Kite and Swallow-tailed Kite plates include an underside view of a soaring figure reaching down to feed. This shows classic in-flight feeding posture in these two species. (Similar aerial feeding occurs with falcons, especially while eating large insects.)

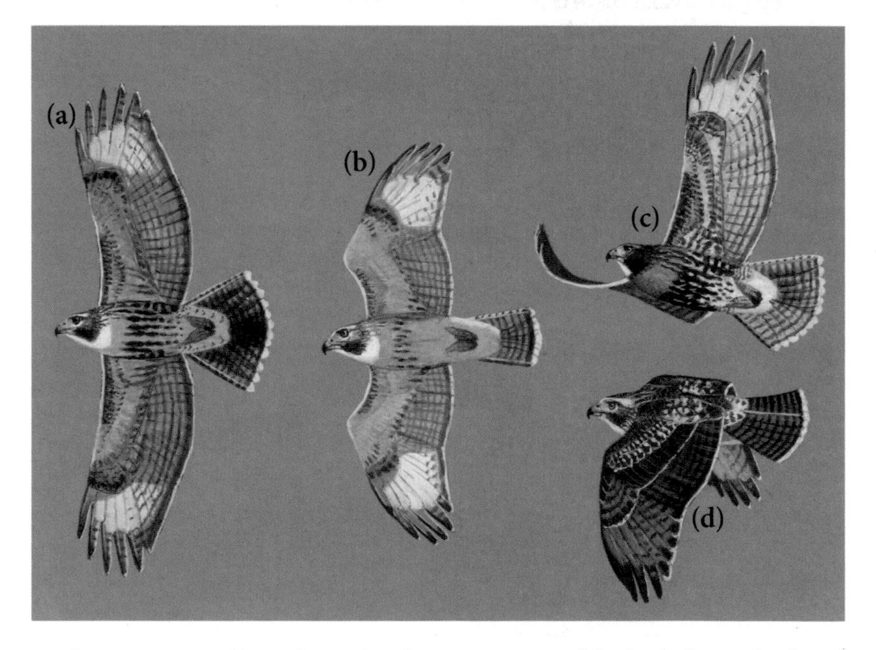

Arrangement.—For each species, the arrangement of the birds depicted is based on age and subspecies. (1) Birds are arranged youngest to oldest, whether in plate order or within a plate. (2) In most species that have subspecies, each subspecies is placed on a separate plate or plates, and youngest to oldest sequence is followed (see "Note," below). (3) Complex species such as buteos, which have many plumage variations, have separate "perching" and "flying" plates in addition to being arranged in youngest to oldest sequence. (4) Figures are arranged from lightest to darkest variants, placed in youngest to oldest order. This includes buteo species with color morphs, which are arranged lightest to darkest in a clinal order, if applicable.

Figures on plates are arranged in a 3- to 5-row linear format. However, this format could not always be followed when arranging perching and flying birds on the same plate. Figures are usually in left-to-right layout, but sometimes a partial vertical or partial diagonal layout was necessary.

Note: The schematic plan for polymorphic species was to put each subspecies (and often each age class of the subspecies) on a separate plate. Exceptions were made with some polymorphic species that did not warrant separate plates because of minor plumage differences or small numbers of figures (such as Sharp-shinned Hawk, Northern Goshawk, American Kestrel, and others). Another exception was "Krider's" and "Eastern" Red-tailed Hawks, which I lumped on the same plate, a decision that took an unexpected twist. At the time I created the Red-tailed Hawk layout, I believed that "Krider's" was a pale color type or pale morph of the variably plumaged "Eastern" Red-tailed Hawk. However, with additional contemplation and study, I changed my opinion and recognized that this pale Red-tailed Hawk was indeed—at least historically—a uniquely plumaged subspecies. (At this period of the book's development, it would have been very difficult to repaint and insert additional plates, so the current format was retained.) "Krider's" limited breeding range on the n. Great Plains is indicative of historic subspecies status. However, its breeding range has been *totally* inundated by the range expansion of the much more adaptive and dominant "Eastern" subspecies over the past 100-plus years. This has resulted in a massive amount of interbreeding and the considerable dilution of the unique, pale plumage traits found in classic "Krider's." Classic-plumaged "Krider's" still exist on their restricted breeding grounds, but they are now a minority and will continue to dwindle with time. Please read the text account for more information and cited material.

Color scheme of plates.—(1) Vultures are on a rich blue background, (2) harriers are on an orange-tan background, (3) accipiters are on a brown background, (4) Osprey, buteos, and falcons are on blue-gray background, (5) eagles are on a blue-green background, and (6) kites and the "Southwestern Specialty Species" section are on a green background.

Neutral-colored backgrounds allow for an easy-on-the-eyes natural-world viewing scenario, but they also help accentuate white tips on feathers, especially on the wings and tail. Such backgrounds also assist in contrast for shadow-side underside gliding and soaring figures with pale-edged feathers and those with pale, translucent windows on their primaries, as noted above.

Plate Descriptions

These are the data pages opposite each plate. The top portion is introductory information on the species, or age of the species if age classes are on separate plates. The following subsections appear in this section, in order. **Ages:** Briefly describes chronology of age classes for the species/subspecies, or of the age featured, and notes how long it takes to attain adult plumage. (1) Species-unique plumage traits are noted 1st and described in *italics*. (2) Each age class is then described, with any age-unique plumage traits noted in *italics*. **Subspecies:** Notes number of subspecies in N. America, and briefly notes any found beyond the scope of the book. **Color morphs:** Only some buteos, Gyrfalcon, and Hook-billed Kite have color morphs. If color morph variation is present, this text notes whether it occurs only in light or dark forms or if there is a clinal variation from light to dark forms. **Size:** Length (L) and wingspan (W) measurements are given. Plates with perching and flying figures give both measurements; perching-only plates give only length. *Length:* Distance from tip of bill to tip of tail. *Wingspan:* Distance between wingtips when in flight. Wing chord, the distance on a folded wing from the wrist to the wingtip, is given for some polymorphic birds. Wing-chord measurements are given in millimeters (mm). **Habits:** Gives general behavior information when it may assist with where the species may be found or help with identification. **Food:** Notes whether a species hunts from a perch or while in flight. Prey sizes are noted, and prey types are listed. **Flight:** Describes flight modes and other information that may assist in identification. "Flight" is not included on "Perching" plates. **Voice:** Any vocalization is described. Many species are vocal only during the nesting season and often only if agitated. Bald Eagle is highly vocal in all seasons, whereas Golden Eagle is silent, even when its nesting territory is entered. Migrants of any species rarely vocalize. **Figure captions:** Each figure on a plate, identified by an alphanumeric designation that corresponds to the plate number and the letter noted on the plate, is described. The age and plumage-variation type for each figure is noted in boldface type. One to 3 (rarely more) lines of descriptive text briefly describe the most important features of a given figure. Species- and age-unique markings for that figure are noted in *italics*.

Natural History Text

Following each species' plates and plate descriptions is additional text about that species. Habitat, status, nesting, and movements are discussed in a condensed manner, followed by comparisons with similar species. Most of these texts are ½–1 page long; more complex species with multiple subspecies or other important data have longer texts. A few species have in-text citations, mostly concerning plumage features but sometimes regarding new biological data. **Habitat:** This text notes areas in which a species can be found in various seasons, especially breeding season and winter. Migration habitat may also be noted. **Status:** For most species, notes whether it is common, uncommon, very uncommon, or rare. Numerical statistics are sometimes given, although they are difficult to assess in most cases. **Nesting:** Notes the span of the basic nesting period from start to finish, nest type, location, and clutch size. If fratricide commonly occurs in a species—as in Golden Eagle and Swainson's Hawk—it is noted, because surviving nestling/fledgling count may be lower than clutch size. Most fledglings remain with their parents for a few weeks after leaving the nest before dispersing on their own. **Movements:** Describes migratory movement, divided into spring and fall, as well as any extralimital movements that a species or age class of a species may seasonally engage in. Span of movement is given, as well as peak periods, often for different latitudes. **Comparison:** Names similar-looking species and describes what to look for to separate them. Comparative species are listed in taxonomic order. Juvenile and adult age classes may be divided so a more direct comparison can be made; as in the rest of the book, these are in younger-to-older order. Direct comparisons are made for each anatomical feature that may be similar or that highlight what is different. These comparisons explain *how* the field mark is different or similar. In these comparisons, the similar species is described 1st, followed by the subject species of the account. The term "*vs.*" separates the field mark description of the similar species from the field mark of the subject species. If a particular age class and/or plumage is compared, that is noted. In the following example, from the Peregrine Falcon text, a comparison is made to Gyrfalcon. In the field mark comparisons, the Gyrfalcon feature is described 1st, the Peregrine feature after "*vs.*"

> **COMPARISON: Gyrfalcon (intermediate morph, both ages).**—Head features, including malar mark, are diffused, *vs.* distinct markings with sharply defined dark malar mark. When perched, wingtips are much shorter than tail tip, *vs.* equal or nearly equal to tail tip. In ventral flight view, flight feathers are paler than coverts, *vs.* same colored. On juvenile, dorsal feathers have white edges, *vs.* tawny edges.

Note about Red-tailed Hawk account: A considerable amount of new data was discovered during the writing of the "Eastern" Red-tailed Hawk account late in the book's development. This information has been compiled, comprising numerous pages of in-depth text on plumages and distribution; basic information appears in the species' natural history text in the "Status" section. Following the natural history text are two separate essays: "Plumage Variation in 'Eastern' Red-tailed Hawk," which lays out most of the new data about plumages and distribution and includes a discussion of the mixing of subspecies; and "Analysis of the Proposed 'Northern' Red-Tailed Hawk, Subspecies (*B. j. abieticola*)," which reconsiders the proposed "Northern" subspecies in light of these new data. The "Eastern" subspecies has previously been taken for granted by all authors. It is actually quite variable—although shares the same plumage traits—throughout its entire range. Lightly marked and more heavily marked resident adults from New York City, New Jersey, or North Carolina can be identical to breeding adults from Labrador and Québec—and even those from Alberta and the Northwest Territories. It was also found that this subspecies extends much farther west and north than previously noted in the literature.

In "Plumage Variation in 'Eastern' Red-tailed Hawk," plumage features are broken down into anatomical parts (e.g., head, breast, belly), and variations are described and accompanied by in-text citations. What is unique is that virtually all in-text citations refer to Internet URLs that depict photographs of birds with the particular plumage features described; some link to typically cited published material. The downside to this approach is that Internet URLs may not be accessible years down the road, as a printed manuscript would be, although a great many will stay live for quite some time. Distribution information based on verified plumage criteria is cited in-text also. (There was not time when preparing this book to take time out to publish an extensive article on this subspecies of Red-tailed Hawk—as is included herein—in a peer-reviewed journal.)

Habitat images.—All species have at least one habitat image that fits between the text and the range map. Complex species with multiple subspecies and multiple range maps often have more than one habitat image; because of space constraints, habitat images are typically not depicted for all subspecies. Images were selected for habitat accuracy as well as aesthetic value, but purely artistic images were not used.

Images depict typical habitat in various seasons; some photographs show habitat during the breeding period; others span migration and winter periods. Breeding habitat images are used for many species, as breeding areas depict optimal habitat. Nest sites were not visited during the critical incubation period; for species such as Ferruginous Hawk, this is necessary to prevent nest abandonment. Some nest images are of unused "alternate" nest sites or were taken during the off-season period. Larger telephoto lenses were used when possible to remain at a minimal-impact distance. Any nest visitations during the nestling period, which occurred only with Common Black Hawk, Hook-billed Kite, and Prairie Falcon, were brief, and nesting success was *never* compromised by my short visitations.

Each image caption notes the county, state, and month the photograph was taken. If not taken by the author, the photographer is credited in the image caption.

A high percentage of images were taken in my "playground" in Weld County in ne. Colorado and in Albany County of se. Wyoming. These areas have an impressive number of very accessible w. breeding species as well as an incredible number and variety of migrating and wintering species. Habitat ranges from rural farmland to remote prairies.

Range Maps

Maps are of the large-size format that was first used in *Raptors of Western North America* (Wheeler 2003b). A few of the maps are reused from that book. However, most are either updated versions of those maps or new maps altogether. The large format allows for maximum accuracy in range plotting. Although topographic features cannot be depicted on this type of map, features such as mountains and rivers are often indicated and become a part of the equation when determining whether a species may inhabit a particular area.

All markings denote specific locations. Every magenta, cyan, or purple mark is there for a reason, indicating an accurately plotted locale. The maps are highly researched and show the latest knowledge. That does not mean ranges will not change or that birds do not stray far away from even current known locations. Case in point: Mississippi Kite and Merlin are both nesting in regions that even a few years ago were unheard of. The kite is extending its range farther north, and the falcon is extending its range farther south.

Maps are also large enough to allow users of the book to make continual adjustments and plotting, and all are *encouraged* to do so.

These maps also incorporate the use of city names, which were first used in a bird book in Wheeler (2003a, 2003b). They assist in even more accurate range plotting and add a unique touch.

Key to Maps

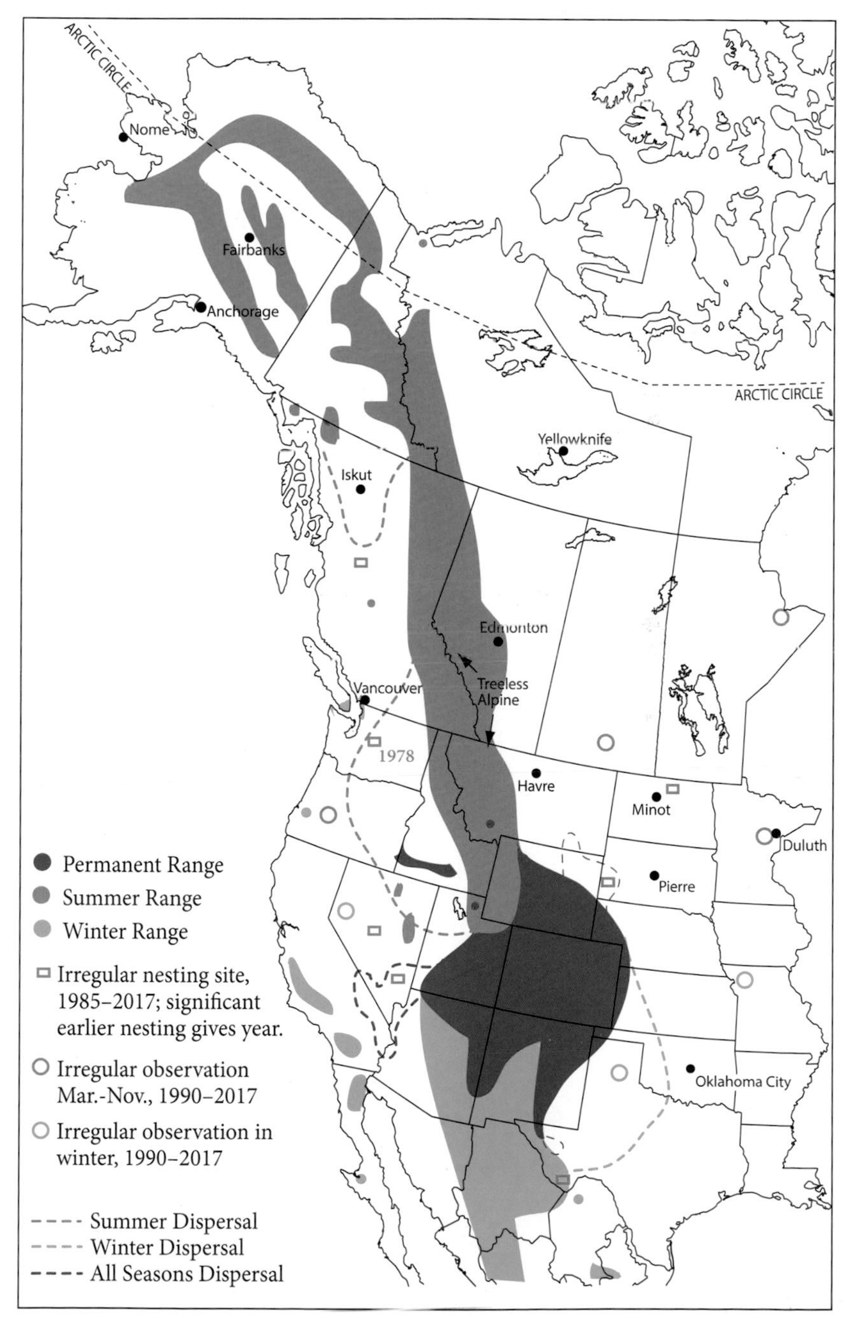

ARCTIC CIRCLE
Nome
Fairbanks
Anchorage
ARCTIC CIRCLE
Yellowknife
Iskut
Edmonton
Treeless
Alpine
Vancouver
1978
Havre
Minot
Duluth
Pierre
Oklahoma City
Permanent Range
Summer Range
Winter Range
Irregular nesting site, 1985–2017; significant earlier nesting gives year.
Irregular observation Mar.-Nov., 1990–2017
Irregular observation in winter, 1990–2017
Summer Dispersal
Winter Dispersal
All Seasons Dispersal

Identifying Birds of Prey

These primarily brown or black birds can initially all seem to look the same, whether at close range or in the distance. And, because most are such muted colors, they can be confusingly similar, even to seasoned bird-watchers. This book deals with close- to moderate-range identification of perching and flying birds where ID markings are readily visible.

For more distant identification by *jizz* (a term sometimes explained as "general impression and shape") of flying birds, which is possible on many species to a certain degree of accuracy once one gains such skills, see *Hawks from Every Angle* (Liguori 2005) and *Hawks at a Distance* (Liguori 2011). With practice, even age class can be distinguished in a few raptors without a view of actual markings or coloration; for instance, a flying adult Red-tailed Hawk can be told from a juvenile based on width of wing and length of tail.

Perching Birds

Look at the head, wings, and tail. Markings on the body mass may be identical on many species, especially on darker morphs. However, whether the undertail coverts are dark or light can sometimes assist in separating some darker buteos. Other characters, such as details of the head, wings, and tail, provide further information. For example, an adult Krider's Red-tailed Hawk can be as rufous on its upperparts as some adult light morph Ferruginous Hawks, but features on the head can help separate the two species.

Head.—All species and age classes of species can be identified by head pattern. (The exceptions are juveniles of some "Eastern" and "Southern" Red-shouldered Hawks and the often similar-looking juvenile light morph Broad-winged Hawk. All three can have a dark mid-throat streak, and rarely "Eastern" lacks throat markings, as do some Broad-wings. Both subspecies of Red-shoulder, however, can have all-dark throat, which never occurs on Broad-wing. Eye color, cere color, and basic head patterns between Red-shoulder and Broad-wing can be identical.) Separation by head pattern is possible even on species with seemingly confusing dark morphs and on both age classes of smaller accipiters. Compare the following: (1) eye color, (2) whether forehead is dark or white, (3) if throat is dark or white or marked, (4) color of lores and whether dark or light (white/gray), (5) thickness of any dark line behind eye, (6) presence of a dark malar mark—and how prominent it is if present, (7) color of cere, (8) presence of pale supercilium, and (9) sometimes color of base of bill in front of cere.

Wings.—(1) In perched bird, check especially the wing-tip-to-tail-tip ratio: distance of tips of wings in relation to tip of tail. This is very important on many species, particularly buteos, but also separates accipiters from buteos. For instance, dark morph of Rough-legged and Ferruginous Hawks can be confusing, but the Rough-legged's wingtips extend to or beyond the tail tip, whereas the Ferruginous Hawk's wingtips are shorter than the tail tip. (2) Look at the markings on the secondary flight feathers and the greater covert tract in front of them. Whether they are plain-colored or barred can often assist in separating species and even ages (e.g., juvenile Rough-legged Hawk has plain feathers, whereas adult has barred feathers).

Tail.—Tail patterns can be shared by species. Use caution and rely more on wingtip-to-tail-tip ratio. Juveniles of dark morph Rough-legged Hawk and Ferruginous Hawk can have identical patterns, as can juveniles of Swainson's Hawk and Red-tailed Hawk, and adult dark morph "Harlan's" Red-tailed Hawk and adult or juvenile dark morph Rough-legged Hawk. Check the tail for thickness of any barring and how broad the subterminal band is. Tail pattern can help separate accipiters from buteos, especially Northern Goshawk, which has 3 or 4 dark bands, *vs.* more than 4 dark bands on many juvenile buteos. Of course the "red" tail of an adult Red-tailed Hawk is one of the most obvious markings on its many color morphs. *Uppertail and undertail coverts:* These feather tracts attach to the tailbone (as do the actual tail feathers) but appear as part of the body feathering. Their color and markings can assist in separating a few species, but coverts are often difficult to see on perched birds. Undertail coverts are typically concealed by a perch structure, and uppertail coverts are covered by the wings, unless the bird is preening and has the wings drooped outwardly. Tail coverts can be good identification features on flying birds (see the next section), depending on angle of view.

Flying Birds

Wings and tail are usually the only parts that need to be studied, although on dark morph buteos it does not hurt to check the color of the undertail coverts, whether dark or light.

Wings.—(1) *Shape:* Every species has a different wing shape, albeit subtle on some. Even the age reflects the shape: juveniles of accipiters and falcons have broader wings than adults, whereas in buteos it is the opposite. What the bird is doing in flight can alter the shape of its wings—especially the outer primaries—affecting whether the wingtips appear rounded or pointed. A bird looks completely different when it is soaring than when it is gliding with wings pulled in closer to its body to reduce drag. The round-shaped wingtips of a Red-tailed Hawk can become quite pointed when it is engaged in a fast glide. (2) *Attitude:* (2a) Check whether the wings are held on a flat plane or are held in a dihedral. If the white uppertail coverts are not seen, an adult female Northern Harrier can look very much like a juvenile Northern Goshawk; however, the harrier holds its wings in a discernible dihedral, while the goshawk holds its wings on a flat plane. (2b) Compare angle of wings to body. A soaring Cooper's Hawk generally can be separated from a soaring Sharp-shinned because its wings are perpendicular to its body, while the Sharp-shinned holds its wings angled forward of a perpendicular line. (2c) How a bird of prey flaps its wings can also assist in identification. The two eagle species present a classic example: A Bald Eagle can be told from a Golden Eagle at great distances when in powered flight because the Bald Eagle raises its wings high above a horizontal plane, compared to an even up-and-down stroke of the Golden Eagle. (3) *Markings:* (3a) Check for any barred pattern on the undersides of the flight feathers and how thick the barring is. (3b) Check for a dark band along the trailing edge of the wing, whether gray or black, and how broad it is. In a few species, such as Mississippi Kite and Aplomado Falcon, it is a white band, of differing widths. (3c) Check wingtips, whether barred or solid dark, and how extensive the dark areas are (e.g., juvenile dark morph Rough-legged Hawk has all-dark primary tips, whereas the virtually identical dark morph Ferruginous Hawk has dark only on the very tips of its outer primaries). (3d) Check for any dark marks on the leading edge of the wing, which is the patagial area. All lighter morphs of Red-tailed Hawks have a dark mark, but similar species do not. (3e) Check for presence and extent of pale areas on the surfaces of the primaries. On the dorsal side these pale areas are called wing panels; on the backlit ventral side they are called windows. A juvenile Ferruginous Hawk has a pale window/panel only on its primary flight feathers, whereas a juvenile Red-tailed Hawk has a larger pale area extending onto its greater primary coverts.

Tail.—As in perching poses, tail patterns can be shared between species, and caution must be used. A juvenile Golden Eagle can have an identical tail pattern to a juvenile Bald Eagle that has extensive white on its tail. Tail shapes can be easily distorted, too, depending on whether the bird is soaring or gliding. On birds with tail banding, the pattern on the outer set of tail feathers may be different from that on the other 10 feathers. This is very common on juvenile accipiters and some juvenile buteos, especially Broad-winged Hawk. *Undertail coverts:* All feathers attach to the tailbone, as do actual tail feathers, but appear as part of the body feathering. Color of the undertail coverts, whether light or dark, can assist in separating a few species (e.g., dark morph Swainson's Hawk, which has light coverts, from dark morph Short-tailed Hawk, with dark coverts). Markings, such as a streaked or barred pattern, can also help separate species (e.g., juvenile "Southern" or "California" Red-shouldered Hawk may have barred undertail coverts, but a juvenile Broad-winged Hawk never has barring on its undertail coverts). *Uppertail coverts:* These 12 feathers are also attached to the tailbone, but each adheres directly on top of a respective tail feather. Look for whether this feather tract is white/light or dark or has markings. A Northern Goshawk flying away with a dorsal view will show distinctly tawny/dark-marked uppertail coverts, whereas the similar-looking adult female Northern Harrier will show white uppertail coverts.

Anatomy and Plumage Glossary

Head

See Anatomy and Plumage plate: *Figure 1.* Red-tailed Hawk, "Eastern" juvenile; *Figure 2.* Sharp-shinned Hawk, adult female; *Figure 3.* Merlin (falcon), "Taiga" juvenile; *Figure 6.* Turkey Vulture, 1-year-old.

Auriculars: Stiffly formed feathers that allow airflow into the ear hole on the side of the head behind and below the eyes.

Auricular disk: A semicircular formation bordering the *auriculars* that adorns harriers and, to an extent, juvenile Northern Goshawk; also called "facial disk." See Northern Harrier, Plate 10.

Bill: Hardened skin that covers the bone of the upper and lower *mandibles*.

Bill notch: An anatomical feature that adorns the distal part of the upper mandible on falcons. It is used, at least by large falcons, to sever the spinal cord of prey birds after they are captured.

Cere: Fleshy area at base of bill in which the nostrils are located. The cere often changes color as a bird matures from juvenile to older ages.

Cheek: Area just below the eye and in front of the *auriculars*.

Crown: Top portion of the head.

Ear: Covered by the feathered *auriculars* on true raptors but highly visible as a hole in the side of the head behind and below the eye on bareheaded vultures.

Eye line: A variable-width, but usually thin, dark line extending from behind the eye and above the *auriculars* along the side of the head.

Forehead: The area of the front top of the head, just behind the cere of raptors and above the eyes on vultures.

Gape: Fleshy "lip" area along the edge of the mouth that connects the upper and lower *mandibles*. It is not readily visible on some species, but on others, such as eagles and large buteos, it can be quite visible.

Hackles: Elongated feathers of the *nape* that are raised in cool weather or when a raptor is excited or alarmed.

Hindneck: The area at the top of the neck that is below the *nape* and above the *back*.

Lores: Short feathers with long hairlike bristles located between the *cere* and eyes.

Malar/Mustache mark: An often long, dark mark attaching to feathers of the lower jaw, sometimes on the cheek area, just below the *auriculars*. The mark may be faint, as on Merlin, or distinct, as on Red-tailed Hawk and Peregrine Falcon.

Mandibles: The bony structure to which the bill attaches. The upper mandible is fixed and the lower mandible is movable.

Nape: Feathers attached to the back of the skull; when erected, show as *hackles*.

Nostril/Nares: Air intake holes behind the bill, within the fleshy *cere* of raptors and falcons, and enlarged see-through cavities in vultures.

Nostril post/baffle: A protrusion within the center of the nostril hole of falcons that may permit air intake at high speeds.

Orbital skin/ring: Fleshy patch on the lores, which in falcons also thinly encircles the eyes. On juveniles of large falcons, it is typically blue or gray flesh and turns yellow within a few months. It is yellow on all ages of small falcons. The yellow is brighter on males.

Supercilium: A pale patch or linear pattern above the eye; may extend from the forehead onto the nape.

Supraorbital ridge: The bony structure above the eyes of most raptors and falcons that creates the "mean" look. It is bare-skinned on most species, but often difficult to see at field distances.

Throat: Region under the lower mandible.

Tubercles: White, wart-like bumps that surround part or all of the eye of most adult "Eastern" race of Turkey Vulture (*C. a. septentrionalis*).

Body (Perching)

See Anatomy and Plumage plate: *Figure 2.* Sharp-shinned Hawk, adult female; *Figure 3.* Merlin (falcon), "Taiga" juvenile; *Figure 5.* Red-tailed Hawk, "Eastern" adult.

Back: Feather tract attaching to and overlapping the inner part of the 2 scapular tracts. Most visible on rear, perched angle of view. It is a broad feather tract coming off the hindneck but becomes narrower on the distal part over the middle scapulars.

Barring: Horizontal dark markings on any or all of the following regions: *breast, belly, lower belly, flanks,* and *leg feathers*; also on *undertail coverts.*

Belly: Middle part of underparts below the *breast* and between the *flanks.*

Bellyband: Darkly marked feathers on the *belly* and the *flanks.* Markings may be spots, blotches, streaks, or a solid band.

Breast/front of neck: Large region below the *throat* and above the *belly.* The feathers are actually the front of the neck, and all feathers attach to a tract on each side of the neck.

Crop: A food-holding pouch on the *breast/front of neck* that is elastic and expands when full of food. Food is initially stored in the crop before entering the stomach.

Feet: Powerful, bare, scaled appendages for grasping prey, except in vultures, which cannot grasp prey. There are 3 toes in front and 2 in the rear; exception is the Osprey, which has 4 toes, 1 of which is an opposable outer toe that can be in front or placed in the rear to form a double 2-pronged appendage for grasping slippery fish.

Flanks: Elongated, fluffy feathers on the sides of the *belly* region. They cover the legs and front of the wings when fluffed out. These feathers may also form the side portions of a dark *bellyband.*

Leg: Upper portion (*thigh/tibia*) is feathered. Lower portion (*tarsus*; pl., *tarsi*) is bare and scaled except on 2 species. More-aerial species have short legs; more-terrestrial species have long legs.

Leg/thigh feathers: Long fluffy feathers attaching to the tibia (drumstick, or thigh) region just above the bare (tarsus) portion of the leg. Often protrude slightly past the somewhat shorter feathers of the *flanks* when bird is perched vertically. This anatomical area is difficult to see in flight because it is mostly covered by the *flanks.*

Lower back: On perched bird, this feather tract is often covered by the 2 scapular tracts, and not visible unless wings are drooped when preening. Visible in dorsal view of flying bird (see *Fig. 8*).

Lower belly: Elongated, fluffy feathers below the main *belly* region that cover the upper part of the *undertail coverts.* This tract is most visible in front-perched angle of view of a bird in a vertical-stance posture.

Rump: Visible on perched bird only if wings are drooped. This tract is the lower part of the *lower back* and covers the upper part of the very separate tract of the *uppertail coverts* (see *Fig. 8*).

Scapulars: Two large feather tracts that fill the void between dorsal body and wings and streamline a bird.

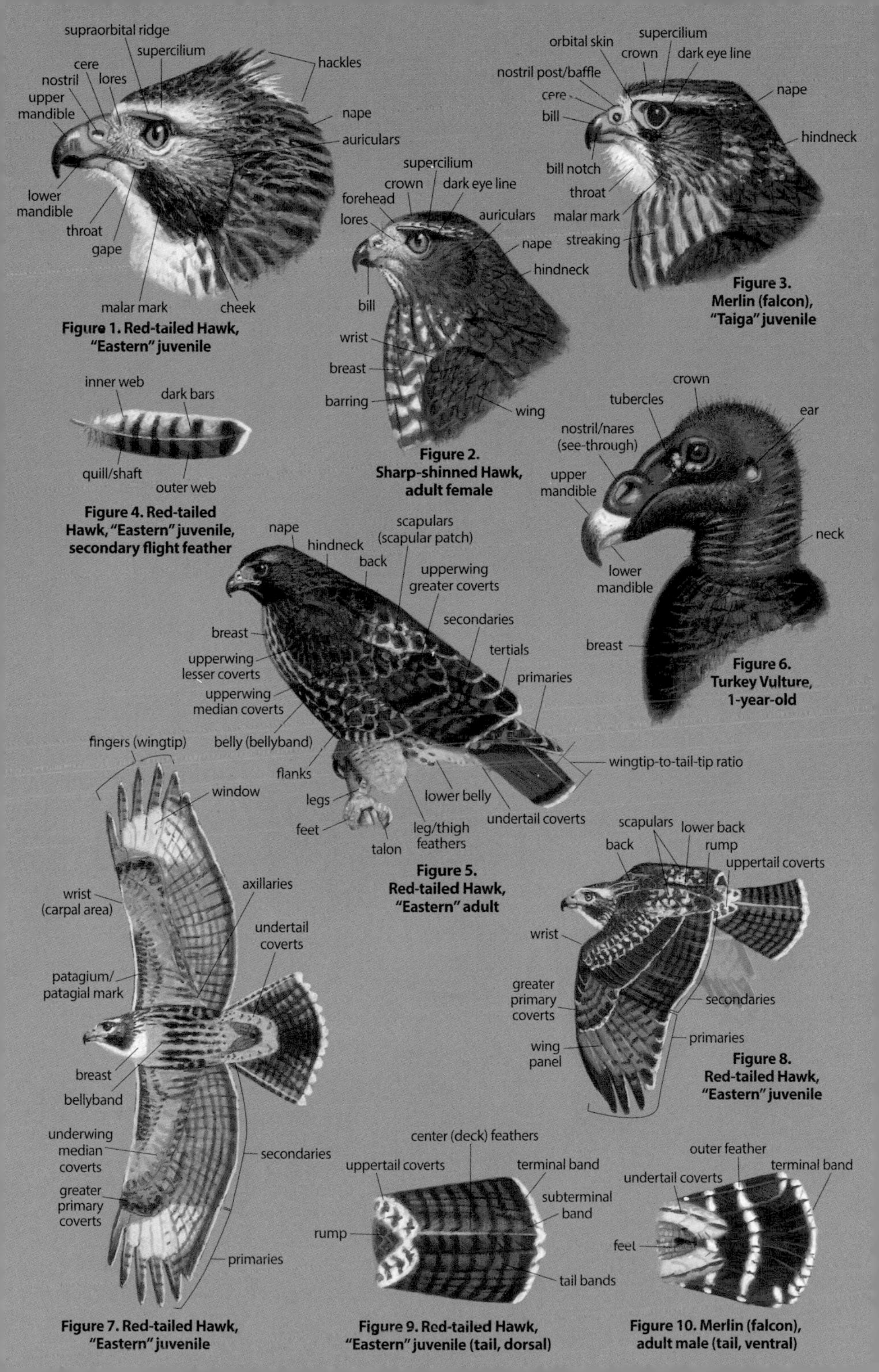

Figure 1. Red-tailed Hawk, "Eastern" juvenile

Figure 2. Sharp-shinned Hawk, adult female

Figure 3. Merlin (falcon), "Taiga" juvenile

Figure 4. Red-tailed Hawk, "Eastern" juvenile, secondary flight feather

Figure 5. Red-tailed Hawk, "Eastern" adult

Figure 6. Turkey Vulture, 1-year-old

Figure 7. Red-tailed Hawk, "Eastern" juvenile

Figure 8. Red-tailed Hawk, "Eastern" juvenile

Figure 9. Red-tailed Hawk, "Eastern" juvenile (tail, dorsal)

Figure 10. Merlin (falcon), adult male (tail, ventral)

Scapular patch: Pale marking, usually on mid-region of scapular tract, on some buteo species.

Streaking: Vertical dark markings on any or all of the following regions: *breast, belly, lower belly, flanks,* and *leg feathers*; also on *undertail coverts.*

Talon: Long, needle-sharp toenail used for grasping prey. Rear and innermost talons are longest. On eagles, the larger talons can be very large.

Underparts: Term collectively referring to *breast, belly,* and *lower belly.*

Upperparts: Term collectively referring to the *back, scapulars,* and dorsal side of the *wings.*

Body (Flying)

See Anatomy and Plumage plate: *Figure 7* and *Figure 8.* Red-tailed Hawk, "Eastern" juvenile.

Back: Visible in dorsal angles of view between the *scapular* tracts.

Belly: Highly visible in ventral angles of view. Species with dark *bellyband* markings show this area well.

Breast: Highly visible in ventral angles of view.

Lower back: Visible to some degree in dorsal angles of view. This anatomical area is covered much of the time by the dominant *scapular* tracts.

Lower belly: Markings, if any, are visible on this tract, which covers the region between the legs, just above the feet.

Rump: Most visible in dorsal flight angles between the distal *scapular* tracts. Often collectively lumped with *uppertail coverts*; however, they are 2 distinct feather tracts attaching to 2 separate anatomical regions.

Wings (Perching)

See Anatomy and Plumage plate: *Figure 4.* Red-tailed Hawk, "Eastern" juvenile, secondary flight feather; and *Figure 5.* Red-tailed Hawk, "Eastern" adult.

Primaries: *Flight feather* group that attaches to the "hand" section of the wing, which is outward of the *wrist.* On perched bird, the 10 primary feathers are folded into a long, slender formation that projects beyond the *secondaries.* Even when folded, these feathers can show field markings (e.g., silver-edged feathers on Ferruginous Hawk, Plates 47 and 49).

Primary projection: Distance the folded *primaries* project beyond the folded *secondaries.*

Secondaries: Flight feather group that is inward of the *primaries.* In perched bird, secondaries sit above the primaries and look to be part of the main body; they are overlapped tightly, forming a large block, and it is difficult to see individual feathers. Only the *outer web* of each feather shows when bird is perched. Markings on this tract can assist in ageing and separating species.

Secondary flight feather: See *Fig. 4.* Markings, such as dark or light bars, or lack thereof (solid colored) can separate ages and species; these feathers also show molt, which can help in ageing a bird. *Quill/shaft.*—Stiff linear structure that attaches to the wing bone (radius). *Outer web.*—Bottom part, below the *quill*; visible in dorsal angles of view. *Inner web.*—Top part, above the quill; visible in ventral angles of view. In dorsal view, inner web is overlapped and covered by the outer web of adjacent secondary. Inner part of the feather is completely covered by *upperwing* and *underwing coverts,* and the *quill* is never visible.

Tertials: Usually the 3 or 4 innermost feathers of the *secondary flight feather* tract. The innermost feather is never visible, since it is always covered by the longer, overlapping *scapulars*. The 2 or 3 longer feathers are visible when bird is perching but are covered by distal scapulars in dorsal flight view. The tertials are usually not visible in ventral flight view. *Note:* Authorities' views of the differentiation between outer tertials and inner secondaries are difficult if not impossible to discern on some species.

Upperwing coverts: Collectively 3 feather tracts: (1) *Greater coverts.*—Attach above respective secondaries, 1 per secondary, and overlap in the same outer-web-showing manner as described for *secondary flight feather.* (2) *Median coverts.*—Attach above respective greater covert, but overlap in inner-web-showing manner. (3) *Lesser coverts.*—A large group in upper part of wing attaching to the elastic *patagium* region. Lesser coverts attach in inner-web-showing manner, as *median coverts*. Coverts (mainly lesser) form "shoulder" region of wing. All upperwing covert tracts are readily visible in perched or dorsal flight views. The *scapulars* cover the inner portion of all covert feathers in perched bird.

Wingtip-to-tail-tip ratio: Distance from tips of primaries to tip of tail in a perched bird. The ratio is best seen in dorsal or profile angles of view. This distance is important in separating certain ages and species.

Wrist: Forward-most part of the folded wing.

Wings (Flying)

See Anatomy and Plumage plate: *Figure 7* and *Figure 8*. Red-tailed Hawk, "Eastern" juvenile.

Axillaries: A small feather tract that fills the void between the underside of the wing and the body. Also called the "armpit" of the wing.

Carpal area: Region on the *wrist*/bend of wing on ventral side. *Carpal patch* is a dark patch on this part of the wrist (e.g., Osprey (Plate 6) and Rough-legged Hawk (Plates 43, 45)).

Fingers: Outermost *primaries* that are thinly formed, very stiff feathers and appear as separate fingers at the end of the wing. Buteos and harriers have up to 5 fingers, accipiters have 6, and eagles have 6 or 7.

Flight feathers: Collectively, the *primaries* and *secondaries*, in layperson's terminology, which is used throughout the book. Also called *remiges* (sing., *remex*).

Patagial mark: Dark mark adorning the front edge of the *patagium* (e.g., Red-tailed Hawk).

Patagium: The elastic tissue that spans the leading edge of the wing between *wrist* and shoulder. The *underwing* and *upperwing lesser coverts* attach to this tissue.

Primaries: Ten *flight feathers* that attach to the "hand" section of the wing, which is outward of the *wrist*. All species have 10 primaries, including falcons and vultures. Molt sequence on these feathers is useful in ageing immature buteos and eagles. Feathers are sequentially numbered innermost (p1) to outermost (p10).

Remiges: Technical term for *flight feathers*, or *primaries* and *secondaries* (sing., *remex*).

Secondaries: *Flight feather* tract that has from 11 to 15 feathers (not including *tertials*), depending on size of bird. Sharp-shinned Hawk and falcons have the minimum amount, and eagles and vultures the maximum amount; 12 or 13 feathers are typical on buteos. Molt sequence on these feathers assists in ageing immature birds, especially buteos and eagles.

Tertials: The innermost shorter feathers of the *secondary flight feathers*. The outermost (longest) feather may be partially visible in a ventral flight angle of view. The tertials are completely covered by the *scapulars* in dorsal flight views on all species.

Trailing edge: "Rear" edge of the *flight feathers* (i.e., *primaries* and *secondaries*). This can refer to the shape of the trailing part of the wing, whether bowed or straight, or to a pattern on the distal end of the flight feathers, such as a dark band or white edging.

Underwing coverts: Collectively, 3 main feather tracts, which are not as visible or separable as *upperwing coverts*. (1) *Greater coverts.*—Barely visible and often hidden on the *secondary* flight feather region because they are overlapped by the long median coverts. The *greater primary coverts* are quite visible on the *primary* flight feather region. (2) *Median coverts.*—A quite visible tract, as long and prominent feathers, on the *secondaries*; visible to a lesser extent on the *primary* flight feather region. (3) *Lesser coverts.*—Highly visible on the forward portion of the underwing, which is the *patagium*.

Window: Translucent square, rectangular, or crescent-shaped area on ventral surface of wing, formed by light shining through a backlit pale *wing panel* on the dorsal surface of the inner *primaries* on some birds. The window is visible only when ventral side of the raptor is in shadow. This mark helps in ageing some buteos and also in separating some buteo species.

Wing panel: A rectangular, square, or crescent-shaped pale patch on dorsal surface of the inner *primaries* and/or *greater primary coverts* that is distinctly paler than the rest of the wing. On shadowed ventral side, the backlit panel forms a pale, translucent *window*. This mark assists in ageing and separating species in genus *Buteo*.

Tail

See Anatomy and Plumage plate: *Figure 7*, *Figure 8*, and *Figure 9*. Red-tailed Hawk, "Eastern" juvenile; *Figure 10*. Merlin (falcon), adult male.

Deck (central) set: Innermost of the 2 symmetrical sets of feathers that compose the tail. In a dorsal view, the outer web of each feather is visible. In a ventral view, these inner feathers are not visible because of the overlapping formation. These 2 feathers, each designated "r1" (*r* is for *rectrix*), are the 1st to commence molt.

Outer feather set: The outermost of the tail feather sets, made up of 1 feather on each of the tail's outer edges. The full surface of each feather in the set is visible in ventral angles of view, especially in flight. These 2 feathers may overlap the rest of the 10 feathers when the tail is closed. These feathers, designated "r6," are typically the 2nd set of feathers to molt.

Rectrices: Technical name for the tail feathers (sing. *rectrix*). It is not used in the book.

Subterminal band: A variable-width dark band adjacent to the pale tip of the tail in some birds.

Tail: 12 long, stiff feathers in 2 symmetrical sets of 6 feathers each.

Tail band: A linear pattern perpendicular to the tail feathers.

Terminal band: A light or dark band on the very tip of the tail.

Undertail coverts: Feather tract at base of tail on ventral side of body. On distal part of the tract, there is an elongate feather for each of the 12 tail feathers; shorter feathers attach inside these longer 12 feathers and can blend with lower body feathers. This feather tract assists in identification by its color (dark or pale) or by markings (barred or streaked). These coverts be difficult to see when bird is perched because they may be hidden by perch structures, especially branches and/or leaves. The tract is most visible in flying bird in a ventral angle of view.

Uppertail coverts: Feather tract on base of tail on dorsal side of body, with 12 feathers that attach to each of the 12 tail feathers. Not always visible on perched bird in a dorsal or profile view because of overlap by the wings; readily visible in dorsal-side flight view. Inner part of the tract is overlapped by the *rump*; these separate tracts are often mistakenly lumped together.

Age Classification and Molt Stages

Age Classification

The terms used for designating birds' age classes and their molt stages are typically synonymous, both indicating the age of a bird.

There are 4 methods of labeling raptor/bird ages: (1) U.S. Bird Banding Laboratory (Canadian Wildlife Service and U.S. Fish and Wildlife Service), (2) Humphrey and Parkes (1959), (3) Humphrey-Parkes-Howell (summarized in Clark and Pyle 2015), and (4) layperson's terminology (used in this book).

Over the years authors have struggled with age labeling. With Bald Eagles, such terms as *white-bellied age* and *transition age* have been used in the quest to label plumage changes. The word *subadult* was also commonly used by all of us at one time or another. Below is a composite of terminology that can be used. Authors writing peer-reviewed articles will use the more technical terminology.

I prefer to keep things as simple as possible, especially since plumage and age variations of raptors can seem so daunting. I am using simplified but logical labels in a manner all can easily understand.

U.S. Bird Banding Laboratory.—Age advances on Jan. 1 of each year, regardless of when the bird was hatched and even though no molt or plumage change occurs in many species to reflect an age-classification change. Hatching date to Jan. 1 is HY (hatch year); thereafter, the bird automatically becomes SY/AHY (second year/after hatch year); the following Jan. the bird becomes TY/ASY (third year/after second year); on large birds such as eagles, it goes further, to FY (fourth year) and FFY (fifth year).

I find this method confusing, since the bird's age advances but the plumage does not advance until several months later, in mid–late spring, long after the "annual" Jan. 1 age-label change. And a bird that is 2 or 3 years old could have at least 2 age labels having nothing to do with its actual birthdate and molt stage. (E.g., a raptor automatically becomes a TY/ASY bird on Jan. 1 of its 2nd year, for no reason other than that fixed date. It is still a 2-year-old and typically months away from actually being 3 years old.)

Humphrey and Parkes (H-P).—Age advancement is correlated with birthdate and molt. This method has been used by a considerable number of authors since its inception (used in Wheeler 2003a and 2003b).

The H-P method denotes that birds overall may have at least 2 seasonal plumages: a summer/breeding plumage and a winter/nonbreeding plumage. In H-P terms, the breeding plumage is the *basic plumage* and the nonbreeding plumage is the *alternate plumage*. Since raptors have but 1 plumage once breeding age is attained, their plumage is labeled *basic*. (*Alternate* designation is for birds such as Western Tanager that have a breeding plumage and a winter plumage.) A full adult plumage is called *definitive basic*. (The age at which adult plumage is attained varies, depending on the bird's size.) So any molt into a given *basic* plumage for a respective age is a "pre-" or interim stage of molt; thus, it is called *prebasic molt*. Molt into the final *definitive basic* adult plumage is called a *predefinitive basic molt*. H-P is a logical method but overly confusing, especially concerning birds entering adult age.

The term *juvenile* is used for a bird in the 1st set of feathers attained. The 1st molt cycle of mainly annual plumage after *juvenile* is *basic I*, the 2nd molt cycle of mainly annual plumage is *basic II*, the 3rd molt cycle of mainly annual plumage is *basic III*, the 4th is *basic IV*, and so on, until the *definitive basic* adult plumage is attained.

Humphrey-Parkes-Howell (H-P-H).—This molt-cycle/ageing method tweaks the H-P *basic* molt sequence and is the recommended ageing method of some contemporary authors (Clark and Pyle 2015). This newer method adjusts the H-P terminology to accommodate a partial body molt, labeled *partial 1st-year molt* (term used in this book) or, more technically, *1st-cycle preformative molt* (both terms noted in Pyle 2005a). This molt stage occurs in kites and many falcons, and occasionally on several species of raptors (but not vultures), when they are only a few months old (Pyle 2005a). The H-P-H terminology seems somewhat complex, but it does address the rather complex partial-age variable molt sequence that occurs in many species that other aging methods cannot address.

In H-P-H terminology, a bird in juvenile plumage would be called *1st-cycle juvenile*; a bird in juvenile plumage but engaging in a partial/full body-only molt in fall (e.g., White-tailed Kite and American Kestrel) is in *partial 1st-year molt* (term used herein), or the more technical *1st-cycle preformative molt* (FPF); a species such as American Kestrel that has completed the FPF molt but has not engaged in its 1st full body and wing molt is in *1st-cycle formative plumage*, which spans winter–spring in most cases, until the 1st actual full wing and tail molt is initiated. Subsequent annual molt changes and plumages are *2nd-cycle molt/2nd basic plumage, 3rd-cycle molt/3rd basic plumage*, etc.

A modern approach to the H-P method (seemingly aligned with H-P-H) is to consider juvenile plumage as *basic I* (although the bird is still called a juvenile), and former basic I as *basic II*, and so forth.

Age and Molt Terminology Used in This Book

Format in this book adheres to a combination of layperson's terminology for annual age classification and P. Pyle's (layperson's) terminology, noted above in the H-P-H discussion, to properly describe the previously difficult-to-label juvenile to 1-year-old stage that some birds of prey undergo. This plumage stage was not addressed in the otherwise tried-and-true H-P format.

Juvenile: The 1st plumage (feather set) attained after hatching, held until any subsequent molt. This plumage may be worn for only a few months—as in kites and American Kestrel—or, in most cases, a full year.

Partial 1st-year molt/stage: This layperson's terminology was coined by P. Pyle for what the H-P-H system terms *1st-cycle preformative molt/1st-cycle formative plumage*. It occurs in American Kestrel, all kites, and some other birds of prey to a lesser extent. Pyle (2005a) found that a great many species molt 10–15% of body plumage in this early molt. This molt begins when a bird is a few to several months old, but well shy of the normal 1st full prebasic body-wing molt that typically starts at about 1 year of age. It is a partial molt on all or part of the body. The only exception is White-tailed Kite, on which it can include some secondaries (but no primaries) and some or all tail feathers (Pyle 2005a). Partial 1st-year molt is never a complete molt and is an interim plumage stage between juvenile and adult plumage.

Prebasic molt: The time-trusted H-P term is used to note the annual birth-date molt sequence after any partial 1st-year molt that may occur (e.g., 1st prebasic molt, 2nd prebasic molt, etc.).

Immature: Dictionary definition is "not fully developed." Used herein in 2 variances: (1) for an interim plumage stage during or after partial 1st-year molt and until the 1st full body-wing molt is initiated at 1 year of age; (2) as a general term that concerns larger raptors (e.g., eagles) that have annual interim plumages and do not fully develop into adult plumage until a few years of age.

Early/Late stage: This wording is used to cover 2 scenarios in a broad but logical application: (1) It denotes progression of body molt and plumage change on species undergoing partial 1st-year molt; (2) denotes progression in bare skin color from juvenile to adult-like character in otherwise juvenile-plumaged vultures.

Annual Age Progression Terminology.—After their 1st year, raptors go through partial or full annual molts that begin around their birth date. Therefore, a simple annual age-labeling method is used, just as we denote human age progression. **1-year-old:** A bird in its 2nd year of life, beginning near its 1st birthday and at or near its 1st annual molt. It aligns with the basic I H-P age class (basic II of H-P-H). **2-year-old:** A bird in its 3rd year of life, beginning near its 2nd birthday and at or near its 2nd annual molt. It aligns with basic II H-P age class (basic III of H-P-H). **3-year-old:** A bird in its 4th year of life, beginning at its 3rd birthday and at or near its 3rd annual molt. It aligns with the basic III H-P age class (basic IV of H-P-H). 4-year-old and 5-year-old, if applicable, can also be noted as such.

In-Flight Field-Visible Molt

To elevate one's identification skills a notch or more higher, knowledge of molt is necessary. This section deals with molt that is visible within a reasonable distance to anyone using binoculars, moderate to long telephoto lenses, or scopes. Birds such as immature ages of Bald Eagle—with their distinctly longer, retained juvenile feather groupings—can be photographed if they fly overhead at a moderate altitude and then reliably aged from the photo, even one taken with a smartphone.

Perching birds of prey present more of a problem, because flight and tail feathers, which are more obvious when spread apart in flight, are squished together and hidden from view when wings and tail are folded. Still, retained juvenile feathers on wing coverts often readily show on a perched bird.

Timing of molt varies with age, diet, breeding season, and migration periods. Breeding females will initiate molt while incubating, but males will wait until nesting is mostly over, needing energy to supply food to their mate and youngsters. Pairs that have undergone a failed nesting may start molt more quickly than successful pairs. Molt is often suspended or slowed during periods of low food supply and migration. Few birds molt during the winter (Pyle 2005a, Bloom and Clark 2001).

Wing Molt Terms.—These terms apply to all birds of prey. **Molt center (node):** The flight feather where molt is initiated. **Molt unit:** A group or block of flight feathers that molts in a particular direction; the unit's molt stops where the next molt unit begins. **Molt wave/Stepwise molt:** A group of feathers within a *molt unit* that molts within a given period, usually each spring–fall. All feathers within a *molt unit* may molt in 1 season on some species. Large species, such as eagles, may have 3 molt waves, with 3 ages of feathers, occurring in a season. **Suspension limit:** Point where a *molt wave* ends for a given season; on adults, sometimes ends for breeding season or in autumn.

On 1-year-old birds, easily identifiable faded and worn retained juvenile feathers signify the end of the *molt wave*.

Accipitriformes (Eagles, Harrier, Hawks, Kites, and Osprey)

Primary molt.—The 10 primaries are numbered from innermost to outermost; innermost is p1 and outermost is p10. (*Note:* All subsequent labeling uses this numerical system.) The 10 primary feathers molt as 1 sequential *molt unit*, with molt always beginning on p1 and ending at p10. Larger species molt in *molt waves* or a *stepwise* manner, with only a portion of the primaries replaced annually. On larger species, the innermost feather, p1, is molted annually regardless of extent of molt on rest of primaries (Bloom and Clark 2001).

Molt is distinct in spring on juveniles approaching the 1-year mark: p1 and p2 (and often more primaries) are dropped before any secondaries are replaced. This is notable at hawk watches in late Apr. and especially in May, as returning juveniles of Broad-winged, Red-tailed, Rough-legged, and Swainson's Hawks show gaps on the inner primaries where new feathers have not yet replaced the dropped ones.

Molt sequence on 1-year-old/immature classes: The inner few or more primaries are molted into new feathers in variable-length *molt waves*. The outer feathers are retained older juvenile feathers, which get frayed and faded, are easy to see at field distances.

Adult molt begins during nesting cycle, earliest on incubating females, and on males typically after young are nearly fledged. Larger species molt in *stepwise* manner (in *molt waves*). Smaller species always molt their entire set of 10 primaries.

Primary feathers of most juveniles are more pointed on their tips than those of older ages.

Osprey.—Stepwise molt can be seen on adults. Molt is visible during fall migration on some adults (Sullivan and Liguori 2009). One-year-olds winter south of the U.S., thus molt sequence is not encountered (*see* Edelstam 1984).

Kites.—White-tailed Kite: (Based on Pyle 2005a.) In *partial 1st-year molt* stage (immature) does not molt any primaries (but molts other areas). Annual wing molt occurs on adults spring–fall. *Mississippi Kite:* One-year-olds molt the inner 5–7 primaries before migrating in late Aug.–Sep. Adult females will replace 3–6 inner primaries, adult males 1–3. One-year-olds begin primary molt in Jun., often losing 3 or 4 inner primaries in quick succession, so there is often a *huge* gap in the wing. Inner 2 primaries appear to molt in unison, with p3 just behind them. Adult females begin annual molt in Jul., males in Aug.; adults may have a large gap of 3 missing feathers. Primary molt is suspended for migration, so birds show new dark feathers adjacent to old primaries with no gap between them. Wing molt is completed on S. American winter grounds. *Swallow-tailed Kite:* Inner primary molt in May near the 1-year mark is noticeable, as a few feathers are dropped. Molt is occurring on wing coverts at this time. No molt is occurring on other areas yet. Molt occurs during summer on adults, and through their late-summer migration (Sullivan and Liguori 2009). *Snail Kite:* One-year-olds begin 1st wing molt at 1 year of age, and annually thereafter.

Accipiters and Northern Harrier.—Northern Goshawk: Almost always replaces all 10 primaries, but may occasionally retain a few outermost feathers; Pyle (2005b) noted that 70% molted all primaries. *Sharp-shinned Hawk:* Always replaces all juvenile primaries before fall. *Cooper's Hawk:* One-year-olds and adults migrating through the Great Basin tend to be in heavy wing molt in Sep., further along than other species for the time of year (Sullivan and Liguori 2009). Outer 1 or 2 primaries may be retained until mid-fall but are always molted into new feathers by late fall. *Northern Harrier:* Replaces all primaries, though p9 and p10 may still be growing in Sep. (pers. obs.).

Buteo types.—Common Black Hawk: One-year-olds may retain outer 1–3 juvenile primaries until the following spring (see Wheeler and Clark 2003, Fig. BH01). *Harris's Hawk:* One-year-old entering 1st prebasic molt will be adorned with 1 or more new, adult-colored, dark gray inner primaries (no molt on the secondaries, body, or tail). Several inner primaries are molted before molt of secondaries begins. *White-tailed Hawk:* One-year-old (based on W. S. Clark, unpubl. data) may replace all primaries or, more commonly, retain juvenile p7–10; also likely to retain p8–10, p9, and p10, or just p10. Molt is usually suspended during winter. *Gray Hawk:* Primary molt begins in May on returning 1-year-old; all juvenile primaries are replaced, in 2nd summer, as 1-year-old. *Red-shouldered Hawk* and *Short-tailed Hawk:* Both species replace all primaries each year. *Broad-winged Hawk:* Has a slower molt than the previous 2 species, possibly due to shorter summer season, and 1-year-old commonly retains juvenile feathers on 1–3 outer primaries (p8–10); *molt wave* is suspended by Oct., in early stage of fall migration, and resumed the following spring. *Zone-tailed Hawk:* May retain old juvenile p8–10. *Swainson's Hawk:* One-year-old and 2-year-old begin primary molt in Apr.–May. One-year-olds may still be molting outer primaries during Sep., often retaining old juvenile p8–10 until the following spring, when molt starts again where it was suspended the previous autumn and continues as 2-year-old. *Red-tailed* and *Rough-legged Hawks:* One year-olds typically retain 1–4 outer juvenile primaries. Red-tailed Hawks and many Rough-legged Hawks from northerly latitudes will retain the most juvenile feathers, due to a shorter summer growing season and/or less prey. Some more-southerly Red-tails may molt all primaries as 1-year-old. *Ferruginous Hawk:* Molts all primaries in the 1st molt, which is unusual for a large raptor. Molt on older ages may not be complete: an adult growing a new p8 during May shows molt starting/progressing, presumably, where it left off in a *molt wave* the previous fall (Sullivan and Liguori 2009).

Eagles.—Both species typically replace 3–6 inner primaries each year. Slow-molting birds, especially Golden Eagles, may replace only the inner 3 primaries in their 1st molt. Three ages of primaries adorn older immatures and adults. *Molt on older ages:* Innermost primary (p1) molts anew each year. Waves of old and new primaries decorate the wing. The more subtle annual differences can be difficult to see in the field. Molt continues the following year where it left off the previous year, and slowly keeps cycling the 10 primaries, with an average of 6 feathers replaced each year. *Note:* For full wing-molt data on all ages of Golden Eagle, *see* Bloom and Clark (2001).

Secondary molt.—Secondary feathers are numbered sequentially from outermost (s1) to innermost (number varies by genus and size of species; see below). Separation between secondaries and tertials is not defined herein, and the innermost visible secondary may actually be the longest tertial, which is often equal in length to or only slightly shorter than the innermost secondary (P. Pyle pers. comm.). Sharp-shinned and Cooper's Hawks, smaller birds with shorter "arms," have 10 or 11 visible secondaries; Northern Goshawk has 11 or 12 visible secondaries. Most species of *Buteo*, Northern Harrier, and kites have 12 or 13 visible secondaries. Large species such as eagles and

Osprey have 14 or 15 visible secondaries. (Feather counts are readily visible on the close, sharp photos in Wheeler 2003a/b.)

There are 3 *molt centers* and 3 *molt units* on the secondaries in the Accipitriformes: s1 (start of outer unit), s5 (start of middle unit), and innermost secondary/tertial (inner unit). Molt from s1 extends inward and stops at the end of the *molt unit* at s4; molt from s5 also extends inward and stops at the end of the molt unit, typically at s9 but can be at s8, depending on genus and molt advancement. Molt from the innermost secondary extends outward to meet inward extension of the s5 unit, at s10 or sometimes s9.

Molt sequence: Retention of old, faded, and often frayed juvenile feathers is best clue to identifying immature ages. Juvenile feathers can be retained on both eagle species and many buteos on the *innermost* feather or feathers of the *outer* and *middle molt units*, or the outermost feather or feathers on the *innermost molt unit*. It can be a large block of retained feathers or perhaps 1 or 2 feathers. On the *outer unit*, the s4 juvenile feather is typically the retained feather on many species (except Golden Eagle).

Molt of secondaries starts long after primaries have been molting. In 1-year-olds/immature birds, it can start as early as spring. Often 3–6 inner primaries are replaced prior to start of secondary molt (P. Pyle pers. comm.). Molt begins on the outermost (s1) feather, or the innermost feather (inner molt unit), before s5 is dropped. However, s5 and the innermost feather may drop in unison or at different times. Which one of these may be dropped 1st may indicate how far the *middle molt unit* extends inward or how far the *inner molt unit* extends outward, whether to s9 or s10 (W. S. Clark pers. comm.). Adults molt in the same sequence, but start later, during the nesting cycle. Females start while incubating, males toward the end of the nesting cycle.

Osprey.—Molt is not noticeable on most birds at field distances in N. America.

Kites.—White-tailed Kite: Juveniles in their *partial 1st-year molt* may replace 2–6 secondaries (but no primaries) in typical sequence at this time, which is bizarre, since all other birds of prey start primary molt long before secondaries begin molting (Pyle 2005a). *Mississippi Kite:* One-year-olds and adult females may drop and sometimes replace the outermost secondary (s1) before migrating south in late Aug.–Sep. No bird has been seen with a dropped s5 of the middle molt unit (pers. obs.). Secondaries are primarily molted on the S. American winter grounds. All ages molt new tertials while in the U.S. *Swallow-tailed Kite:* It is not known when this species' secondary molt begins.

Accipiters and Northern Harrier.—Northern Goshawk: One-year-olds may replace all juvenile feathers, or slow-molting birds may retain juvenile secondaries at s3, s4, and s7–9. More typically, if they retain juvenile feathers, they are at s4 and s7 and/or s8 locations. *Note:* Having molt stop *regularly* at s8 rather than at the more typical point of s9 on the *inner molt unit* may indicate that this species has 1 less secondary (usually 12) than species of *Buteo* (usually 13). However, the s7 and s8 combination also sometimes, but not regularly, occurs on the large buteo-type raptors (e.g., seen on 1-year-old White-tailed Hawk; W. S. Clark, unpubl. data). *Sharp-shinned* and *Cooper's Hawks:* Replace all juvenile flight feathers (but may retain upperwing coverts). *Northern Harrier:* May retain s4 and either s8 and s9, or s7 and s8.

Buteo types.—Common Black Hawk, Harris's Hawk, Gray Hawk, Red-shouldered Hawk (all subspecies), *Zone-tailed Hawk*, and *Ferruginous Hawk:* All tend to replace all juvenile secondaries as 1-year-olds. Red-shouldered Hawk, however, may retain s4 (W. S. Clark, unpubl. data). *Broad-winged Hawk*, *Red-tailed Hawk*, and *Rough-legged Hawk:*

May retain s3 and s4 or only s4 in the *outer molt unit* and s8 and/or s9 in the *middle molt unit* as 1-year-olds. Some Broad-winged Hawks will replace all secondaries as 1-year-olds. Red-tailed Hawks of southerly latitudes and low elevations may replace all juvenile secondaries; birds of northerly latitudes and high elevations are likely to retain some or much juvenile feathering, because of shorter summer growing season.

Eagles.—One-year-olds: Typically retain a vast number of juvenile secondaries, with new, shorter, older-age feathers coming in at *molt centers* and nearby locations. *Two-year-olds:* Have replaced most juvenile feathers, but often retain last feathers of *molt units* (e.g., s4, s9, s10). Some may replace all secondaries with shorter, more rounded adult-length feathers. *Three-year-olds:* All secondaries are usually replaced, except on slow-molting Golden Eagles, which may retain last feathers of molt units, as noted above. The slowest-molting birds, especially Golden Eagles, will molt only the inner 1 or 2 secondaries but no outer or especially middle secondaries.

A difference between Bald and Golden Eagles is that Bald Eagles quickly molt s1, then s5 and s6, leaving the typical s4 of the outer molt unit to be the very last of the long juvenile secondaries to be replaced as a 2-year-old (if an eagle is seen with only s4 projecting out, or s4 and s9 and/or s10, it is very likely a Bald Eagle). Golden Eagles, by contrast, molt s1, s5, then especially s2–4 more quickly than Bald Eagles, leaving them more likely to have 2 spikelike juvenile feathers retained at s9 on *middle molt unit* and s10 of *inner molt unit* as 2- or sometimes 3-year-olds (if an eagle has s9 and s10 as a single 2-feather group of retained juvenile spikelike feathers, it is very likely a Golden Eagle; pers. obs.).

Note: Eagles, Northern Harrier, and Northern Goshawk have longer juvenile secondaries than the new older-age incoming feathers, and such retained feathers on 1-year-old and older age classes are easy to detect, as they project distinctly beyond the new, shorter feathers on the wing's trailing edge. Species of *Buteo* have shorter juvenile feathers than newer adult-type feathers, and they show as indents along the rear edge of the secondaries.

Tertial molt.—Starts on all species after primary molt has been in progress and before molt on other secondary flight feathers; usually begins after primaries 1–3 or more have molted (P. Pyle pers. comm.). New tertials are most easily seen on perching birds, since they are covered by the scapulars during flight.

Molt sequence on most species, counting from innermost location, is 2, 1, and 3. Most species have a 4th tertial, but it is always covered by the scapular tract.

Upperwing covert molt.—One-year-olds of many species, from Sharp-shinned Hawk to eagles, retain some or many juvenile feathers on the upperwing coverts. These are generally faded, more brownish feathers. These feathers can be visible at a distance when bird is in flight but are more easily seen at closer range in perched bird viewed through a spotting scope.

Tail molt.—In all species, the tail comprises 12 feathers in 2 symmetrical sets. Innermost (deck set) is r1 (*r* for *rectrix*), and feathers are counted outward on each side to outermost, r6. Tail molt is easiest to see on the *1st prebasic molt* from juvenile plumage to older age class. Tail molt on most species does not begin until primary flight feather molt is well underway, though on some species of *Buteo* it can begin just after primary molt has begun (easy to see in May on springtime Broad-winged, Red-tailed, and

Swainson's Hawks with "split" tails, showing a gaping hole in the mid-tail due to the loss of the r1 set). Molt *always* begins on r1 set on all birds of prey (and on most birds). Molt proceeds to r6 before or after r1 set is replaced. Molt then goes to r3 and then r4, with r2 and usually r5 the last to be replaced. On some species, r2 may molt just after r1 (noted on White-tailed Hawk and Golden Eagle). Tail molt is quick and is typically completed by mid-fall. Slow-molting birds may retain 1 or more juvenile tail feather sets, usually r5, or r5 and r2, throughout the winter until molt begins again in the spring.

One-year-olds of Northern Goshawk, Red-tailed Hawk, and Rough-legged Hawk are likely to retain juvenile feathers on r5 or r2 and r5 throughout the winter. One-year-old Northern Goshawks can still be simultaneously growing r2 and r5 (which appear half-grown) in late Oct. (pers. obs. of migrants passing Duluth, Minn.).

White-tailed Kite.—Immatures in the *partial 1st-year molt* may replace some or all of their tail feathers, usually at least the 2 deck feathers (Pyle 2005a).

Eagles.—Golden Eagle: May exhibit an irregular molt sequence (based on Bloom and Clark 2001). Molt begins at r1, then goes onto r6 or even r2; only 1 feather of the set may be replaced in many cases. Middle feathers are irregularly molted, but often r4 is the last to be replaced. This species will retain some tail feathers for up to 3 years. *Bald Eagle:* This species has a rapid tail molt, replacing all 12 feathers in its 2nd year, as a 1-year-old.

White-tailed Hawk.—One-year-olds of this species have a unique fall 1-year-old partial tail molt. This partial molt was noted in independent observations/studies by the author and W. S. Clark. Data below are based on findings by W. S. Clark (unpubl. data) of 48 in-hand birds. One-quarter of the birds did not engage in a partial tail molt and retained all typically patterned 1-year-old feathers until the typical 2nd prebasic molt began in spring. Nearly half of the birds, however, engaged in a partial molt that attained 1 or more adult-like feathers; 4% replaced all but 1 juvenile tail feather. None of the birds molted all of the 12 feathers.

Body molt.—A partial to full molt, depending on species, occurs annually. *Partial 1st-year molt* occurs on all kites, but especially White-tailed Kite (and American Kestrel of Falconiformes); this early molt occurs to a much lesser degree on many other species, which replace up to 10–15% of their plumage (Pyle 2005a). This molt usually starts when birds are only a few months old; it is suspended during winter, then merges with annual *1st prebasic* body-wing molt in spring and continues until molt is complete in all anatomical areas.

Molt begins on the breast and back, then continues downward. The head is the last area to molt. On most species, wings and tail have been engaged in extensive molt before body molt starts. On larger species, most notably eagles, only a portion of the body plumage is replaced annually; this is why eagle plumages can be mottled with old faded feathers and new darker feathers. (On Golden Eagle this helps separate juveniles, which are in uniformly dark plumage, from mottled 1-year-olds that may otherwise look very similar, with retention of most of their juvenile-age flight and tail feathers.)

One-year-olds of smaller species of birds of prey, such as Sharp-shinned and Cooper's Hawks, may retain brown juvenile feathers on the lower back and rump. Sometimes they are large blocks of brownish feathers, at other times scattered brown feathers among adult-colored feathers. These feathers can be viewed easily in flight when birds are seen at dorsal angles of view, when banking or if flying below the observer at ridgetop hawk watch locations.

Cathartiformes (Vultures)

Primary molt.—There are 10 primaries, which molt as a single *molt unit* in *molt waves*, as in larger Accipitriformes. *Black Vulture:* Molt pattern is unknown but undoubtedly occurs in molt wave pattern, with only some of the primaries molting in any given year. *Turkey Vulture:* Primary, secondary, and tail molt information for this species is based on Chandler et al. (2010). This species has a unique molt in that it has a nearly continuous cycling of *molt waves*, with feathers being replaced constantly each year. Molt can begin in late Jan., prior to the bird's attaining 1 year of age, and completed by late Jul. Then, as a 1-year-old, bird replaces p1–3, or p1–4, for the 2nd time within the year, in Sep.–Nov., and as a 2-year-old replaces p3–10 or p4–10 in Jan.–Mar. Cycling continues with the 2 groups (p1–4 and p5–10), over and over, so 1 of the 2 *molt waves* molts twice each year. *California Condor:* Molt is slow, occurring, as on eagles, in a *stepwise* sequence. It may take 3 years to replace all primaries.

Secondary molt.—*Turkey Vulture:* As with Accipitriformes, molt sequence starts at *molt centers* at the innermost secondary and the tertials, then at s1 and s5. Molt extends outward from the *innermost molt center* (s14/15) and meets at an irregular mid-wing area. One-year-olds can retain juvenile s4, s7, s8, and s10 (but not s9, which is so typical of larger Accipitriformes). Tertial molt begins before molt begins on the secondaries.

California Condor: Slow progression in *stepwise* formation from *molt centers* at s1, s5, and innermost secondary, similar to Accipitriformes.

Tail molt.—*Turkey Vulture:* Starts with typical r1 and r6 sequence, then r2, r4, r3, and r5 (once past r6, this sequence is a bit different from tail molt on Accipitriformes). *Black Vulture:* Undoubtedly follows the same sequence as Turkey Vulture.

Body molt.—Begins on neck and goes down the body after wings and tail have begun molting. Change in fleshy head and neck color and loss of natal downy feathers on head and neck of juveniles occur gradually during the 1st year, long before any type of molt occurs.

Falconiformes (Falcons)

Primary molt.—There are 10 primaries. As noted in "Taxonomy" segment, falcons' molt sequence is much different from that of Accipitriformes and Cathartiformes. Molt begins at 1 *molt center* but extends in 2 *molt units*. Molt begins on p4 or sometimes p5, and is bidirectional, to p1 on the *inner molt unit* and to p10 on the *outer molt unit*. Molt expands more quickly on the *inner molt unit* than on the *outer molt unit*. This is easy to see on 1-year-old Peregrine Falcons in their 1st prebasic molt in summer: p3–5 are new, p6 is dropped, and p7–10 are old juvenile feathers. Molt is completed in a direct manner, except on "Arctic" Peregrine, which will suspend molt during migration and resume it on winter grounds. In falcons, primary molt begins before secondary molt starts.

Secondary molt.—Information based on Pyle (2005b, 2013). There are 10 or 11 visible secondary feathers on birds of genus *Falco*, 12 or 13 on genus *Caracara* (Crested Caracara). There are 2 *molt centers* and 3 *molt units*. The 1st *molt center* is on the mid-wing, at s5 location; molt proceeds in a bidirectional fashion, to s1, forming the *outer molt unit*, and to a variable point on the inner secondaries on the *middle molt unit*. The 2nd *molt center* starts on the innermost feather and molts outward to a variable point between s7 and s10 of the *middle molt unit* (Pyle 2005b). Molt is typically complete, but older small falcons (American Kestrel and Merlin) may rarely retain an old s1 (pers. obs.; Pyle 2005b).

Tail molt.—Tail molt is much different from that of Accipitriformes. Its sequence is identical to that of parrots and passerines: Molt begins on the typical r1 set, but then follows a sequential order to the outermost set, r6; however, r6 is sometimes replaced prior to r4 and r5 (P. Pyle pers. comm.). Tail molt on large falcons begins soon after wing molt starts but long after body molt has been in progress. It is most obvious on 1-year-old Peregrine Falcons, with the 2 deck feathers sprouted among otherwise juvenile feathering (and body 80–90% adult feathering).

Body molt.—Molt begins on back, breast, and head. A large percentage of body plumage is replaced on large falcon species before wings and tail proceed to molt. *American Kestrel:* All juveniles engage in the *partial 1st-year molt* on the body. All body feathers are replaced during fall; molt is suspended in winter, and full molt of wings and tail begins in late spring (Pyle 2005a). Molt is most obvious on males, because plumage change is more obvious. Thereafter, full molt on kestrels is an annual summer-through-fall affair. *Merlin: First prebasic molt* is visible in Apr. on "Taiga" males, with new, blue back feathers growing in on migrating 1-year-olds that are in otherwise juvenile plumage; molt probably begins in Feb.–Mar (pers. obs.). This molt precedes any wing or tail molt and may be a *partial 1st-year molt.* One-year-old "Richardson's" male can still be undergoing some body molt in mid-Sep. (pers. obs.). Annual molt on adult Merlin occurs during summer and fall: "Taiga" female in Oct. shows a distinct blotchy pattern of faded old and dark new body feathering. *Gyrfalcon:* No doubt because it is a large bird and lives in regions with a short summer, molt is slow. Younger birds may take at least 2 or 3 years to replace juvenile plumage, especially on wing coverts and lower back and rump areas. Adult molt is also likely to be slow, with only a portion of body feathering being replaced annually. *Aplomado Falcon:* This species may regularly engage in the *partial 1st-year molt* on the body. First-year birds in Texas in May can show adult feathering on the head, back, and breast; but full juvenile characters on rest of lower body, wings, and tail (also seen on a juvenile male from Chiapas, Mexico, in early Jan.; pers. obs.). *Peregrine Falcon:* "Anatum," more than any other subspecies, may engage in a *partial 1st-year molt*; birds with mostly adult-colored heads and body have been noted in Dec. (Pyle 2005a; pers. obs.). This early molt may be suspended during winter or continue into the normal *1st prebasic molt* of the whole body, wing, and tail, beginning in spring. Molt thereafter is annual in spring–fall. One-year-old Peregrine Falcons show a few scattered juvenile feathers on all plumage areas in autumn.

Boulder Co., Colo. (Mar.)
Bald Eagle, 1-year-old. A classic molt stage for this age class is shown on the wings of this bird (shows best on right [lower] wing). On the primaries, outer 5 feathers (p6–10) are frayed, retained juvenile feathers. On the secondaries, an outer group of 3 feathers (s2–4) and a middle group of of 5 longer, retained juvenile feathers (s7–11) project beyond the new, shorter feathers (s1, s5 and s6, and innermost, s12–14/15).

In-Flight Field-Visible Wing and Tail Molt

Knowledge of wing and tail molt takes bird of prey identification to a more detailed level. Many species have at least 1 interim age between juvenile and adult, and large birds such as eagles have multiple interim plumage stages. A basic understanding of wing molt, and to a lesser extent tail molt, greatly assists in ageing birds of prey. All data presented herein apply to birds seen at close and moderate distances, and to eagles sometimes at great distances. This page contains a summary of data presented in the preceding text, and the facing plate depicts typical examples of molt patterns of eagles, hawks, and kites, of the order Accipitriformes, and for comparison, molt patterns in the Peregrine Falcon, of Falconiformes. The graphics on this plate illustrate the preceding detailed descriptions of *molt units* and *molt waves*.

Most figures illustrated in this guide that depict interim molt patterns are of plumages seen in fall and winter, when molt is typically suspended. Summertime molt stage is usually not shown, because most birds of prey are in exceedingly ratty-looking plumages. Exceptions are of a few species of *Buteo* illustrating the earliest stage of 1st prebasic molt in spring; kites, which undergo a partial 1st-year molt or, as with Mississippi Kite, as 1-year-olds, are in the midst of 1st prebasic molt during their summer occupation in the U.S.; and the Peregrine Falcon, which illustrates the falcon molt strategy that occurs mainly during summer.

Figure 1. Mississippi Kite, 1-year-old (early stage, Jun.): The earliest stage of molt of Accipitriformes, with p1 dropped. A small gap occurs with loss of this juvenile primary; new ones do not grow until more have dropped.

Figure 2. Mississippi Kite, 1-year-old (early stage, late Jun.–Jul.): Primary molt has advanced, and juvenile p1–4 are dropped; new adult primaries have sprouted at p1 and p2 locations, which may grow in unison.

Figure 3. Mississippi Kite, 1-year-old (late stage, Aug.–Sep.): Primary molt is at final stage before bird leaves the U.S. (Molt suspended for migration.) 6 primaries (p1–6; can be to p7) are new. Outer secondary (s1) can be dropped. New (black) r1 and half-grown r6 adorn tail. Tail molt begins well after primary molt has been in progress.

Figure 4. Broad-winged Hawk, 1st prebasic molt (May): 1st stage of wing and tail molt of genus *Buteo*, in which the innermost primaries start molting and tail set r1 is close behind. A small gap occurs where p1 and p2 are dropped but not yet replaced by adult feathers. The loss of r1 is most visible when tail is fanned.

Figure 5. Bald Eagle, 1-year-old: Inner 5 primaries (p1–5) are new (darker); p6–10 are retained worn juvenile feathers. Secondaries s1, s5 (*molt centers*) and s13, 14/15 are new; s2–4, s6–12 are old juvenile feathers. A slow molt stage

Figure 6. Bald Eagle, 2-year-old: Molt continues from previous year's molt (*Fig. 5*); p6–9 are now new, darker feathers; molt stopped before reaching p10, which now is a *very* worn juvenile feather. P1 and p2 are new again (p1 is replaced annually). S4, s9, and s10 are the last of their respective *molt unit* feathers; these juvenile feathers are longer and worn.

Figure 7. Bald Eagle, 3-year-old: P10 is finally replaced and is new and dark. Though difficult to see in the field, many inner primaries have also been replaced again (darker). All juvenile feathers have now been replaced.

Figure 8. Red-tailed Hawk, "Eastern" 1-year-old (Nov.): Retains old, faded juvenile outer 3 primaries (p8–10) and s4, s8, and s9 secondaries; old secondaries are shorter. Tail sometimes retains juvenile r2 and/or r5.

Figure 9. Broad-winged Hawk, 1-year-old (mid-Sep.): Retains old, faded juvenile outer 2 primaries (p9, p10) and 2 secondaries (s4, s9). Primary p8 could still be growing in mid-Sep. Tail is typically molted into adult pattern.

Figure 10. Northern Goshawk, 1-year-old female (late Oct.): All primaries are new adult feathers. Juvenile secondaries adorn s4, s7, and s8 locations. Inner *molt unit* may extend molt outward past s10, to s9, as depicted by the arrows. This often occurs on larger species that replace feathers in *molt waves*. Juvenile secondaries are longer.

Figure 11. Peregrine Falcon, "Anatum," 1st prebasic molt (Jul.): Depicts bidirectional molt of primaries and secondaries, with new p3–5 and half-grown p6, and new s5–7. Much of body is molted into adult feathering before wing molt occurs. Tail has new adult r1; rest of tail is juvenile.

Figure 12. Golden Eagle, 1-year-old (tail, ventral): Depicts 1st stage of tail molt, with the new mostly concealed immature set at r1 and new immature set at r6. *Note:* Outer feathers overlap inner feathers on the ventral side. This molt pattern is attained in autumn and held until spring. Immature feathers are shorter and darker.

Figure 13. Golden Eagle, 1-year-old (tail, dorsal): Same stage of 1st molt as on *Fig. 12*, with new immature r1 set visible, but new r6 set mostly concealed. *Note:* Inner feathers overlap outer feathers on the dorsal side.

p10
p2
s1
no molt yet on secondaries or tail
1
p10
p5
p2
p1
s1
no molt yet on secondaries or tail
2
p10
p6
p1
s2
3
r6
r1
p10
p3
s1
no molt yet on secondaries
r1 set missing due to molt
4
s10
s9
s4
s1
p1
p10
p5
p6
5
s10
s9
s4
s1
p1
p2
p10
p5
p6
6
molt is difficult to see once juvenile secondaries are molted
7
p1
p10
p5
p6
p8
p10
p1
s1
s4
s8
s9
8
p9
p8
p10
p1
s1
s4
s9
9
s9
s8
s7
s4
s1
p1
p10
10
11
r1
s5
s1
p1
p4
p5
p10
r6
r5
r4
r3
r2
r1
r1
12
r6
r5
r4
r3
r2
r1
r1
13

Plate 1. BLACK VULTURE (*Coragyps atratus*)
Typical (*C. a. atratus*) and "Southwestern" (*C. a. brasiliensis*)

Ages: Juvenile, 1-year-old (1st prebasic molt stage), and adult. Head is featherless. Feathers on neck can be raised (when cool) or lowered (when warm). Bill is long and thin. When bird is perched, wingtips barely protrude past secondaries, and wingtips equal short tail tip. White legs are long. In flight, outer 6 primaries show as a white patch, and white toes reach tail tip. *Juveniles.*—Bill tip is black. Head and neck skin is smooth, with fuzzy, buffy, downy feathers on crown and nape, which disappear quickly, though darker fuzzy feathers remain throughout age class. Neck slowly acquires bumps and wrinkles on lower part. Bill turns slightly yellow by end of stage. Plumage is uniformly dull black, with no molt. *1-year-olds.*—Transitional stage of 1st prebasic molt and change of fleshy head/neck from juvenile to adult characters that occurs during 2nd year. Bill is mainly dusky yellow. Head skin is smooth, but neck has adult-like grayish bumps and wrinkles. Plumage is a mix of dull blackish-brown juvenile feathers and new iridescent adult feathering. There is considerable wing molt, beginning with acquisition of darker, iridescent adult tertials and inner primaries. *Adults.*—Bill tip is yellow. Head is gray, with bumps and wrinkles. Plumage is iridescent black. **Subspecies:** Typical is in East and much of West; *C. a. brasiliensis* in Ariz. (subspecies not separable in the field); 1 other subspecies in S. America. **Color morphs:** None. **Size:** L: 23–28" (58–71cm), W: 53–63" (135–160cm). **Habits:** Tame. This species is gregarious. Found in pairs or in flocks. Terrestrial; prefers open perches. Individuals are aggressive toward each other and other scavengers. **Food:** Carrion. Food is located by sight while in flight. **Flight:** Short burst of snappy wingbeats between glides. Wings are held in a shallow dihedral when soaring and flat when gliding. Regularly dips wings in a downward motion. **Voice:** Grunts and hisses when agitated; *wuff* when alarmed.

1a. Juvenile (recently fledged): *Black tip on thin black bill. Smooth-skinned black head and upper neck skin is covered with fuzzy, buffy, downy feathers; buffy, downy feathers on nape disappear within weeks of fledging.*

1b. Juvenile (early stage): *Black tip on thin bill. Smooth black head and upper neck skin is covered with fuzzy, downy feathers* (late spring–midsummer). Lower neck and ruff feathers are non-iridescent brownish black.

1c. Juvenile (late stage): *Partial yellow areas on black bill tip on thin black bill. Smooth black head skin with fuzzy, downy feathers on nape; wrinkles on lower part of upper neck.* Lower neck feathers are non-iridescent brown.

1d. 1-year-old: *Mostly yellow bill tip on thin bill* (sometimes similar to *1c*). *Smooth, black head skin has scattered fuzzy, downy feathers; distinct gray bumps and wrinkles on much of upper neck.* Lower neck feathers are partly iridescent.

1e. Adult: *Yellow bill tip on thin bill. Gray head and upper neck skin are wrinkled and bumpy.* Black lower neck feathers are iridescent green.

1f. Juvenile (early stage): *Smooth-skinned black head and neck.* Skin of upper neck is lowered (compare *1g*). Brownish-black plumage lacks iridescence and has faint brownish feather tips, which quickly wear off. *Legs are long and white.*

1g. Juvenile (early stage): *Black head.* Skin of upper neck is raised, making neck appear fully feathered.

1h. 1-year-old: Plumage is mix of old, faded, brownish-black juvenile feathers and new, iridescent, greenish-black adult feathers. Adult plumage features are most notable on wing, especially the new, glossy black tertials and respective inner greater coverts. *Long legs and toes are white. Note:* Tail is cocked upward when walking and trotting. *Note:* This is the early stage of 1st prebasic molt.

1i. Adult: *Large, white wing panel on outer primaries on all age classes.* Black tail is short and square.

1j. Flight silhouette: *Soaring.* Wings held in shallow dihedral.

1k. Adult: *Very wrinkled gray head skin; upper neck skin is raised.* Black plumage has greenish iridescent cast. *White legs are long. Note:* Head-bow posture is shown.

1l. Adult: *Very wrinkled gray head and upper neck skin; upper neck skin is lowered. Black plumage has greenish iridescent cast. Note:* Alert, high-body stance is shown.

1m. Adult: Upper neck skin is raised, making neck appearing nearly fully feathered.

1n. Adult: *Large white panel on outer 6 primaries. Wings are broad and square-shaped. White legs and toes extend to tip of short, square-shaped black tail.*

a
b
c
d
e
f
g
h
i
j
k
l
m
n

BLACK VULTURE (*Coragyps atratus*)
Typical (*C. a. atratus*) and "Southwestern" (*C. a. brasiliensis*)

HABITAT: *Year-round.* **Arizona and West Texas.** Semi-open, arid lowlands up to mid-elevation montane areas. Habitat encompasses riparian zones and/or rocky terrain in lowlands and regions of moderate tree growth and density and/or rocky terrain in higher elevations. **East Texas and eastward.** Semi-open, humid lowlands, including partially wooded areas of mesquite (*Prosopis* spp.), or tall trees, pastures, coastal savannas, and towns. **STATUS:** Population is stable but greatly reduced from historic times due to vastly different farming, ranching, and refuse practices in years past. **Arizona.** Fairly common species in its restricted range. Range extension occurred from Sonora, Mexico, into s. Ariz. in 1920s. **East Texas and eastward.** Common; appears to be stable. **NESTING:** Begins Jan.–Mar., depending on latitude. No nest is built; eggs are laid on ground. Nest sites are in protected dark locations in thick brush, hollow logs, and small caves; sometimes in tree cavities and abandoned buildings. Clutch size is typically 2 eggs. **MOVEMENTS:** Populations in n. Ark., Mo., and Okla. migrate south for winter. Other areas may have dispersal movements in quest of food during this time. *Extralimital.*—Records for Colo., Iowa, Kans., and Minn. for summer and fall; Calif. also for fall–winter. Canada has recorded this species in Alta., s.-cen. B.C., se. Sask., and sw. Y.T. in spring and summer. *Spring.*—Jan.–Apr., though actual migration is often difficult to assess. *Fall.*—Sep.–Nov. Peak movement in s. Tex. occurs from late Oct. to early Nov. **COMPARISON:** **Turkey Vulture (juvenile).**—Head is gray and smooth-skinned with pink around nasal area, and base of bill is ivory, *vs.* smooth black skin and black bill or gray wrinkled skin and whitish bill. When perched, primaries extend far past tertials, *vs.* primaries barely extending past tertials. In flight, all flight feathers are pale gray, *vs.* large white patch on outer primaries, and tail extends far beyond toes, *vs.* toes equaling tail tip. Plumage is dark brown, *vs.* black. **Eagles (flight).**—On underwing, white only on inner 2 primaries (Bald Eagle) or on base of inner primaries (Golden Eagle), *vs.* outer 6 primaries all-whitish. Tail extends far beyond feet in eagles, *vs.* equaling toes. Wingbeats are slow, *vs.* quick and snappy.

Figure 1. Refugio Co., Tex. (Sep.)
Coastal cattle pasturelands in humid climate.

Figure 2. Duval Co., Tex. (Aug.)
Inland mesquite brush in dry climate.

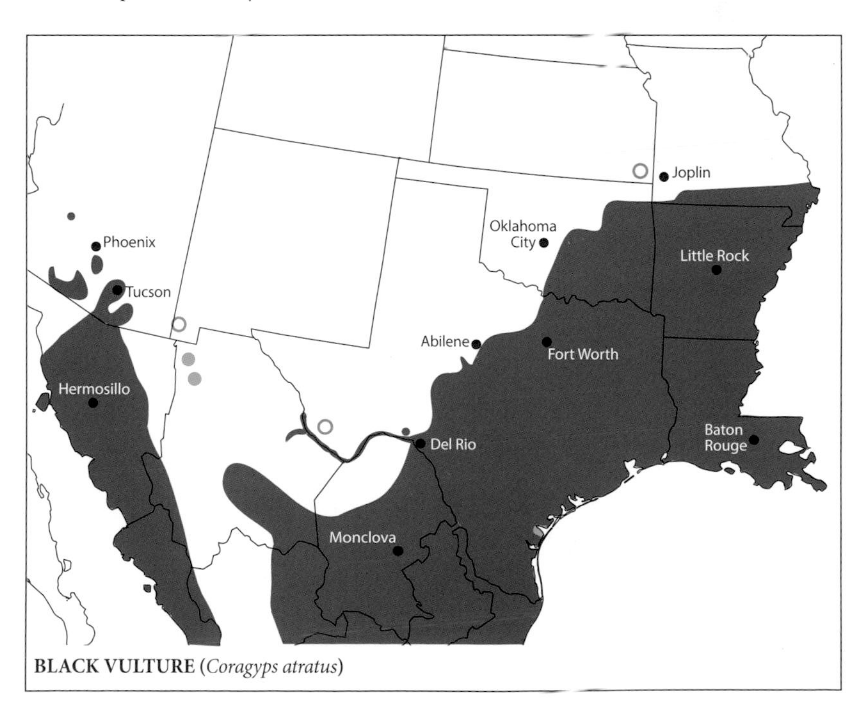

BLACK VULTURE (*Coragyps atratus*)

Plate 2. TURKEY VULTURE (*Cathartes aura*) "Eastern" (*C. a. septentrionalis*), "Western" (*C. a. meridionalis*), "Southwestern" (*C. a. aura*)

Ages: Juvenile, 1-year-old (1st prebasic molt stage), and adult. *Head and neck are bare-skinned, except on juveniles, which have some downy feathering. Nostrils (nares) have large, see-through holes.* When bird is perched, wingtips equal tail tip, with long primary projection beyond the tertials. *Underside of wing is 2-toned, with pale, unmarked, gray flight feathers contrasting with black coverts. Tail is wedge-shaped. In flight, short legs reach only to rear edge of undertail coverts. Legs and toes are reddish or pink but become white from excrement. Juveniles.*—Plumage is held for 1st year. *Crown of head and nape are covered with downy feathering; head slowly changes from gray to reddish.* Underside of flight feathers is darker gray than on older ages. Dark brown plumage has some iridescent quality; pale edges wear off. Age class is divided into 3 stages. *1-year-olds.*—Transition from juvenile to adult features, especially to glossier plumage. *Head is adult-like red, but ivory bill has dark tip. Note:* Only head is depicted. *Adults.—Head is red, and bill is solid ivory.* Plumage is iridescent. **Subspecies:** "Eastern" extends only into very e. part of West; "Western" in much of West; "Southwestern" in Southwest (and Greater Antilles); 3 more subspecies in Cen. and S. America. **Color morphs:** None. **Size:** L: 24–28" (61–71cm), W: 63–71" (160–180cm). *C. a. aura* is smallest. **Habits:** Tame in South, wary in North. Found singly, in pairs, or in flocks. Uses open perches, or perches and walks on ground. **Food:** Carrion. Food is located by smell. **Flight:** Mainly soars and glides; rocks back and forth when gliding. Powered flight is with deep, labored, equidistant up-and-down wingbeats. Regularly dips wings in downward motion to gain speed. **Voice:** Grunts and hisses when agitated.

2a. Juvenile (recently fledged): *Black bill with narrow ivory base. Nostril area and mandibles are pink, and bare head and neck skin is gray. Fuzzy, downy feathers on crown and hindneck. White downy feathers on base of upper neck disappear after fledging. Lower neck feathers are brownish black.*

2b. Juvenile (early stage): *Black bill has thin ivory base. Nostril, lores, eyelids, and front of neck are pinkish; rest of head skin is gray, with downy feathers on crown and hindneck.* Lower neck feathers are brownish black.

2c. Juvenile (late stage): *Distal ½ of bill is black, basal ½ is ivory. Head and neck are pink, with downy feathers on crown and hindneck.* Lower neck feathers are brownish black.

2d. 1-year-old: *Distal ⅓ of bill is blackish, basal ⅔ is ivory. Head and neck are red. Small to moderate amount of white tubercles in front of and/or below eyes. Lower neck feathers are iridescent black.*

2e. Adult (moderate tubercles): *Ivory bill. Head and neck are red.* Moderate amount of white tubercles in front of and below eyes. *Note:* Fairly common on birds east of Rocky Mts. and Great Plains; uncommon to rare westward.

2f. Adult (extensive tubercles): *Ivory bill. Head and neck are red.* Large amount of white in front of and below eyes. *Note:* Fairly common on birds east of Rocky Mts. and Great Plains; uncommon to rare westward.

2g. Adult (very extensive tubercles): *Ivory bill. Head and neck are red.* Very large amount of white tubercles encircling eyes and on forehead. *Note:* Found only in very e. part of West (and eastward).

2h. "Western"/"Southwestern" adult: Red head *lacks* white tubercles on all *aura* and on most *meridionalis* west of Rocky Mts.

2i. Juvenile (early stage): Pink-gray head and neck. Pale edges adorn dark brown feathers of dorsal plumage. Legs and feet are red.

2j. Juvenile (early stage): Upper neck skin is raised, and neck appears fully feathered.

2k. Juvenile (early stage): *Small gray head.* Brownish-black body and underwing coverts contrast with pale gray flight feathers. Tail is medium gray and wedge-shaped.

2l. Adult (moderate tubercles): *Bill, head, and neck as on 2f.* Black, iridescent lower neck feathers are raised, making head look small. Wing coverts are iridescent, with broad brownish edges. *Note:* Moderate tubercles on some "Western" birds.

2m. "Western"/"Southwestern" adult: Head as on *2h*; feathered neck lowered. Feet white from excrement.

2n. Adult: *Small red head; ivory bill. Black plumage contrasts with pale gray flight feathers (paler than on 2j). Tail is long, gray, and wedge-shaped. Note: Bare skin of crop shows after feeding; reddish on adults and pinkish or grayish on younger birds.*

2o. Flight silhouette: *Gliding.* Wings held in modified dihedral; tipsy side-to-side motions.

2p. Flight silhouette: *Soaring.* Wings held in high dihedral; makes tipsy motions.

a
b
c
d
e
f
g
h
i
j
k
l
m
n
o
p

TURKEY VULTURE (*Cathartes aura*)
"Eastern" (*C. a. septentrionalis*), "Western" (*C. a. meridionalis*), "Southwestern" (*C. a. aura*)

HABITAT: *Summer.*—Found in wide variety of habitats, but rarely in heavily agricultural areas. East and south of Great Plains found in semi-open, humid, flat and hilly regions. On Great Plains and west, where climate is arid and semi-open, occupies open, flat, hilly, and moderate montane areas. Absent from open areas if topography is flat and lacks riparian zones or rocky habitat. *Winter.*—Similar to summer but found at elevations and latitudes below and south of regular snowfall. May occupy extensive agricultural areas. **STATUS:** Very common. Fall migrant counts in e. Mexico alone often exceed 2 million, with untold numbers migrating or wintering elsewhere in s. U.S. and Mexico. *Note:* Demarcation between subspecies is not shown on the range map because subspecies are difficult to separate in the field. This is a highly adaptable species. It has expanded its range and increased its numbers in s. Canada in recent decades. **NESTING:** Begins in Feb. in s. areas and Apr.–May in n. areas; can end as late as Sep. in n. areas. Pairs stay together as long as mate survives. No nest is built. Eggs are laid on bare surface. Nest sites are in sheltered and protected areas in dense underbrush, hollow logs and stumps, or in small caves or rocky outcrops in hilly and montane areas. Also nests in old abandoned homestead buildings (especially noted in Sask.; C. S. Houston pers. comm.). **MOVEMENTS: "Eastern."** Migratory; winters as far south as S. America. Of 2 marked Sask. nestlings, 1 wintered in Costa Rica and 1 in Venezuela; of 3 marked Canadian 1-year-olds, 2 returned to Sask. and 1 to Nebr. Migration is at leisurely pace, with periodic 1–4 day stops (Houston et al. 2006, 2011). Accidental in spring to s. Y.T. (Sinclair et al. 2003). **"Western"/"Southwestern."** Can be sedentary, short-distance migrant, or long-distance migrant. Some "Southwestern" birds winter into cen. S. America. *Spring.*—Movement detected in early Mar. in s. Tex., with peak later in the month. In cen. Rocky Mt. region, movement is seen in early–mid-Mar. and peaks in early–mid-Apr., with some moving through May. *Fall.*—Begins in mid-Aug. in n. and cen. regions, with peak in late Sep.–early Oct., and ends by late Oct. Marked Sask. birds begin heading south late Sep. (Houston et al. 2006, 2011). Migrants show up in s. Tex. in mid-Sep., peak in late Oct., with movement through Nov. (Migration is virtually identical to that of Swainson's Hawk; both species are often in same mixed flocks.) **COMPARISON: Black Vulture.**—Thin, black bill and black head (juvenile) or white bill and gray wrinkled head (adult), *vs.* white-based bill and pale gray head (juvenile) or white bill and red head (adult). In flight, large white patches on primaries, and feet extend to tail tip, *vs.* all-gray flight feathers, and feet far short of tail tip. Dark brown plumage, *vs.* black. **California Condor.**—White axillaries and whitish to white portion of underwing coverts, *vs.* all-dark underwing. Darker-headed immatures are similar to gray-headed juvenile vulture. Flight is very stable, without flapping, *vs.* tipsy manner and occasional flapping. Wings are mainly flat, *vs.* strong dihedral. **Rough-legged Hawk (juvenile dark morph).**—2-toned underwing, but flight feathers are white, *vs.* pale gray flight feathers. Flight is steady, *vs.* tipsy manner. Often hovers, *vs.* does not hover. **Bald Eagle.**—Soars with wings held on flat plane, *vs.* wings held in high dihedral. All immatures have white patches on axillaries and inner primaries, *vs.* pale gray flight feathers. **Golden Eagle.**—Soars with wings in similar dihedral, and adults and some juveniles can have similar 2-toned underwing; flight is steady, *vs.* tipsy. Tawny nape is often visible, *vs.* small gray/red head.

Figure 1. Larimer Co., Colo. (Jun.) Classic nesting habitat in rocky areas.

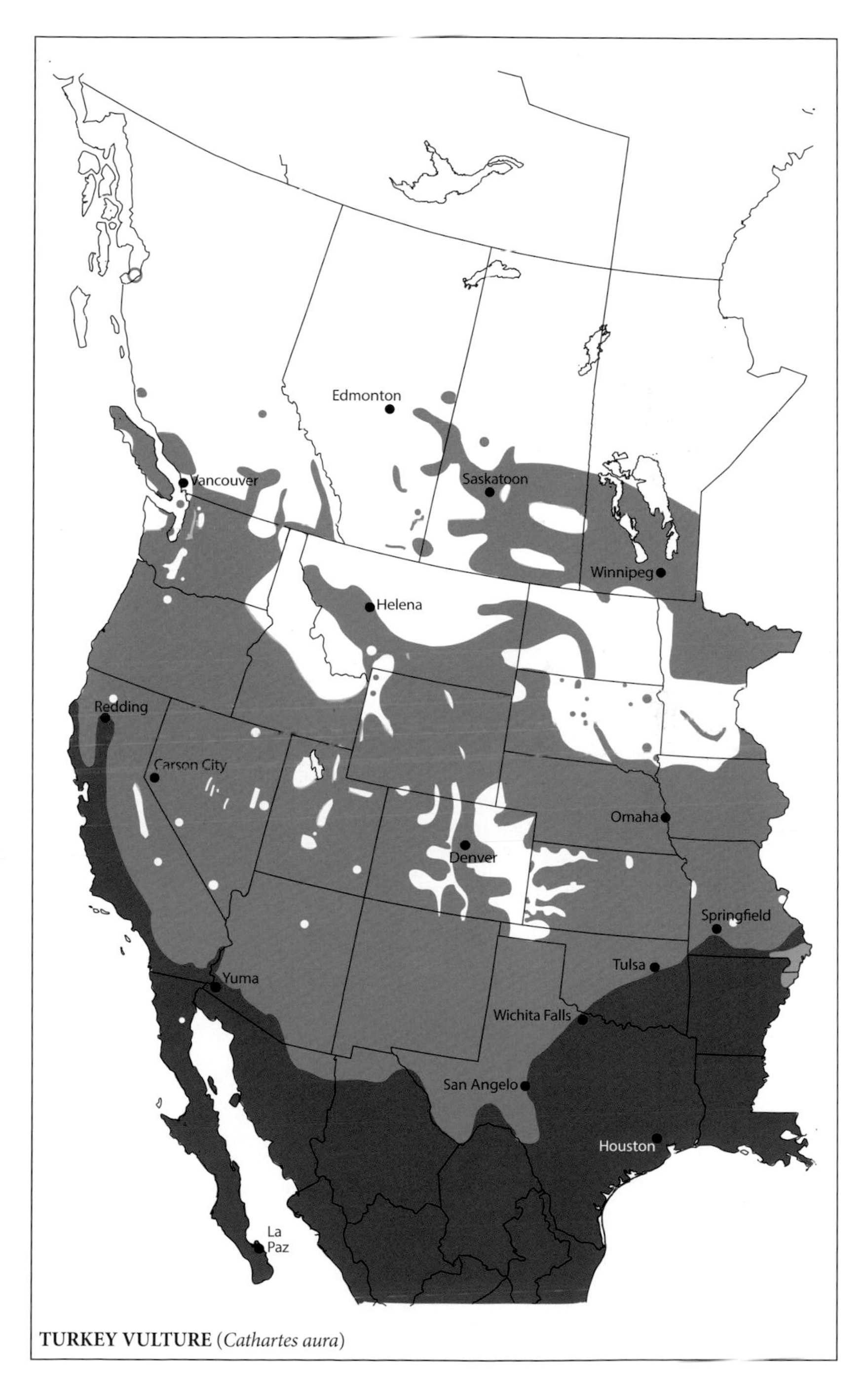

TURKEY VULTURE (*Cathartes aura*)

Plate 3. CALIFORNIA CONDOR (*Gymnogyps californianus*) Head Portraits

Ages: Juvenile, immature (ages 1–4), and adult. *Patch of short, stiff black feathers adorns forehead and area in front of eyes.* Spike-shaped feathers of lower neck can be raised and often conceal all of upper neck. Puffy, inflatable throat region, which begins developing after age 1, can be inflated or deflated at will. There are no sexual differences in any age class. *Juvenile.*—Head features held until nearly 1 year of age. The sclera, a small tissue area behind eye, is pale pink; all older ages have a red sclera. *Immatures.*—Ages 1–4. Progression of bill, head, and eye features to full adult characters is slow and variable. Variation of head characters can span at least 2 years of age classes (T. Hauck pers. comm.). *Adult.*—Full adult plumage is attained when 5–7 years old. Basic adult characters that are visible at field distances are generally attained by age 5, which includes head color but very rarely pale bill color. Head features, such as lack of black freckles and acquisition of pale blue bill, and dorsal wing features, which include broad white bar on secondary greater coverts and white edges on inner secondaries and tertials, are usually not attained until age 6 or 7. Head and bill characters are visible only on perched birds at close range; plumage features are visible only on perched birds when viewed in profile or dorsal angle; wing features are visible in flight only when dorsal surface is shown. Progression toward adult-colored head and neck starts on lower neck, next shows on throat and nostril region, and then slowly expands onto rest of head. **Subspecies:** Monotypic. **Color morphs:** None. **Size:** L: 43–50" (109–127cm). **Habits:** Tame and inquisitive. **Food:** Scavenger of ungulates. **Voice:** Hisses and grunts during feeding confrontations with other condors, Golden Eagles, Turkey Vultures, and Coyotes (*Canis latrans*). *Note:* Illustrations are based on close photographs of known-age birds from Grand Canyon Nat'l. Pk., Ariz., by the author and Ned Harris; figure *3b* (#85) was photographed at Pine Mountain, Calif., by Ned Harris. Juvenile figure *3a* is based on images in Osborn (2007).

3a. Juvenile (less than 1 year old): *Black bill. Bare-skinned gray face. Downy, grayish-brown feathers adorn crown, nape, all of neck, and base of throat.* Eye is brown, and sclera is pale pink. *Note:* Only wild-born birds would be seen in this age class.

3b. 1- and 2-year-olds: *Black bill. Bare-skinned gray face. Downy, grayish-brown feathers adorn crown and much of neck, except lower neck.* Eye is brown, and sclera is red. *Note:* #85 at 1 year, 5 months old (hatched Apr. 1998, photo Sep. 1999); also #59 at 1 year old (hatched Jun. 2007, photo Jun. 2008) and #EO at 2 years old (hatched Apr. 2006, photo Jun. 2008).

3c. 2- and 3-year-olds: *Black bill. Bare-skinned gray face. Downy, grayish-brown feathers adorn nape and upper hindneck, sparsely on front of neck. Visible neck skin is pale pink.* Eye is brown. *Note:* #C7 at 3 years old (hatched May 2005, photo May 2008).

3d. 3- and 4-year-olds: *Black bill. Bare-skinned gray face has small yellowish patch on front of chin. Gray strip on hindneck has very sparse downy feathers. Bright pink neck has lost much of downy feathering.* Eye is brown. *Note:* #A6 at 4 years old (hatched May 2004, photo Jun. 2008).

3e. 3- and 4-year-olds: *Neck ruff raised. Black bill. Bare-skinned gray face has large orange-yellow patch on chin and gape; yellow lower eyelid. Bright pink neck as on 3d but hidden by raised neck feathers.* Eye is brown or reddish brown. *Note:* #80 at 4 years old (hatched May 2002, photo May 2006).

3f. 4- and 5-year-olds: *Black bill. Head and neck are bare-skinned; gray-masked appearance on crown and face. Nostril and throat regions are orange; gray neck with rosy spot on front is adult-like.* Eye is reddish brown. *Note:* #99 at 4 years old (hatched Apr. 2003, photo Jun. 2007) and #16 at 5 years old (hatched May 2003, photo Jun. 2008).

3g. 5- and 6-year-olds: *Mainly black bill. Eye is reddish brown. Orange head has gray crown and gray freckles on much of head. Gray neck has rosy spot on front. Note:* #97 at 5 years old (hatched Apr. 2003, photo Jun. 2008).

3h. 5- and 6-year-olds: *Bill blackish but turning paler. Orange head has sparse freckles on crown and nostril regions. Throat region is deflated. Gray neck has rosy spot on front. Note:* #-3 at 5 years old (hatched Apr. 2003, photo Jun. 2008) and #74 at 6 years old (hatched Apr. 2002, photo Jun. 2008).

3i. 6-year-old and older: Pale bill. *Head is bright orange. Eye is reddish brown. Gray neck has rosy spot on front.* Breeding birds have richly colored heads. *Note:* #75 at 6 years old (hatched Apr. 2002, photo Jun. 2008) and #46 at 7 years old (hatched Apr. 2001, photo Jun. 2008).

a
b
c
d
e
f
g
h
i

Plate 4. CALIFORNIA CONDOR (*Gymnogyps californianus*) Perching

Ages: Juvenile, immature (ages 1–4), and adult. *Patch of short, stiff black feathers adorns forehead and area in front of eyes.* Spike-shaped feathers of lower neck can be raised and can conceal all of upper neck. Puffy, inflatable throat region, which begins developing after age 1, can be inflated or deflated at will. There are no sexual differences in any age class. *Juvenile.*—Features held until nearly age 1. *Immatures.*—Ages 1–4. Progression of head, body, and wing features to full adult characters is slow and variable. Variation of head characters, in particular, can span at least 2 years of age classes (T. Hauck pers. comm.). Head color, but very rarely bill color, can attain full adult character at age 5. *Adult.*—Head features may not be attained until age 6 or 7. Wing features, which show full adulthood, may occur at age 6, but often not until age 7 or 8. **Subspecies:** Monotypic. **Color morphs:** None. **Size:** L: 43–50" (109–127cm). **Habits:** Tame and inquisitive. "Suns" with spread wings. **Food:** Scavenger of mainly ungulates. After feeding, bare-skinned crop is visible as large, distended sac. **Voice:** Hisses, grunts during feeding confrontations with other condors, Golden Eagles, Turkey Vultures, and Coyotes. *Note:* Illustrations are based on close photographs of known-age birds from Grand Canyon Nat'l. Pk., Ariz., by the author and Ned Harris. Most birds have highly visible vinyl wing tags with alphanumeric markings on 1 or both wings. A small radio transmitter and antenna used for telemetry tracking adorn dorsal side of 1 or both wing markers. (Wing markers, which are different for every individual, were painted *exactly* as on respective individuals they adorned.) A few birds, however, lack alphanumeric designation on 1 or both tags or lack wing tags altogether.

4a. 1- and 2-year-olds: *Black bill. Gray head has brownish downy feathers on nape and upper ½ or more of neck. There may be faint pale pink on very lower neck, especially on 2-year-olds. Dark brown body and wings; thin, pale tawny bar on greater upperwing coverts.* Distended crop is gray. Little if any body or wing molt shows. *Note:* #59 at 1 year old (hatched Jun. 2007, photo Jun. 2008). Juvenile is similar, except neck has more downy feathers and pale wing bar may be wider; crop is gray.

4b. 2- and 3-year-olds: Black bill. Gray head has downy feathers on nape and upper neck. Much of neck is pale pink. Thin, pale tawny wing bar. Distended crop is pale grayish pink. Wing and body molt will show. *Note:* #C7 at 3 years old (hatched May 2005, photo May 2008).

4c. 3- and 4-year-olds: *Black bill. Gray head has very little downy feathering. Bare-skinned neck is bright pink. Considerable molt shows on body and especially wing; new wing feathers are paler gray. Greater secondary coverts have narrow whitish bar.* Distended crop is pink but can be grayish pink on some birds. *Note:* #A6 at 4 years old (hatched May 2004, photo Jun. 2008). This bird did not have an alphanumeric designation on its left wing tag.

4d. 4- and 5-year-olds: *Black bill. Gray face mask on orange head.* Lower neck ruff is fully raised and conceals upper bare neck. Crop does not show; if it did, it would be pink. *Note:* #-4 at 5 years old (hatched May 2003, photo Jun. 2008). #99 at 4 years old (hatched Apr. 2003, photo Jun. 2007) is similar but wing as on *4c*.

4e. 5- and 6-year-olds: *Blackish bill. Orange adult head is lightly freckled; freckling can be absent on many birds this age. White wing bar on greater secondary coverts begins getting more prominent at this age but is still fairly thin.* Secondaries are all-dark. Crop does not show; if it did, it would be pink. *Note:* #-3 at 5 years old (hatched Apr. 2003, photo Jun. 2008). (Antenna can partially overlap alphanumeric designation.) #74 at 6 years old (hatched Apr. 2002, photo Jun. 2008) is similar.

4f. 6-year-old and older: *Pale bill. Head is bright orange. Greater secondary coverts have broad white bar. Inner secondaries and tertials have thin white edges. Broad, white secondary bar and white-edged secondaries signify full adult. Note:* #22 at 13 years old (hatched May 1995, photo Jun. 2008).

4g. 6-year-old and older: *Pale bill. Head is bright orange. Distended crop is deep pink. Underwing has large, white triangular mark.* Bird is sunning. *Note:* #95 at 9 years old (hatched Feb. 1999, photo Jun. 2008). This bird was missing tag on its right wing.

a
59
b
C7
d
c
-4
-3
e
g
95
f
22

Plate 5. CALIFORNIA CONDOR (*Gymnogyps californianus*) Flying

Ages: Juvenile, immature (ages 1–4), and adult. *On each underwing, axillaries and coverts form long, whitish or white triangular region. 8 fingers on outer primaries, unless affected by molt.* Condors soar for extended periods without flapping. Most flapping occurs when taking off, landing, or when flying close to canyon walls or trees. *Juvenile.*—Features are held until nearly age 1. *Immatures.*—Ages 1–4. Progression of head, body, and wing features to full adult characters is slow and variable. Variation of head characters, in particular, can span at least 2 years of age classes (T. Hauck pers. comm.). Head color, but very rarely bill color, can attain full adult character at age 5. *Adult.*—Head features may not be attained until age 6 or 7. Wing features, which show full adulthood, may occur at age 6, but often not until age 7 or 8. There are no sexual differences in any age class. Molt is slow and variable. Figures are painted *exactly* in molt stage as photographed. *Note:* Most viewing is in spring through fall, when wing molt is active. **Subspecies:** Monotypic. **Color morphs:** None. **Size:** L: 43–50" (109–127cm), W: 8.1–9.8' (2.5–3m). **Habits:** Tame and inquisitive. **Food:** Scavenger of ungulates. After feeding, bare-skinned crop is visible as large, distended sac. **Flight:** Soars for extended periods, often for hours, without flapping wings. Wings are held flat or in very shallow dihedral when soaring, with fingers splayed widely. Powered flapping mode of slow, labored wingbeats used only for short period when getting airborne. **Voice:** Hisses and grunts during feeding confrontations with other condors, Golden Eagles, Turkey Vultures, and Coyotes. *Note:* Illustrations based on close photographs of known-age wing-tagged birds from Grand Canyon Nat'l. Pk., Ariz., by the author and Ned Harris. Most birds have highly visible vinyl wing tags with alphanumeric markings on front edge of 1 or both wings. Small radio transmitter with antenna used for telemetry tracking adorns dorsal side of 1 or both wing markers. Wing markers, which are different for every individual, were painted *exactly* as on respective individuals they adorned. A few birds, however, lack alphanumeric designation on 1 or both tags or lack wing tags altogether. *Note*: Primary molt sequence on Figures *5d*, *5e*, and *5g* illustrate *stepwise/molt wave* feather replacement. Innermost primaries are replacing feathers as well as a 2nd *molt wave* on mid-primaries, where molt stopped the previous year (reason for the isolated fully-grown p5 on Fig. *5d*, and irregular lengths of primaries on Figs. *5e*, *4g*). (All primaries are molting in an outwardly direction.)

5a. Juvenile to 2-year-old: Gray head and neck. *Axillaries and coverts show as long, dirty-white triangle.* Innermost primary is missing due to molt (slight notch in mid-wing), which signifies 1st stage of 1-year-old. Juvenile would lack wing molt; 2-year-old would be molting many inner primaries. *Note:* #59 at 1 year of age (hatched Jun. 2007, photo Jun. 2008).

5b. 3- and 4-year-olds: *Gray head with thin, pale pink neck ring. Dirty-white underwing triangle is similar to that on younger ages.* Inner primaries show considerable molt as large gap of missing feathers. Usually 5 or 6 old, worn juvenile outer primaries show. *Note:* #71 at 3 years of age (hatched Apr. 2005, photo Jun. 2008). #72 at 4 years old (hatched Apr. 2002, photo Jul. 2006) is similar but has brighter pink neck with wing molt similar to *5c*.

5c. 3- and 4-year-olds: *Gray head with thin, bright pink neck ring. Wing has thin, whitish bar.* Wing molt highly variable, but often retains 3 or so very old juvenile primaries. *Note:* #A6 at 4 years old (hatched May 2004, photo Jun. 2008). Left wing tag is blank, and only radio and antenna show.

5d. 5- and 6-year-olds: *Orange head. Whitish underwing triangle exhibits remnant dirty-white features of younger ages; primary greater coverts are black with white outer edges. Note:* #-3 at 5 years old (hatched Apr. 2003, photo Jun. 2008). #97 at 5 years old (hatched Apr. 2003, photo Jun. 2008) has dirty-white underwing markings similar to younger-age birds. #-7 at 6 years old (hatched May 2002, photo Jun. 2008) has underwing markings similar to #-3.

5e. 6-year-old and older: *Orange head. Underwing triangle is pure white. Note:* #22 at 13 years old (hatched May 1995, photo Jun. 2008). #75 at 6 years old (hatched Apr. 2002, photo Jun. 2008).

5f. 5-year-old and older: *Orange head. Thin, white wing bar. Note:* #6 at 6 years old (hatched Apr. 2002, photo Jun. 2008). #97 at 5 years old (hatched Apr. 2003, photo Jun. 2008) similar but has gray-crowned head. #46 at 7 years old (hatched Apr. 2001, photo Jun. 2008) is similar.

5g. 6-year-old and older: *Orange head. Wide, white wing bar on coverts and thin white edges on inner secondaries.* Wing markings depicted signify full adult plumage traits, mainly on birds at least 7 or 8 years old. *Note:* #4 at 8 years old (hatched May 2000, photo Jun. 2008).

5h. Flight silhouettes: *Top:* Wings on flat plane or slight dihedral. *Bottom:* Legs down when landing.

a
59
59
b
71
71
c
f
6
d
-3
-3
c
22
g
4
h

CALIFORNIA CONDOR (*Gymnogyps californianus*)

HABITAT: California and Baja California, Mexico. *Year-round.* Seasonally found along seacoasts up to elevation of 9,000' (2750m) in open, semi-open, and forested hills and rugged montane topography. Climate is semi-arid to arid. Nests in rugged areas with high cliffs and tall dead trees. Regularly breeds up to 4,500' (1370m), but has nested at 6,500' (1980m). **Arizona and Utah.** Occurs year-round in Ariz., and recently year-round in Utah, where peak usage is summer. Nests in areas with deep, rugged canyons and high cliffs. Forages in adjacent open, semi-open, or wooded montane and plateau regions at about 4,500' (1370m) elevation. Climate is arid. **STATUS:** Federally listed as endangered species since 1967. All wild condors were taken into captivity for captive breeding; the last was captured in mid-Apr. 1987. 1st captive-bred birds were released into the wild in Calif. in Jan. 1992, in Ariz. in Dec. 1996, and in Baja Calif., Mexico, in 2002. Release and monitoring of captive-raised stock occur in 3 separate populations: (1) U-shaped region of Calif. in Coast Ranges and adjacent Pacific coast, Tehachapi Mts., and Sierra Nevada. (2) Coconino Co., along and north of Colorado River, in Ariz., with recent expansion along the river to se. Nev.; and Kane and Washington Cos. in s. Utah. (3) Sierra de San Pedro de Mártir Nat'l. Pk. in Baja Calif., Mexico. 1st-release-era nesting: Calif. in 2001 (successfully in 2002); Grand Canyon Nat'l. Pk., Ariz., in 2001 (successfully in 2003); Zion Nat'l. Pk., Utah, in 2014 (successfully in 2016); Sierra de San Pedro Mártir Nat'l. Pk. in 2007 (successfully in 2009). At the end of 2016 there were 276 condors in the wild: 166 in Calif., 76 in Ariz./Utah, and 34 in Baja Calif. All condors are wing-tagged and carry telemetry devices for GPS tracking. Lead poisoning from ingesting fragments of lead bullets in carcasses and entrails of big game is a *major* threat to the species' survival. From 45% to 95% of Ariz./Utah population is annually contaminated by lead poisoning (Ariz. Game and Fish Dept. 2016). Lead-poisoned birds are captured and given chelation therapy (Starin 2016). Calif. has banned lead bullets for big-game hunting in condor range since Jul. 1, 2008, and will totally ban lead ammunition for all hunting by Jul. 1, 2019 (Calif. Dept. of Fish and Wildlife 2016). Ariz. has had a *voluntary* non-lead-ammunition program within condor range that "encourages" use of nontoxic (e.g., copper alloy) bullets (and retrieval of lead-embedded entrails of big game) since 2005, and Utah has had a similar program since 2011 (Ariz. Game and Fish Dept. 2016; Utah Division of Wildlife Resources 2016). 2016 was the 1st year that more condors were produced in the wild than were lost (Little 2016). Condor population is *not* self-sustaining yet—due to lead poisoning—and requires constant human intervention to remain viable (Little 2016). *Authors note*: To greatly enhance the survival of this prehistoric relic in our modern world—and survival of other predators and scavengers, which include both eagle species—lead hunting ammunition must be unilaterally banned. *Range map.*—Range and status are based on U.S. Fish and Wildlife Service (USFWS) data as of end of 2016. Nests in very low density within most solid purple regions. Linear dash pattern in Ariz. is a region of fairly regular sightings, based on eBird (2017). **NESTING:** Jan.–May; eggs are laid early Feb.–mid-Mar. Nest is scrape in soil in sheltered cave on high cliffs. Cavities in standing Giant Sequoia (*Sequoiadendron giganteum*) were historically used. Now regularly nests in tall, standing, broken-off, dead Ponderosa Pine (*Pinus ponderosa*) and Coast Redwood (*Sequoia sempervirens*) in coastal Calif. (Little 2016). 1 egg is incubated by both sexes. Breeding occurs every other year, or a pair may nest in 2 consecutive years, then skip a year due to late nesting in 2nd year. Pairs remain mated as long as members survive. 1st breeds when 5–8 years old. **MOVEMENTS:** Mainly sedentary. Seasonal shifting occurs from s. Utah into n. Ariz. in winter and to lower elevations in all areas in winter. Younger ages of Ariz. and Utah population are highly prone to far ranging eastward excursions throughout the rugged terrain of w. Colo., n. N. Mex., and sw. Wyo. Populations of Calif. and Mexico are mainly resident. **COMPARISON: Turkey Vulture.**—In adults, red head, *vs.* orange head. In younger ages, gray head with white base on bill, *vs.* gray head and all-dark bill. In flight, underwing coverts and axillaries are uniformly blackish, *vs.* whitish. Wings are held in dihedral with tipsy flight mannerism and occasional flapping, *vs.* wings held in low dihedral with steady flight and rarely flapping. **Bald Eagle (immatures).**—In flight, both show white axillaries and underwing coverts. Eagle soars with wings held on flat plane, *vs.* low dihedral. Legs are short and yellow, *vs.* long and white. **Golden Eagle.**—Golden-colored, feathered nape is visible perching or flying. If white mark adorns wing, it is on basal area of inner primaries, *vs.* white on axillaries and/or underwing coverts. Flaps wings regularly, *vs.* rarely flaps wings.

Figure 1. Tehachapi Mts., Kern Co., Calif. (Aug.) Area of last natural wild population; also current habitat.

Figure 2. Grand Canyon Nat'l. Pk. (South Rim), Coconino Co., Ariz. (Jun.)
Current year-round habitat.

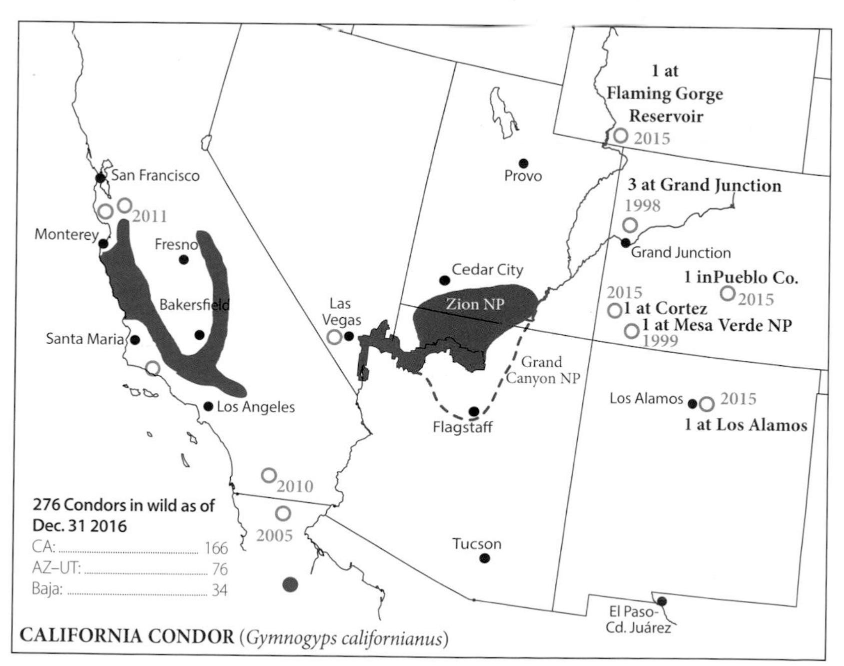

CALIFORNIA CONDOR (*Gymnogyps californianus*)

Plate 6. OSPREY (*Pandion haliaetus*)

Ages: Juvenile and adult. Juvenile plumage held for 1 year (plumage seen only in late summer and fall, since juveniles winter south of U.S.). Nape feathers are long and bushy when raised. *There is a black lore patch in front of eyes and a long black stripe behind eyes that connects to hindneck.* Bushy nape feathers are raised or lowered depending on mood and temperature. *When perched, wingtips extend far past tail tip. Underwing has large, rectangular black patch on wrist (carpal patch).* Wings are held in M-shaped arch when gliding and soaring. *Juvenile.—In fresh plumage, dark, blackish-brown upperparts show white-edged, scalloped pattern on all feathers* (wears off mid-autumn). Underwing coverts have rusty-tawny wash. *Adult.*—Uniformly dark brown upperparts are paler than juvenile's, due to sun bleaching and wear. Breast markings are variable: Those with no markings are males; lightly to moderately marked can be either sex; heavily marked are females. **Subspecies:** *P. h. carolinensis* is in much of U.S. and Canada. *P. h. ridgwayi* is in Caribbean and sparingly in s. Fla; 2 other subspecies elsewhere in world. **Color morphs:** None. **Size:** L: 20–25" (51–64cm), W: 59–67" (150–170cm). Females are larger than males. **Habits:** Tame to fairly tame where acclimated to humans, otherwise wary. Perches on open, natural or human-made structures. **Food:** Aerial hunter of live fish caught near surface of water. **Flight:** Glides with wings angled in M shape, with wrists sharply bent. When viewed head-on, wings are held in arched position when gliding. Powered flight is with slow, deep wingbeats. Dives in shallow or steep angle when capturing fish. **Voice:** Highly vocal at all seasons. Main call is loud, piercing, clear *chee-urp*, as single note or repetitive series of notes. Also emits loud *eeep* note.

6a. Juvenile male (old nestling/recently fledged), unmarked breast: *Eye is reddish orange or orange.* Black lore patch. Tawny-rufous nape and hindneck; crown often heavily streaked. Tawny breast fades quickly fades to white once fledged. Nape feathers are lowered.

6b. Adult, lightly/moderately marked breast: Yellow eye. Black lore patch. *Head is white with dark spot on forehead and on top of nape.* Light-headed type (shown) has isolated dark spot on top of nape. Auricular stripe connects to hindneck. Nape feathers are partially raised.

6c. Adult female, heavily marked breast: Yellow eye. Black lore patch. *Head is white with dark spot on forehead and on top of nape.* Dark-headed type (shown) has auricular stripe connecting to dark nape spot and dark hindneck. Nape feathers are fully raised.

6d. Juvenile, lightly/moderately marked breast: Tawny-rufous nape; nape feathers raised. *Dorsal feathers edged with white.* Breast is lightly marked. Wingtips extend far past tail tip. Tail has broad, white tip.

6e. Flight silhouette: *Gliding.* Wings are held in M-shaped arch. Fish are carried facing forward.

6f. Juvenile, lightly marked breast: Black auricular stripe. *Large, black rectangular carpal patch (on wrist) is mottled with tawny. Secondaries are pale gray with dark barring.* There is a tawny wash on underwing coverts. Trailing edges of wings and tail have broad white tips.

6g. Juvenile female, heavily marked breast (worn plumage): Tawny-rufous nape; nape feathers lowered. White feather edges on upperparts are mostly worn off. *Note:* Late fall. Shown with captured fish (Lake Trout, *Salvelinus namaycush*).

6h. Adult male, unmarked breast: Dark-headed type with auricular stripe that connects to dark nape spot. Tail has thin, white tip. Outer toe is rotated front to back (2 in front, 2 behind); this helps it grasp fish better.

6i. Adult, lightly marked breast: Black auricular stripe. *Each wrist has large, solid black, rectangular carpal patch. Secondaries are dark gray with partial dark barring.* Tail has thin, white tip.

a
b
c
d
e
f
g
h
i

OSPREY (*Pandion haliaetus*)

HABITAT: *Summer.*—Locations with fresh, brackish, or saltwater rivers, lakes, artificial impoundments, seacoasts. Waterways may be near human-inhabited areas or in remote locations, from sea level up to near tree line. *Winter.*—Low-elevation inland waterways, but most inhabit lower-latitude seacoast regions. During migration, can temporarily be found far from water, especially birds crossing Great Plains. **STATUS:** Locally fairly common. Population continues to naturally expand at fairly rapid rate since ban of DDT (organochlorine pesticide) in 1972 in U.S. and final ban in 1990 in Canada. Colo., Iowa, and Kans. engaged in release programs in 1990s. Many states still erect predator-proof nest poles/platforms near waterways to assist breeding success. **NESTING:** Begins in Mar. in southerly latitudes and even in mid-latitude lower elevations. Higher-elevation and northerly latitude populations begin in Apr.–May. Nesting ends Jun.–Aug., depending on elevation and latitude. Breeding does not occur until birds are 3 years old (remain on winter grounds until 2 years old). Pairs may reunite to nest but do not remain mated year-round as eagles do (data gathered by telemetry show birds of mated pairs migrate and winter in vastly different areas). Both sexes bring nest material, but females do most of construction. New nests can be shallow platform of sticks, but nests are typically reused and can become several feet deep. Nests are placed on various structures, both natural and artificial, and in undisturbed areas even on the ground. Nest sites are usually near water but can sometimes be several miles away. Tree nests are often in dead trees or, if in live tree, in dead exposed upper section above the foliage. Buoys, utility poles, and platforms on top of poles are commonly used artificial structures. Clutch is 2 or 3 eggs. **MOVEMENTS:** Highly migratory. *Spring.*—Breeding adults begin leaving winter grounds in Mexico and Cen. and S. America in Feb. Early migrants can be seen in late Feb. in s. areas and even inland mid-latitudes but usually not until Mar.; Canadian birds are 1st seen in mid–late Apr. Peak movement along w. Great Lakes is in late Apr.–early May. Nonbreeders may move until mid-Jun. *Fall.*—Adult females tend to move before males and begin heading south in late Aug.–Sep. Juveniles follow soon after in Sep.–Oct. Movement is often rapid, and wintering grounds are reached in 2–3 weeks. **COMPARISON: Bald Eagle (immatures).**—Share similar whitish heads and necks, dark stripe behind eye, and pale eye. Bill and cere are yellowish or brownish, *vs.* all-black bill with blue cere. When perched, wingtips fall short of tail tip, *vs.* extending beyond tail tip. In flight, axillaries are white, *vs.* mainly white with black carpal patch on wrist. **Large gulls (*Larus* spp.).**—In flight, arched or bowed, crooked wing attitude is similar, but gulls' underwing is uniformly pale and lacks dark carpal patch. Straight bill, *vs.* hooked bill.

Figure 1. Grand Lake, Grand Co., Colo. (Sep.) Nest on top of dead tree.

OSPREY (*Pandion haliaetus*)

Plate 7. WHITE-TAILED KITE (*Elanus leucurus*)

Ages: Juvenile, immature (partial 1st-year molt stage), and adult. All have narrow black bill with yellow on cere and base of lower mandible. There is a large, black lore spot. *Upperwing coverts form large, black "shoulder" patch. Underwing has small, black carpal (wrist) spot. When perched, wingtips extend nearly to tail tip. Tail is white with 2 gray center feathers. Juvenile.*—Eyes are brown to orange-brown. *Feathers of brownish upperparts and most wing feathers are scalloped or edged with white.* Tail has thin, dusky subterminal band. Plumage is held for 3½–4 months. *Immature.*—Age class is split into early and late stages denoting molt progression (molt based on Pyle 2005a). A partial to full body molt (partial 1st-year molt) produces adult-like plumage by winter. Central feather set or entire tail is molted; sometimes 2–6 secondaries are also molted (but no primaries) by late stage. Little additional molt occurs until spring, when 1st prebasic molt begins, including all flight feathers. Eyes are orange and progress to more reddish with age. *Adult.*—Eyes are orange-red or red. Upperparts are pale gray (females darker). **Subspecies:** *E. l. majusculus* in U.S., south to Cen. America; 1 other subspecies in S. America. **Color morphs:** None. **Size:** L: 14–16" (36–41cm), W: 37–40" (94–102cm). Females average larger. **Habits:** Tame to wary. Perches on exposed branches, posts, and wires but does not perch on cross arms of utility poles. *Tail is bobbed up and down as a display to other kites intruding into territory.* Talon grappling occurs in confrontations with other kites. **Food:** Aerial hunter of small rodents. **Flight:** Wings are held in high dihedral when soaring and gliding. Regularly hovers when hunting. Kites dive headfirst for prey, with legs partially lowered and wings held high above body. When landing, kites lower their legs and gently glide or drop down to a perch. **Voice:** Soft, high-pitched chirps or grating, raspy *kree-aak* when agitated. Emits harsh *grrkk* when chasing intruders.

7a. Juvenile: *Eye is medium brown or orange-brown.* Dark lore spot. Dark eye line is short. Crown, nape, and hindneck are tawny. Breast is tawny, often with brown streaks. *Broad white feather tips on back and scapulars.*

7b: Immature (early stage): *Eye is orange or sometimes reddish orange.* Dark eye line is short. Crown, nape, and hindneck are mainly gray adult feathers, with a few retained tawny juvenile feathers. White adult breast feathers are mixed with remnant tawny juvenile feathers. Upperparts are mix of gray adult feathers and retained white-tipped juvenile feathers.

7c. Immature (late stage): *Eye is orange to orange-red. Crown, nape, and upperparts are pure gray adult plumage. Note:* Head and body as adult's, but eye is often not as reddish. Male (shown) is paler gray than female.

7d. Adult female: *Eye is red but may be orange. Crown, nape, and upperparts are brownish gray.* Male is paler gray.

7e. Juvenile: Tawny crown, nape, and hindneck. Breast is tawny. *Brownish back and scapular feathers are broadly edged with white.* All wing feathers are edged with white. Tail has narrow, dusky subterminal band; 2 central feathers are gray.

7f. Flight silhouette: When hovering, soaring, or gliding, wings are held in high dihedral.

7g. Courtship flight: Fluttering flight with wings held in high, V-shaped dihedral and moved in quivering motion; legs extended.

7h. Immature (early stage): Head, back, and scapulars are mix of scalloped tawny/brown juvenile and gray adult feathers. White edges of feathers in dorsal areas are narrower than on *7e* due to wear; white edges have worn off greater upperwing coverts. Wing and tail are still retained juvenile feathers; tail is as on *7e*.

7i. Underwing, alternate type: A few individuals of any age lack black carpal spot on underwing.

7j. Juvenile: Breast has tawny wash. *Black carpal spot. Primaries are dark gray, and secondaries are pale gray, both with thin, white trailing edge. White tail has darker central feather set and thin, dusky subterminal band.*

7k. Immature (late stage): Pale gray crown and dorsal surface, as on adult; white edges on wing feathers have mainly worn off or are very thin. Tail is mostly juvenile, as on *7e*, but center feather set has molted into adult-like plumage, without dusky band.

7l. Immature (late stage): White underparts. *Black carpal spot.* Primaries are dark gray, and secondaries are paler gray (retained juvenile wing feathers). Tail is as on *7e* but can be adult-like. *Note:* Legs often lowered when hovering.

7m. Adult male: Crown and upperparts are gray. *Black shoulder patch adorns lesser upperwing coverts.* Underparts are white. *White tail has 2 gray central feathers.* Female has more brownish-gray tone on dorsal surface.

7n. Adult female: Crown and upperparts are gray with brownish tinge. *Black shoulder patch.* White tail has 2 gray central feathers.

7o. Adult: White underparts. *Black carpal spot.* Primaries are dark gray, and secondaries are very pale gray; wing lacks pale trailing edge of younger birds. *White tail has 2 gray central feathers.*

a
b
c
d
e
f
g
h
i
j
k
l
m
n
o

WHITE-TAILED KITE (*Elanus leucurus*)

HABITAT: *Year-round.* Dry and humid semi-open and open flat or moderately hilly areas with scattered or small tracts of tall bushes and trees. Occupies natural marshes, meadows, grasslands, and vegetated coastal dunes as well as human-altered pastures and agricultural areas with ample unfarmed edges for foraging. In arid areas of Ariz., Calif., and Tex., also found in irrigated low-impact alfalfa and grass-hay agricultural areas. Elevation varies from sea level to 2,000' (600m) along w. foothills of Sierra Nevada of Calif. and up to 4,000' (1220m) on flat savannas in parts of s. Ariz. **STATUS: Arizona.** Very uncommon and local. 1st seen in the state in 1972 in Cochise Co.; breeding documented in 1983 in Pinal Co. These birds are an extension of Sonora, Mexico, population. **California.** Fairly common and widespread. This state hosts largest population of this species. Population is tied to cyclic trends of California Vole (*Microtus californicus*), a primary food source. **Oregon.** Uncommon and local. This species was 1st seen in the state in mid-1920s, then not again until 1933. Population increased dramatically in 1960s and 1970s, with individuals dispersing northward from Calif. Previously a late summer–fall visitor (from Calif.), has been regular breeder and year-round resident since 1976. **Washington.** Rare and very local. This state hosts northernmost extension of populations originally from Calif. 1st documented in the state in 1975; 1st breeding in 1988 in Pacific Co. A few pairs now nest regularly. Recent spring sighting records for nw. counties. **Texas.** Fairly common to locally common. There are small, isolated breeding pockets in interior locations. Population was disrupted in early–mid-1900s by massive habitat alteration but has adapted and increased. **Louisiana.** Very local. Year-round resident in Calcasieu and Cameron Parishes and probably in Beauregard Parish. **NESTING:** Very long nesting season, often with double or even triple clutches (Calif.), extends from Jan. through fall. Pairs remain together year-round. Most nests are built Jan.–Aug. Moderate-size, deep, compact stick nest is placed 10–60' (3–18m) high in top section of large bush or tree. Female builds nest and incubates. Male hunts and guards territory. Clutch size is 3–6 eggs, average 4. Pairs engaged in 2nd nesting stop feeding 1st brood once 2nd brood hatches. **MOVEMENTS:** Sedentary, but engages in periodic northward excursions (and breeding). From w. populations, w. Nev. and s. B.C., including s. Vancouver Island, have spring and summer records. From presumably Tex. population, 1 or more spring–summer records each exist for Kans., Minn., Mo. (1 Dec. record), N.Dak., Okla., and Wyo. **COMPARISON: Northern Harrier.**—All ages have large, distinct white patch on uppertail coverts *vs.* gray uppertail coverts. Adult male has all-gray upperparts, *vs.* gray with large black patch on upperwing coverts. Adult male has white underwing with sharply defined black primaries and wide black bar on secondaries, *vs.* underwing with black wrist spot, dark gray primaries, and very pale gray, unmarked secondaries. Juvenile harriers are similar to recently fledged kites, with brownish upperparts and rusty underparts, but have all-brown upperwing, *vs.* gray with black wing patch. Flight feathers are barred on underwing, *vs.* pale gray, unmarked flight feathers and black wrist spot. **Gulls.**—All-gray upperparts, black wingtips, and short tail, *vs.* gray upperparts with large, black shoulder patch and long tail. All-white ventral wing, *vs.* underwing with gray flight feathers and small black wrist spot.

Figure 1. Cameron Co., Tex. (Aug.) Coastal savanna.

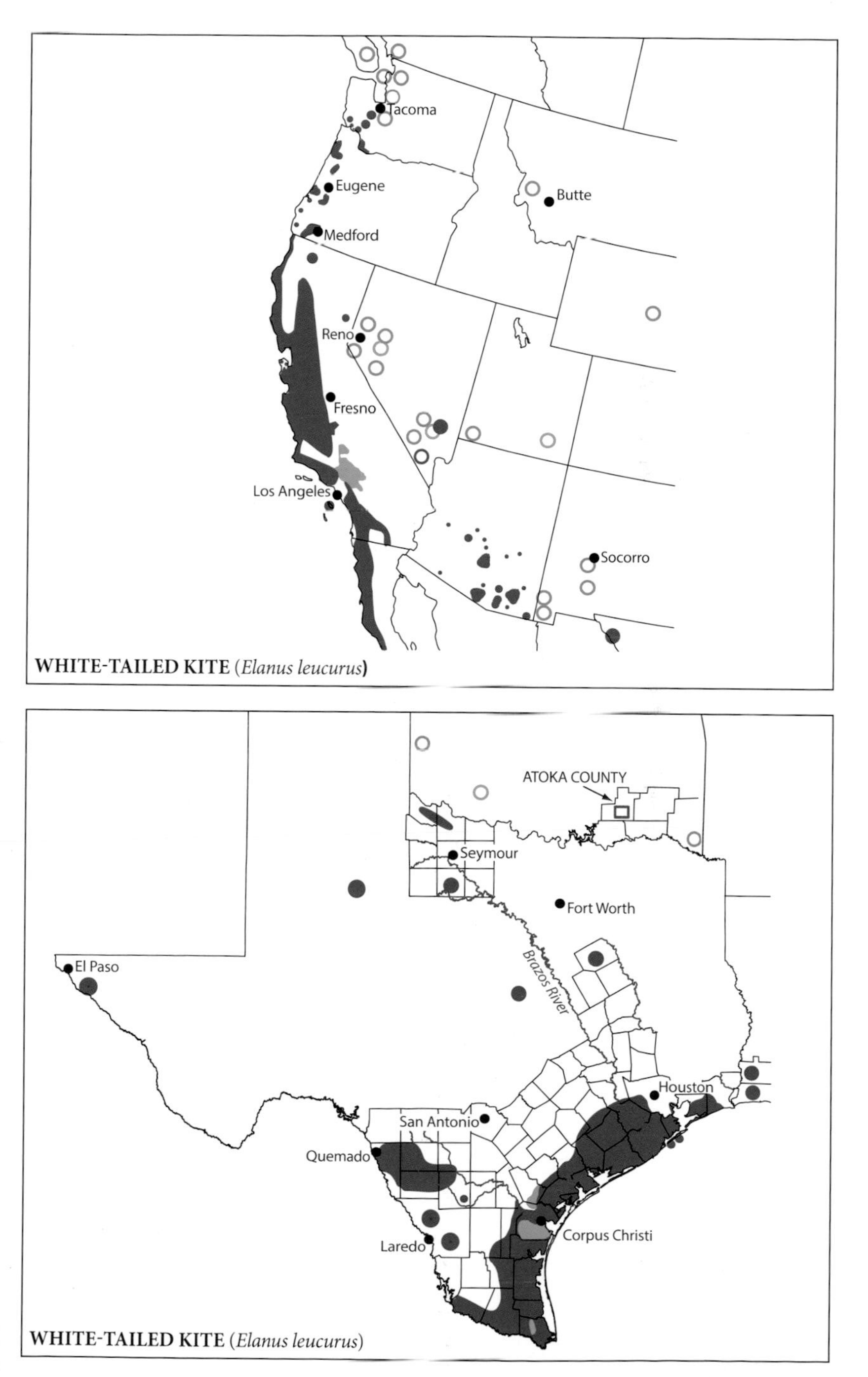

WHITE-TAILED KITE (*Elanus leucurus*)

WHITE-TAILED KITE (*Elanus leucurus*)

Plate 8. MISSISSIPPI KITE (*Ictinia mississippiensis*) — Juvenile

Ages: Juvenile body plumage is retained for 4–6 months, then begins slow partial 1st-year molt, starting on winter grounds in S. America. Head features separate juvenile from other raptors: *short, thick, all-black bill with yellow cere; black lore spot and thin black ring around eye; dark brown eye; short, broad, white supercilium; and lack of dark malar mark. On flying bird, outermost primary is much shorter than longest feathers on wingtip. Secondaries are gray, with wide, white bar on trailing edge.* On perching bird, wingtips equal or extend just past tail tip. Scapulars show variable amount of white spotting. **Subspecies:** Monotypic. **Color morphs:** None. **Size:** L: 12–15" (30–38cm), W: 29–33" (74–84cm). Females are slightly larger. **Habits:** Tame. Acclimates well to humans and human-altered areas. This is a gregarious species, and it roosts, feeds, and migrates in flocks. Often nests communally. Uses exposed branches and wires for perches but also perches inside shaded, heavily foliaged areas to escape heat. *Kites tip forward awkwardly when landing.* **Food:** Aerial hunter of flying insects and of tree-inhabiting amphibians, lizards, and nestling birds. Rarely captures ground-dwelling rodents. **Flight:** *Soars with wings held on flat plane and wingtips bent slightly upward*; glides with wings on flat plane. This kite is acrobatic in flight, especially when hunting. Wingbeats are moderately fast. Often dives to chosen perch. **Voice:** Highly vocal. Loud, piercing, 2-syllable, whistled decrescendo *phee-toooooo*; greeting calls are *phee-too-too-too* and short *phee-too*.

8a. Juvenile, all types: *Large, black lore spot, white supercilium, and gray auriculars and cheeks. Note:* Head may be paler on lightly marked types and darker on heavily marked types.

8b. Lightly marked type: Underparts have thin rufous (shown) or brown streaking. *Secondaries are gray with white bar on trailing edge.* Wingtips equal tail tip. *Note:* Holding a cicada in its right foot.

8c. Moderately marked type: As *8b*, but underparts have moderate-width rufous or brown (shown) streaking.

8d. Heavily marked type: As *8b* and *8c*, but streaking is very heavy and can appear solid rufous or brown.

8e. Juvenile, all types: Dark brown upperparts are spotted. *Secondaries are gray with white bar.* Tail is dark and unbanded on dorsal side.

8f. Banded tail, fanned (ventral): 3 or 4 thin, complete white bands. Outer feather set is paler on basal ¾ of each feather. White terminal band is very thin. *Note:* Common tail type.

8g. Banded tail, closed (ventral): 2 thin white bands; band is incomplete or absent on outer feather set.

8h. Partial-banded tail, fanned (ventral): 1 or 2 thin white bands on all but outer feather set. *Note:* Common tail type.

8i. Partial-banded tail, closed (ventral): Tail appears solid dark with 2 rows of small white spots. *Note:* Can be all-dark and similar to *8k* when seen at a distance.

8j. Unmarked tail, fanned (ventral): *All-dark, except paler on basal ¾ of outer feather set. Note:* Uncommon tail type.

8k. Unmarked tail, closed (ventral): All-dark. *Tail is dark gray with darker terminal band.*

8l. Lightly marked type: Thin rufous streaking. Underwing is very extensive white type: *large white area on all of primaries connects to narrow white strip on basal secondaries.* Tail banded as on *8f.*

8m. Moderately marked type: Moderately wide rufous ventral streaking. Underwing is extensive white type: *large white area on all primaries.* Tail is banded as on *8f.*

8n. Underwing, moderate white type: *Small white patch on outermost primary and moderate-size whitish area on inner primaries.*

8o. Moderately marked type: Moderately wide, brown ventral streaking. Underwing is minimal white type: *small white patch on outermost primary and fairly small whitish patch on inner primaries.* Tail as on *8h* but has 1 thin, white basal band.

8p. Heavily marked type: Nearly solid rufous (shown) or brown underparts. Underwing is all-dark type: *solid dark gray.* Tail as on *8j*; may have partial-banded pattern as on *8h.*

8q. Flight silhouette: *Soaring.* Wings are held on flat plane with wingtips bent slightly upward.

a
b
c
d
e
f
h
g
i
j
k
l
m
n
o
p
q

Plate 9. MISSISSIPPI KITE (*Ictinia mississippiensis*) **1-Year-Old and Adult**

Ages: 1-year-old and adult. *Black bill with gray cere; gray head with black lores and black encircling eyes; red eyes. Legs are yellow or orange with gray toes. On flying birds, outermost primary is much shorter than longest feathers on wingtip. On perching birds, wingtips extend past tail tip.* Males have paler heads than females. *1-year-old.*—Partial 1st-year molt begins on body at 4–6 months of age on winter grounds; molt continues the following summer and fall, as 1st prebasic molt, when rest of body, wings (coverts and up to 7 inner primaries), and tail (r1, r6 sets) continue to molt. *Adult.*—Molts into adult plumage when 1½ years old, while on winter grounds. Males have paler body plumage than females. *Males have pale gray dorsal patch on secondaries; both sexes have white bar on trailing edge of wing. Primaries, especially on males, have rufous on inner feathers.* Wing molt is noticeable in Jun. and later on 1-year-olds, Jul./Aug. and later on adults. **Subspecies:** Monotypic. **Color morphs:** None. **Size:** L: 12–15" (30–38cm), W: 29–33" (74–84cm). Females are slightly larger. **Habits:** Tame. Acclimates well to humans and human-altered areas. A gregarious species, it roosts, feeds, and migrates in flocks. Often nests communally. Uses exposed branches and wires for perches; also perches inside shaded, heavily foliaged areas to escape heat. *Kites tip forward awkwardly when landing.* **Food:** Aerial hunter of flying insects and of tree-inhabiting amphibians, lizards, and nestling birds. Rarely captures ground-dwelling rodents. **Flight:** *Soars with wings held on flat plane and wingtips bent slightly upward*; glides with wings on flat plane. This kite is acrobatic in flight, especially when hunting. Often dives to a perch. **Voice:** Highly vocal. Loud, piercing, 2-syllable, whistled decrescendo *phee-toooooo*; greeting calls are *phee-too-too-too* and short *phee-too*.

9a. 1-year-old male/adult male: *Pale gray. Bill is black; cere is gray. Black patch adorns lores. Eye is red.*

9b. 1-year-old female/adult female: *Medium gray, with whitish forehead, supercilium, and throat.*

9c. 1-year-old female (very early stage, May–Jun.): Head, breast, and scapulars as on *9b*, *9l*. Belly, flanks, thighs are juvenile. Wings, tail are juvenile. White spots are molting, missing feathers.

9d. 1-year-old female (early stage, May–Jun.): More adult-like than *9c*. A few juvenile feathers are on underparts. Wing coverts are mainly adult-like. Flight feathers, tail are juvenile.

9e. 1-year-old male (late stage, Aug.–Sep.): Adult-like, as on *9k*, but all secondaries and several outer primaries are juvenile feathers (2 tertials and inner primaries are new adult). Partially grown, black middle feather set (r1) on tail.

9f. 1-year-old male (early stage): Adult-male-like head and upperparts. Greater wing coverts, flight feathers, and tail are all juvenile.

9g. 1-year-old male (very early stage): *Pale gray head.* Breast is adult-like gray; much juvenile streaking on belly. Wing is juvenile, very extensive white type. Juvenile tail is banded type (see *8f*).

9h. 1-year-old male (early stage): *Underparts are adult-like gray, with some juvenile markings; white spots due to missing feathers. Extensive white underwing is missing (molting) innermost primary (p1).* Tail is banded type (see *8f*).

9i. 1-year-old female (early stage, Jun.): *Adult-like underparts.* Molting, missing inner 4 primaries; 2 innermost primaries are partially grown adult black feathers. *Note:* Eating.

9j. 1-year-old female (late stage): *Adult-like, with white lower belly, white-blotched undertail coverts.* Outer secondary (s1) dropped. Primaries are as on *9g*, *9h*, but with 6 new (black) feathers. On tail, r6 set is ½ grown, r1 set is grown.

9k. Adult male: *Pale gray head. Ventral side of body is medium gray, and dorsal side is dark bluish gray. Secondaries are whitish.*

9l. Adult female: *Medium-gray head with paler supercilium and throat. Medium-gray ventral side of body and dark gray dorsal side are more brownish than on male. Secondaries are medium gray but can be as pale as on male.*

9m. Adult male tail (ventral): *All-black; white only on basal area of outer quills.* Gray undertail coverts (all males).

9n. Adult female tail, black (ventral): *All-black; white outer quills.* All-gray undertail coverts type.

9o. Adult female tail, pale (ventral): Pale outer feather set with white quills. Spotted undertail coverts type.

9p. Adult female tail, banded (ventral): Similar to *9o*, but partially banded and like 1-year-old. White undertail coverts type (lower belly also white). *Note:* Very uncommon to rare tail type.

9q. Adult male: *Pale gray head and secondaries contrast with blue-gray dorsal surface of body. Tail is black. Primaries show rufous.*

9r. Adult male (Apr.–Jul.): *Pale gray head and medium-gray underparts.* No wing molt.

9s. Adult female (Jul.–Sep.): *Medium-gray head and underparts. White lower belly/white undertail coverts type. Underwing coverts are mottled whitish.* Inner 2 primaries are newly molted and dark; new p3 is partially grown. Tail is as on *9n*.

a
b
c
d
e
f
g
h
i
j
k
l
m
n
o
p
q
r
s

MISSISSIPPI KITE (*Ictinia mississippiensis*)

HABITAT: *Summer.*—East of Great Plains found in humid, hot, lowland riparian floodplains, swamps, and lakes with adjacent tall, semi-open tracts of mainly deciduous trees or mix of deciduous and coniferous trees. On s. and cen. plains uses semi-arid to arid, hot, semi-open and open areas of flat topography with large bushes and short or tall deciduous trees. Elevation can be as high as 4,600' (1400m) on high plains of s. Colo. In less arid regions of s. plains, kites use semi-open and open areas that lack water and have minimal tree growth and size. Kites extensively use shelterbelt plantations, especially double-row plantings, in s. Kans., w. Okla., and n.-cen. Tex., since they were 1st planted during Dust Bowl era of 1930s. In arid range of se. Colo., w. N. Mex., and Ariz., they inhabit riverine areas with tall Plains Cottonwood (*Populus deltoides*) trees, including areas within canyons. Except in Ariz., kites have readily adapted to residential areas, city parks, school campuses, and cemeteries. **STATUS:** Locally common, but very uncommon and sporadically distributed north and west of core breeding areas. Corpus Christi, Texas tallied 35,000 in late summer-fall of 2016; largest 1-day tally was on Aug. 31, 2007 with 12,261 seen. (At Cardel, Veracruz, Mexico, fall counts have ranged from 275,000 to 369,000 birds since 2010.). Range has been expanding north and west of core areas since 1970. 1st documented nesting in Ariz. in 1970, Colo. in 1971, and Nebr. in 1991. 1st breeding record for Canada was in Winnipeg, Man., in 2014 (Swick 2014). **NESTING:** May–mid-Sep., and completed mid-Jul–Aug. Nests singly or in loosely formed colonies of up to 20 pairs. Small stick nest can be poorly made or sturdy, compact structure; it is placed among dense foliage in upper portion of large bush or tree. Nests can be as low as 4' (1.2m) in undisturbed areas, or as high as 135' (41m) in tall tree. Old nests can be reused. Both sexes build nest and care for young. Clutch size is 2 eggs. Male does most of feeding of fledglings. **MOVEMENTS:** Long-distance migrant to and from winter quarters in s. Bolivia, Paraguay, and n. Argentina. Travels singly or in small to large flocks; larger flocks, often numbering in the thousands, are encountered in fall (flocks of 5,000 have been seen in fall in se. Tex.). *Spring.*—Breeding adults return 1st, with peak numbers in s. Tex. in early Apr. Northern breeding areas of Colo. are reached by early–mid-May. 1-year-olds straggle through s. Tex. in mid-May and reach n. areas by Jun. *Fall.*—An early-season migrant, its movement spanning Aug.–Oct., with peak in s. Tex. in late Aug.–early Sep. All ages move together, but most late records comprise late-fledged juveniles. *Extralimital.*—1-year-olds, juveniles, and some adults are prone to wandering Mar.–Oct. Records exist for Calif., Nev., S.Dak., Minn., Mont., Wyo., and Sask. **COMPARISON: White-tailed Kite.**—Black shoulder patch on upperwing and black spot on wrist area of underwing, *vs.* uniform underwing or dark underwing with white patches (juvenile/1-year-old). Tail is whitish, *vs.* black or dark with white bands. **Sharp-shinned Hawk (juvenile).**—Has similar streaking as juvenile kite on underparts. Eye is yellow, *vs.* dark brown. When perched, wingtips are much shorter than tail tip, *vs.* equal to tail tip. Wingtips are rounded when in flight, *vs.* pointed. **Merlin.**—Plumage and dark eye are similar to those of juvenile kite. Orbital skin is yellow, *vs.* black spot on same area. Wingtips are shorter than tail tip, *vs.* equal. Similar shape in flight, but outermost primary is barely shorter than wingtip, *vs.* much shorter than wingtip. **Peregrine Falcon.**—Dark head pattern with dark malar mark is distinct, *vs.* indistinctly marked head lacking malar mark. Similar long wings, which reach tail tip when perched. Very similar shape in flight, but outermost primary is nearly same length as wingtip, *vs.* much shorter outermost primary.

Figure 1. Lamar, Prowers Co., Colo. (Sep.) Nesting site in Chinese Elms (*Ulmus parvifolia*) in residential area.

Figure 2. Ellis Co., Okla. (Jun.)
Colonies of kites nest site in Osage-orange (*Maclura pomifera*) shelterbelt groves.

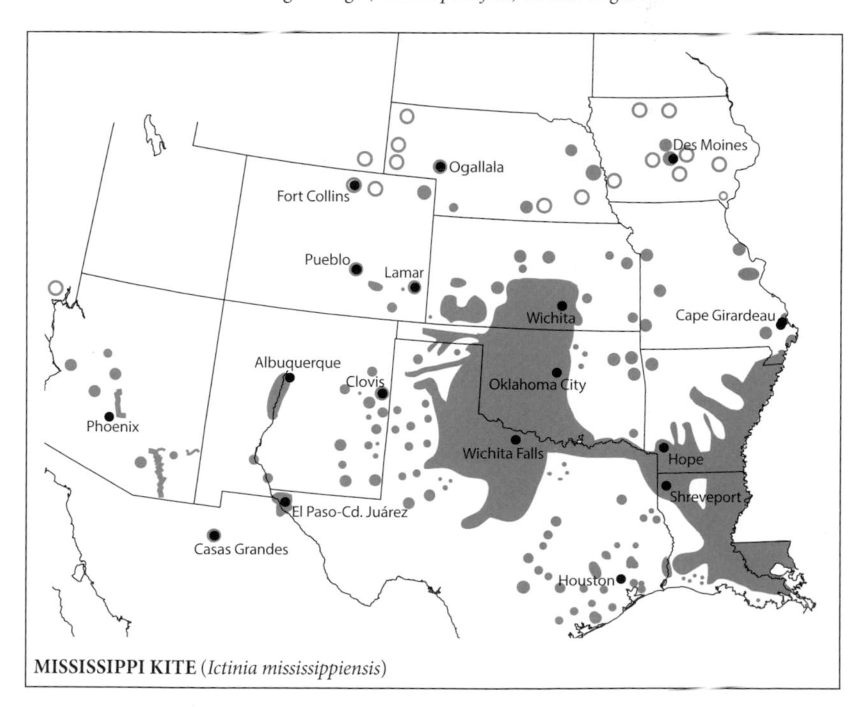

MISSISSIPPI KITE (*Ictinia mississippiensis*)

Plate 10. NORTHERN HARRIER (*Circus cyaneus*) Perching

Ages: Juvenile, 1-year-old, and adult. *Auricular (facial) disk is owl-like in shape and function.* Pale, thin rim surrounds disk on juvenile and female. Pale "spectacle" encircles eye on all ages and sexes. On juvenile and female, central 1 or 2 tail-feather sets are darker than rest of tail. Square-shaped patch formed by white uppertail coverts, which may be thinly streaked, often does not show on perched bird. Plumages, which are often variable for each sex and age, fade by spring. *Juvenile.*—Plumage is held for 1st year and is similar for both sexes, but eye color is sexually dimorphic and always paler on males. Dorsal surface of secondaries and greater wing coverts is solid brown. *1-year-old.*—Plumage and eye color are sexually dimorphic and similar to respective adult sex. Plumage features are usually held for 1 year, but some individuals may retain a few juvenile feathers, and eye color changes slowly to more adult-like colors. 1-year-old males typically have rufous-brown auriculars and are usually more heavily marked than adult males. 1-year-old females have barred, grayish dorsal surface of secondaries and gray bar on upperwing greater coverts. *Adult.*—Plumages are sexually dimorphic. Females may take 3 years to attain pure yellow eye. In adult female, dorsal markings on secondaries and upperwing greater coverts are as on 1-year-old females. **Subspecies:** *C. c. hudsonius* is subspecies in N. America; *C. c. cyaneus* breeds in Europe and Asia. **Color morphs:** None. There are a few records of melanistic adults. **Size:** Male L: 16–18" (41–46cm). Female L: 18–20" (46–51cm). Sexually dimorphic. **Habits:** Wary. Can be gregarious. Perches on ground or on very low open objects. **Food:** Aerial hunter. Prey consists of rodents, hares, rabbits, and small to midsize birds up to large ducks. Amphibians, reptiles, insects are also consumed; carrion is readily eaten, especially in winter. Male takes small prey, female small and large prey. **Voice:** Rapid, chattered *cheh-cheh-cheh* and squeeze-toy-like *squee-aah.*

10a. Juvenile male (fall): Eye is pale brown or pale gray. *Auricular disk and hindneck often streaked rufous.*

10b. Juvenile male (winter–spring): Eye is pale yellow. Plumage is faded, paler than on *10a.*

10c. Juvenile female (fall): Eye is dark brown. *Auricular disk and hindneck are often darker than on males.*

10d. Juvenile female (winter–spring): Eye is medium brown. Plumage is faded, paler than on *10c.*

10e. 1- to 2-year-old female: Eye is pale or medium brown. *Auricular disk and hindneck are streaked with tawny.*

10f. 1- to 2-year-old female: Eye is yellow and flecked with brown. *Auricular disk and hindneck are streaked.*

10g. Adult female: Eye is pale to medium yellow. *Note:* 2 or 3 years old.

10h. Adult male: Eye is orange-yellow. *Head and hindneck are pale gray, and auricular disk lacks pale rim.*

10i. 1-year-old male: Eye is pale yellow or pale orange-yellow. *Auricular disk and crown are rufous brown.* Nape and hindneck are grayish brown. *Note:* Some are adult-like and similar to *10h.*

10j. Juvenile flank feather: Dark brown streak along middle part of feather.

10k. Juvenile male (fall): Eye as on *10a. Underparts are solid orange-rufous, with streaking on neck and flanks.*

10l. Juvenile male, streaked type (fall): Some juveniles are thinly streaked on all of underparts.

10m. Juvenile male (spring): Plumage as on *10k* but fades to pale tawny by spring.

10n. 1-year-old female/adult female flank feather: Dark brown heart-shaped mark near tip and inner dark bar.

10o. 1-year-old female, moderately marked type (fall): Some 1-year-old females are quite rufous on underparts and have minimal pale markings on hindneck, scapulars, and wing coverts. Underparts moderately streaked; barred flanks.

10p. 1-year-old female, lightly marked type (fall): Streaked hindneck; lightly marked underparts with barred lower flanks. *Note:* Typical plumage of many adult females as well.

10q. Adult female, heavily marked type (spring): Broadly streaked underparts, with barring on lower flanks; mottled upperparts. Plumage fades by spring. *Note:* 1-year-old females can be similar, except eye is darker.

10r. 1-year-old male, heavily marked type (fall): *Rufous (shown) or white breast is streaked. Upperparts are dark brown. Secondaries and base of greater wing coverts are gray. Note:* Some are more adult-like, as *10s.*

10s. Adult male (fall): Upperparts are medium brown, with gray head and bases of secondaries and upperwing coverts.

b
d
c
a
f
h
e
g
i
m
l
k
j
s
r
p
q
o
n

Plate 11. NORTHERN HARRIER (*Circus cyaneus*) Flying

Ages: Juvenile, 1-year-old, and adult. *Auricular disk is owl-like in shape and function.* Juveniles and females have thin, pale rim surrounding disk. Pale "spectacle" encircles eye. *Wide black bar on trailing edge of secondaries.* On juveniles and older females, central 1 or 2 tail-feather sets are darker than rest of tail. Square-shaped white patch formed by uppertail coverts; may be thinly streaked. Plumages, which are often variable for each sex and age, fade by spring. *Juvenile.*—Plumage is held for 1st year and is similar for both sexes, but eye color is sexually dimorphic and paler on males. Dorsal surface of secondaries and greater wing coverts is solid brown. *On underwing, secondaries, median coverts, and axillaries are darker than rest of wing. 1-year-old.*—Plumage and eye color are sexually dimorphic and similar to respective adult sex. Plumage features are usually held for 1 year; eye color changes slowly to more adult-like hues.A few juvenile secondaries can be retained in early fall; usually replaced by late fall. Some underwing greater coverts can be held into spring. 1-year-old males are typically more heavily marked than adult males and often have rufous-brown auriculars. 1-year-old females have barred grayish dorsal surface of secondaries and a gray inner bar on upperwing greater coverts. *Adult.*—Plumages are sexually dimorphic. Female may take 3 years to attain pure yellow eye. Dorsal wing surface on adult female is as noted for 1-year-old females. **Subspecies:** *C. c. hudsonius* is subspecies in N. America; *C. c. cyaneus* breeds in Europe and Asia. **Color morphs:** None. There are a few records of melanistic adults. **Size:** Male L: 16–18" (41–46cm), W: 38–43" (97–109cm). Female L: 18–20" (46–51cm), W: 43–48" (109–122cm). Sexually dimorphic. **Habits:** Wary. This species can be gregarious. It perches on ground or on very low open objects. **Food:** Aerial hunter. Prey consists of rodents, hares, rabbits, and small to midsize birds up to large ducks. Amphibians, reptiles, insects are also consumed; carrion is readily eaten, especially in winter. Male takes small prey, female small and larger prey. **Flight:** Courses low over ground when hunting, with head pointing downward, in intermittent sequences of labored flapping and gliding, with wings held in dihedral. When prey is detected, it banks, flips around, and drops to the ground. Will fly at high altitudes, especially when migrating. Wings are held at sharp angle when gliding. **Voice:** Rapid, chattered *cheh-cheh-cheh* and squeeze-toy-like *squee-aah.*

11a. Juvenile (fall): *Orange-rufous underparts. Secondaries are dark gray; median coverts and axillaries are dark.*

11b. Juvenile (fall): White uppertail coverts. Primaries are paler than inner wing; secondaries are solid dark.

11c. Juvenile (spring): *Secondaries are dark gray; median coverts and axillaries are dark.* Underparts very faded.

11d. Juvenile (spring): Plumage as on *11b* but paler due to fading.

11e.1-year-old female, heavily marked type (fall): Underparts are rufous in fresh plumage. Underparts are broadly streaked. *Some have dark gray secondaries and dark median coverts and axillaries, as on juvenile.*

11f. 1-year-old female/adult female, moderately marked type (fall): Tawny underparts are moderately streaked. *Secondaries are medium gray; dark median coverts and axillaries.*

11g. 1-year-old female/adult female, lightly marked type (spring): Lightly streaked or spotted underparts. Secondaries are pale gray and only slightly darker than primaries; sometimes same color as primaries.

11h. 1-year-old female (early fall): Medium brown or gray flight feathers and base of greater coverts, with distinct barring. This bird retains 3 faded brown juvenile secondaries on mid-wing (s4, s8, s9). White uppertail coverts. *Note*: Juvenile secondaries usually replaced by late fall.

11i. 1-year-old female/adult female (spring): Flight feathers are pale gray and barred. White uppertail coverts may have dark streaks.

11j. 1-year-old male, heavily marked type: Heavily marked underparts and wing coverts. Often has rufous breast. *Note:* Some are adult-like, as on *11k* and *11l*, but with 1 or more juvenile underwing greater coverts showing.

11k. Adult male, moderately marked type: Underparts, including axillaries and coverts, are moderately marked.

11l. Adult male, lightly marked type: Forward part of white underparts and axillaries are lightly marked.

11m. 1-year-old male (fall): Back and scapulars are dark brown. Tail is banded. Flight feathers are often barred.

11n. Adult male (fall): Back and scapulars are medium brown. Tail is banded. Inner flight feathers are unmarked.

11o. Adult male (spring): Back and scapulars are pale brown. Paler types lack banding on middle tail feathers.

11p. Flight silhouettes: *Soaring (left):* high dihedral. *Gliding (right):* modified dihedral.

a
b
c
d
e
f
g
h
i
j
k
l
m
n
o
p

NORTHERN HARRIER (*Circus cyaneus*)

HABITAT: *Summer.*—Open marshy areas among freshwater ponds and lakes and brackish or saltwater inlets. Also found in semi-open and open bogs and open, moist or dry upland meadows. To a lesser extent it inhabits dry, long- or short-grass prairies, often those with sagebrush. Breeding habitat is often based on prey availability; marginal habitat often used if prey is abundant. *Winter.*—Same as summer, but also more readily found on dry prairies of various grass heights and fallow and harvested agricultural areas, especially those with grass hay and alfalfa. Adult males tend to utilize more open and less vegetated regions. Rural areas are readily used. **STATUS:** Fairly common and stable. **NESTING:** Begins mid-Mar.–mid-May, depending on latitude and elevation, and ends in Jul.–Aug. Harriers breed as monogamous pairs, often in colonial groups. In areas of high density and good habitat, males may engage in polygyny and mate with as many as 5 females. 1-year-olds, especially females, may breed, but most do not nest until 2 years old. Medium-size nests are built of grassy material or thin sticks and placed on ground surrounded by protective cover or in tall vegetation above waterline in wet areas. Clutch consists of 1–7 eggs, but up to 10 eggs can be found in nest shared by 2 females. Female feeds youngsters; male hunts. **MOVEMENTS:** *Spring.*—For adults, with males moving before females, spans late Feb.–Apr., with most movement in Mar.–Apr. Returning juveniles move later, mainly Apr.–May but into Jun. *Fall.*—Mid-Aug.–Nov. for most areas, but continues into Dec. in very southerly latitudes. All ages move together, but juveniles predominate in early part, older birds, especially males, in latter part. No peak periods of movement. **COMPARISON: Northern Goshawk (juvenile).**—Perching birds similar to adult female harrier, with facial disk, pale eye, and streaked underparts. Flying birds soar with wings on flat plane and have 6 fingers at wingtips, *vs.* wings held in dihedral and 5 fingers. Uppertail coverts are brownish, *vs.* white. **Rough-legged Hawk (light morph, flight).**—Similar in flight, sharing dihedral wing attitude and white uppertail coverts. Underwing has black patch on wrist, *vs.* lack of black patch. **Peregrine Falcon (flight).**—Similar, especially when gliding overhead. Short angular distance from body to wrist and snappy, stiff wingbeats, *vs.* long distance to wrist and slow, floppy wingbeats. Ventral flight feathers are uniformly dark or pale, *vs.* wide black bar on trailing edge.

Figure 1. Bear River Migratory Bird Refuge, Box Elder Co., Utah (Aug.)
Year-round wet habitat. Wasatch Mts. in distance.

Figure 2. Weld Co., Colo. (Mar.)
Year-round dry habitat.

NORTHERN HARRIER (*Circus cyaneus*)

Plate 12. SHARP-SHINNED HAWK (*Accipiter striatus*) Juveniles
Typical (*A. s. velox*), "Queen Charlotte" (*A. s. perobscurus*), "Sutton's" (*A. s. suttoni*)

Ages: Juvenile plumage is held for 1st year. Sexes are similar. Eyes are ocher yellow in fall and turn orange by midwinter. *Most have either short black eye line or eye encircled with black and short eye line.* Very thin legs and toes. Tail tip varies: notched (male-only), square or somewhat rounded (both sexes), or very rounded/wedge-shaped (female-only). Tips of tail feathers are slightly rounded or square-edged (all tail types). Dorsal tail has 3, sometimes 4 equal-width black bands; outer feather set pattern does not align with inner sets on ventral side. Underparts are variably marked with streaking or partial barring. **Subspecies:** "Typical" in most of West (and all of East); "Queen Charolotte" in coastal w. Canada and se. Alaska; "Sutton's" is casual visitor/nester in Ariz.; 4 more subspecies in Caribbean and Mexico. **Color morphs:** None. **Size:** Male L: 9–11" (23–28cm), W: 20–22" (51–56cm). Female L: 11–13" (28–33cm), W: 23–26" (58–66cm). Sexually dimorphic (all subspecies) and separable in the field (with practice). **Habits:** Tame. Except during migration, this hawk is very secretive. It perches on concealed and open branches; does not perch on wires or utility poles. **Food:** Aerial and perch hunter of songbirds. **Flight:** *Wings are angled forward of perpendicular angle to body when soaring.* Regular series of very quick wingbeats interspersed with short glides; soars. **Voice:** Soft, high-pitched *kee-kee-kee* or *kyew-kyew-kyew* when alarmed; soft *chirps* when curious. *Note:* Unless indicated, figures show typical subspecies, *A. s. velox*. "Queen Charlotte" are from Haida Gwaii (Queen Charlotte Islands), B.C. in RBCM; "Sutton's" are from San Luis Potosí, Mexico in AMNH.

12a. Female, lightly marked type (fall): *Ocher-yellow eye. Supraoribital ridge skin is yellow. Dark area encircles eye; short, dark eye line.* White supercilium is moderately wide; pale auriculars.

12b. Female, moderately marked type (fall): As on *12a* but darker, with thin white supercilium and thicker streaking on underparts.

12c. Female, heavily marked type (mid-winter): Orange eye. Head is darker than on *12b*, with only hint of supercilium.

12d. "Queen Charlotte" female: Dark brown head with little or no pale supercilium. In spring, eye color as on *12c*.

12e. Notched tail (ventral): *Notched tip* (males only).

12f. Squared tail (ventral): *Square-shaped tip* (both sexes, but mainly males).

12g. Rounded tail (ventral): *Rounded tip* (both sexes, but mainly females).

12h. "Queen Charlotte, female, "wedge-shaped (ventral): *Some females of this subspecies have streaks on undertail coverts. Wedge-shaped tip.*

12i. Tail (dorsal): Brown with 3 or 4 equal-width dark bands; thin white terminal band. *Uppertail covert tips rufous.*

12j. Female, lightly marked type (fall): Thin brown or rufous streaking on underparts. Pale tawny edges on dark brown feathers of upperparts.

12k. Female, moderately marked type (fall): Moderate-width rufous or brown streaking on breast and belly; barred flanks.

12l. Female, heavily marked type (spring): Broad rufous streaking on breast; barred flanks and belly. White supercilium is worn off. White-blotched bases of scapular feathers are exposed when fluffed.

12m. "Sutton's" female, moderately marked type (fall): Head and dorsal body as on typical birds, but flanks and leg feathers are solid rufous. Tail tip is rounded type.

12n. "Sutton's" male, lightly/moderately marked type (fall): Head and dorsal body as on typical, but flanks and leg feathers are solid rufous. Tail tip is notched type.

12o. "Queen Charlotte" male (all seasons): Dark brown head and upperparts. Thinly or broadly (shown) streaked, dark brown or rufous underparts; flanks are barred. Leg feathers nearly solid brown or rufous. *Note:* Darkest variant.

12p. "Queen Charlotte" female (all seasons): Dark brown head and upperparts. Very broadly streaked rufous underparts; flanks are broadly barred. Leg feathers nearly solid rufous. *Note:* Darkest variant.

12q. Female, lightly marked type: Thin brown or rufous streaking; unmarked lower belly. *Flight feathers have gray band on trailing edge.*

12r. Female, moderately marked type: Fairly wide rufous streaking. Black band on trailing edge of wing adorns some heavily marked birds (as on *12t*, bottom).

12s. Female, heavily marked type: Broad rufous streaking often becomes barred on belly. *Wings held forward of perpendicular angle to body.*

12t. Secondary feathers: *Top:* Rear edge has typical gray band. *Bottom:* Rear edge is black only on some heavily marked birds.

a
b
c
d
e
f
g
h
i
j
k
l
m
n
o
p
q
r
s
t

Plate 13. SHARP-SHINNED HAWK (*Accipiter striatus*) Adults
Typical (*A. s. velox*), "Queen Charlotte" (*A. s. perobscurus*), "Sutton's" (*A. s. suttoni*)

Ages: Adult dorsal plumage is sexually dimorphic (no overlap): males are grayish blue; females are gray or gray with bluish tinge. *There is either a short black eye line, or eye encircled with black and a short eye line. Legs and toes are very thin.* Tail tip varies: notched (male-only), square or somewhat rounded (both sexes), or very rounded/wedge-shaped (female-only). Tips of tail feathers are slightly rounded or square-edged (all tail types). Gray dorsal tail has 3, sometimes 4 equal-width black bands. *1-year-olds.*—Plumage is identical to respective sex. May retain some juvenile feathering on rump (rarely visible) and sometimes on upperwing coverts; eye is always orange. Plumage is attained by 2nd fall; all adult flight feathers are attained. *Adults.*—Eye color varies from orange to dark red, darkest on males. Full plumage attained by 3rd fall when 2 years old. **Subspecies:** "Typical" in most of West (and all of East); "Queen Charolotte" in coastal w. Canada and se. Alaska; "Sutton's" is casual visitor/nester in Ariz.; 4 more subspecies in Caribbean and Mexico. **Color morphs:** None. **Size:** Male L: 9–11" (23–28cm), W: 20–22" (51–56cm). Female L: 11–13" (28–33cm), W: 23–26" (58–66cm). Sexually dimorphic (all subspecies) and separable in the field (with practice). **Habits:** Tame. Very secretive, except during migration. Uses mainly concealed branches for perches but will use open branches; does not use utility poles or wires. **Food:** Aerial and perch hunter of songbirds. **Flight:** *When soaring, wings are held forward of perpendicular angle to body.* Regular series of very quick, snappy wingbeats interspersed with short glides; soars to high altitudes. **Voice:** Soft, high-pitched *kee-kee-kee* or *kyew-kyew-kyew*; also soft *chirps*. *Note:* Unless otherwise indicated, figures show typical subspecies, *A. s. velox*. "Queen Charlotte" from coastal se. B.C., s.Calif., in AMNH; "Sutton's" from Coahulia, San Luis Potosí, Sonora in AMNH, LSM.

13a. 1-year-old female/adult female (fall): Orange eye. Crown, nape, and upperparts are gray. Many females have thin, whitish supercilium and thin dark eye line. Auriculars are rufous.

13b. Female (fall): Pale red eye. As on *13a*, but lacks whitish supercilium.

13c. Female (spring): Medium red eye (darkest for a female). Crown, nape, and upperparts fade to pale gray by spring; auriculars are pale rufous. *Black area encircles eye of some females.* Supercilium is whitish, as on *13a*.

13d. Male (fall): Pale red eye. *Black ring encircles eye of most males. Crown, nape, and upperparts are gray-blue; auriculars are rufous.*

13e. Male (spring): Dark red eye (darkest for a male). *Black encircles eye. Grayish-blue crown, nape, and upperparts fade by spring, especially auriculars.* Some males have bluish on part of auriculars.

13f. Female (fall): Barred flight feathers. Wing covert markings are blackish. Tail tip is rounded.

13g. Tail (ventral): Undertail coverts are flared and retracted during some courtship flights.

13h. Female (spring): *Wings held forward of perpendicular line to body.* Flight feathers are barred. Auriculars and ventral barring are pale rufous. Tail tip is squared or rounded type.

13i. Underwing median covert: *Coverts are barred with gray, brown, or black.* (Rufous-barred on adult Cooper's Hawk; *15n*.)

13j. Female (fall): *Uniformly gray crown and upperparts.*

13k. Female tail, rounded type (ventral): Rounded tip. Feathers are square-edged.

13l. "Queen Charlotte" female tail, rounded type (ventral): Rounded tip. Feathers are square-edged. Undertail coverts have rufous or gray markings.

13m. Female tail, all subspecies: Gray with 3 or 4 equal-width black bands; thin white terminal band.

13n. Male (fall): *Grayish-blue crown, nape, and upperparts.* Underparts are uniformly rufous-barred.

13o. Female (fall): *Gray crown, nape, and upperparts.* Underparts are uniformly rufous-barred.

13p. Female, blue-backed type (fall): Dark bluish-gray crown, nape, and upperparts.

13q. 1-year-old female (fall): As on *13o*, but has a few faded brown juvenile wing coverts.

13r. "Sutton's" male (fall): Dorsal color as on *13n*, but breast, flanks, and leg feathers are solid rufous.

13s. "Sutton's" male (spring): Faded pale, grayish-blue upperparts (springtime typical male has identical upperparts). Breast fades to grayish pink, and flanks and leg feathers are rufous.

13t. "Sutton's" female (fall): Dorsal color as on *13o* (shown) or *13p*, but breast, flanks, and leg feathers are solid rufous.

13u. "Queen Charlotte" female (all seasons): Dorsal color as on *13p*. Underparts are broadly barred with rufous and can be solid rufous on breast; very rarely, underparts are all-rufous. Leg feathers are rufous with thin white tips; can be all-rufous. Undertail coverts can be as on *13l*.

a
b
c
d
e
f
g
h
i
j
k
l
m
n
o
p
q
r
s
t
u

SHARP-SHINNED HAWK (*Accipiter striatus*)
Typical (*A. s. velox*), "Queen Charlotte" (*A. s. perobscurus*), "Sutton's" (*A. s. suttoni*)

HABITAT: Typical. *Summer.*—Inhabits thick stands of coniferous or mixed woodlands with young and mid-aged trees. To a much lesser extent, found in dense stands of deciduous growth with thickly foliaged understory. Favors Black Spruce (*Picea mariana*) bogs in n. lowland areas. Bulk of breeding area is montane elevations up to tree line and in boreal forest. In se. part of w. U.S., inhabits upland and lowland pine woodlands and, to lesser extent, deciduous (Osage-orange) or coniferous shelterbelts and small coniferous woodlots and plantations; pine woodlands managed for Red-cockaded Woodpecker (*Picoides borealis*) are favored areas. A quiet stream, river, or pond is in most breeding habitat. *Winter.*—Inhabits areas identical to summer, including high-elevation woodlands near timberline. Also found in rural or suburban semi-open wooded locations. Many migrate through and a few winter in open areas of Great Plains with very limited vegetation. **"Queen Charlotte."** *Summer.*—Coniferous and mixed rain forest with stands of young and mid-aged trees from sea level to near tree line at montane elevations. Much of breeding area is on coastal islands and coastal mainland. *Winter.*—Found within most breeding habitat, but also rural and suburban locales in breeding range. Many females and juveniles opt for similar moist climate farther south on Pacific coast, including wooded and semi-open areas, often in rural and suburban locales. Southernmost females and juveniles inhabit dry climate of near-coastal s. Calif. in wooded and semi-open remote, rural, and suburban areas. **"Sutton's."** *Summer.*—Inhabits montane elevations in coniferous and mixed woodlands within dense stands of younger and mid-aged trees. (1st U.S. nesting record was in deciduous tree.) *Winter.*—There are no verified records in U.S. **STATUS: Typical.** Common but secretive and difficult to find during breeding season. **"Queen Charlotte."** Uncommon and secretive; encountered mainly in nonbreeding season. **"Sutton's."** Accidental in mountains of se. Ariz., with 1st documented nesting in 1993. There is probable nesting record for Davis Mts., Jeff Davis Co., Tex., in 2000 and 2001; may occur in mountains of sw. N.Mex. **NESTING: All subspecies.** Most breed when 2 years old. 1-year-old females in mainly juvenile plumage often mate with adult males; pairs of 1-year-olds also may nest. Nesting occurs May–Aug. Males arrive on breeding grounds prior to females and establish territories. Territories are often used in consecutive years, but a new nest is built each year. Small nests are built of thin sticks and placed next to trunk, mainly in densely foliaged conifers but sometimes in densely foliaged portions of deciduous trees, 16–70' (5–21m) high. Nests are often built on top of leafy squirrel or crow nests (Northwestern Crow [*Corvus caurinus*] nests for "Queen Charlotte") or mistletoe. Both sexes bring material, but female builds nest. No greenery is added. Female incubates the 2–6 eggs and tends to young; male hunts. **MOVEMENTS: Typical.** Sedentary to long-distance migrant, with females moving farther south than males. Movements are notable in fall when large concentrations occur along cen. Calif. coast, Great Basin ridges, and Lake Superior shoreline. Southernmost winter to w. Panama. *Spring.*—Mar.–early Jun. Adults move before returning juveniles (1-year-olds) and peak late Apr.–early May. 1-year-olds peak in mid-May but straggle northward into Jun. *Fall.*—Mid-Aug.–Nov.; a few move later. Juveniles move 1st, and females precede males; females peak in mid-Sep., males peak in mid–late Sep. In adults, females migrate before males, and younger birds before older; females peak in late Sep., males peak in early–mid-Oct. A few juvenile females migrate with adult male segment. **"Queen Charlotte."** Sedentary or moderate-distance migrant. Females also move farther than males, to s. Calif. Exact winter range is unknown (museum examples from s. Calif.); some may move east into or through mountains of Great Basin (photo of adult in fall from ne. Nev., in Liguori 2005). **"Sutton's."** Lacks data. Probably sedentary in its normal Mexican range but may move to lower elevations in winter. **COMPARISON:** ***Sharp-shinned Hawk female.*** **Cooper's Hawk (male).**—*All ages:* Voice separates the species well: nasal *kek-kek-kek*, *vs.* soft, high-pitched *kee-kee-kee*. Male Cooper's is at least 3" (7.6cm) longer than female Sharp-shinned. Head is often square-shaped, *vs.* small and rounded. Region around eyes is pale, *vs.* often having dark ring around eye or short dark eye line. Soars with wings perpendicular to body angle, *vs.* angled forward. Some female Sharp-shins have rounded to wedge-shaped tail similar to tail of Cooper's. *Juveniles:* Eye is pale yellow/pale gray in fall, *vs.* bright yellow; eye is bright yellow in winter/spring, *vs.* orange (small percentage have bright yellow eye in fall, orange in winter/spring). Black band on trailing edge of underwing *vs.* black band only on heavily marked Sharp-shinned, which are more heavily marked than any Cooper's. *Adults:* Black crown and pale gray/rufous nape, *vs.* medium-gray crown *and* nape. Males may have rufous auriculars *and* nape, *vs.* rufous *only* on auriculars. Upperparts of males are blue-gray, *vs.* gray of females. Pale supercilium, if present, is gray/rufous patch, *vs.* white, thin, linear pattern on females.

Sharp-shinned juvenile. **American Kestrel.**—2 black stripes on face and dark brown eye, *vs.* no discernible head marks and yellow/orange eye. Upperparts are rufous, *vs.* dark brown. Wingtips are pointed in flight, *vs.* rounded. **Merlin.**—Eye is dark brown, *vs.* yellow/orange. Wingtips are pointed, *vs.* rounded. Dark underwings have small pale spots, *vs.* pale underwing with strong black barring. Dark tail has thin pale bands, *vs.* pale brown with 3 or 4 equal-width dark bands.

Figure 1. Medicine Bow Mts., Carbon Co., Wyo. (Jun.) Year-round montane habitat.

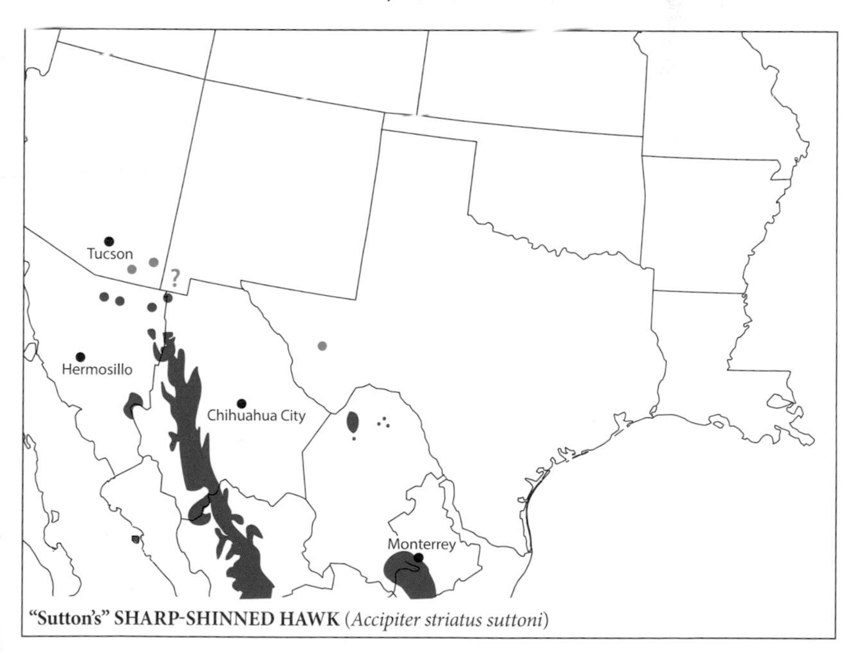

"Sutton's" SHARP-SHINNED HAWK (*Accipiter striatus suttoni*)

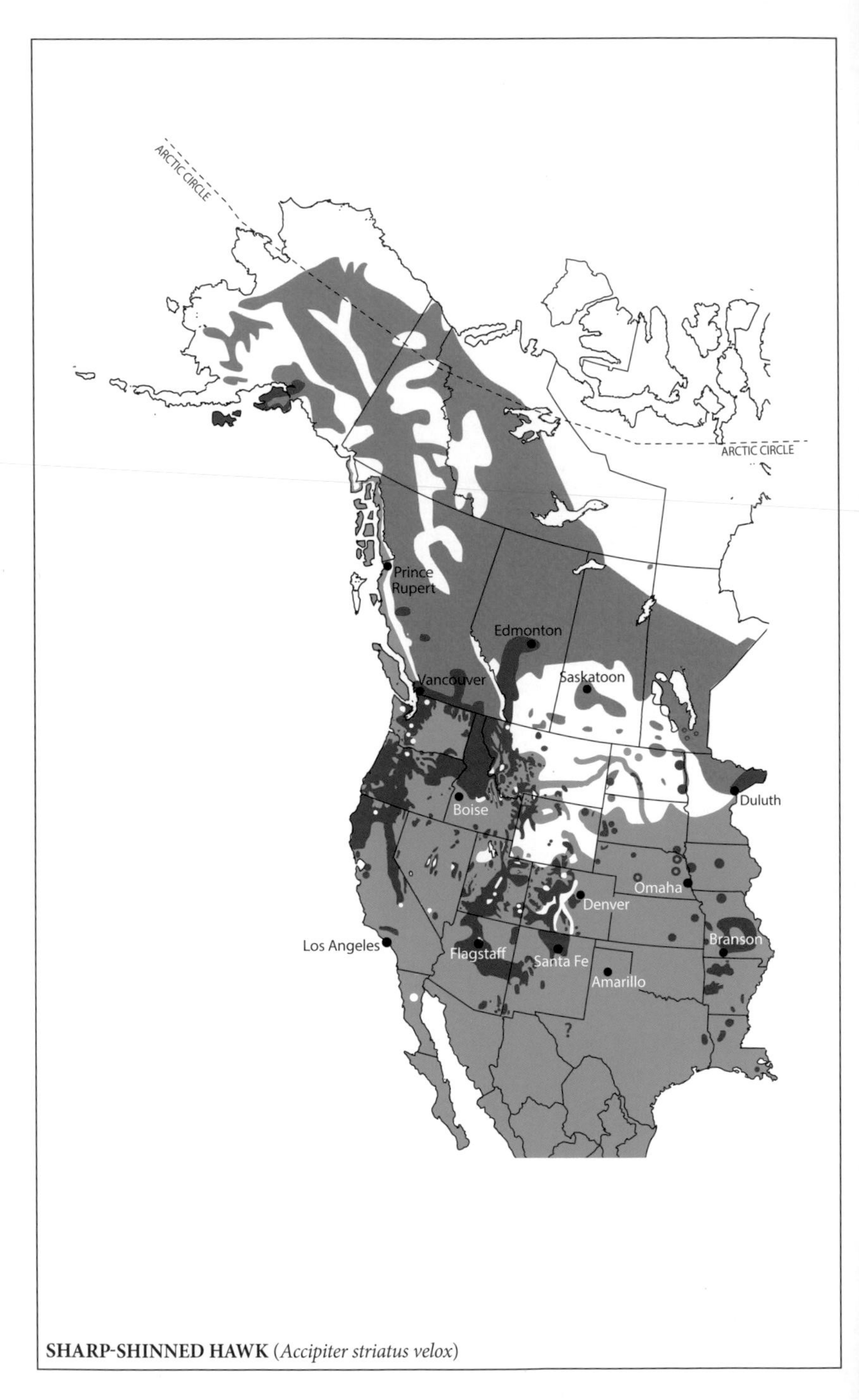

SHARP-SHINNED HAWK (*Accipiter striatus velox*)

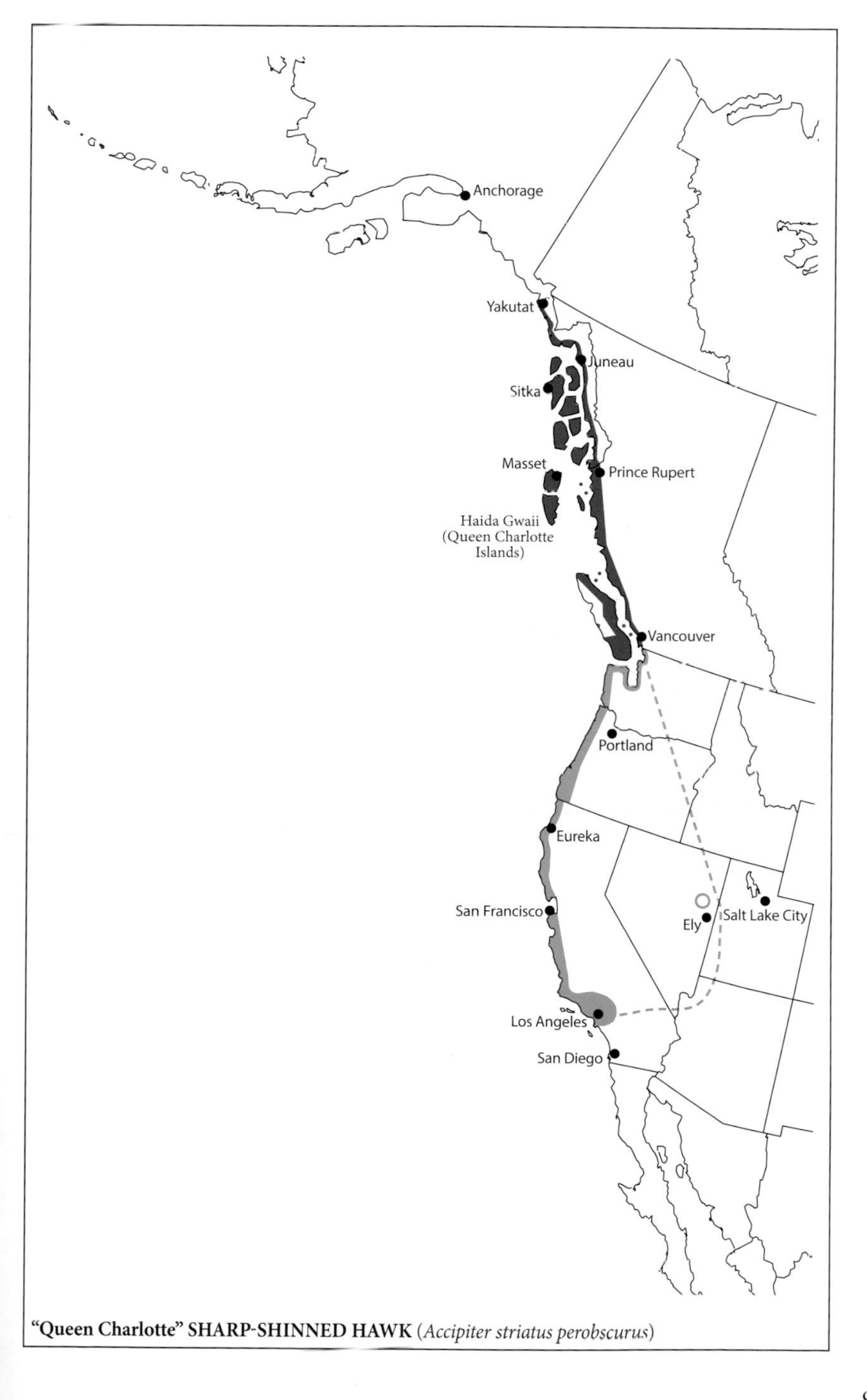

"Queen Charlotte" SHARP-SHINNED HAWK (*Accipiter striatus perobscurus*)

Plate 14. COOPER'S HAWK (*Accipiter cooperii*) — Juvenile

Ages: Juvenile plumage is held for 1st year. Sexes are similar. *Long nape feathers are raised or lowered at will (raised in cool temperatures or when alarmed); if raised, create square-headed appearance.* Eyes are pale gray or pale yellow in fall; in small percentage bright yellow. By midwinter–spring, eyes turn bright yellow on birds born in n. regions and orange on those born in s. regions, especially males. There is variation in amount and thickness of streaking on underparts. Some birds w. of Rocky Mts. can be heavily marked. *On ventral side, trailing edge of flight feathers has distinct black band* (separates from juvenile Sharp-shinned Hawks). *Tail is very long. Tip of each tail feather is rounded. Tail shape is very rounded, with outer feather sets progressively much shorter than inner sets.* **Subspecies:** Monotypic. **Color morphs:** None. **Size:** Male L: 14–16" (36–41cm), W: 28–30" (71–76cm). Female L: 16–19" (41–48cm), W: 31–34" (79–86cm). Sexually dimorphic and separable in the field; smaller w. of Rocky Mts. **Habits:** Moderately tame to tame. A secretive species, but is obvious in many areas. Perches on open poles, posts, and branches, as well as concealed branches. **Food:** Aerial and perch hunter of songbirds up to small upland game birds, rodents, and lizards. **Flight:** Fairly quick wingbeats interspersed with short glides; soars. **Voice:** Nasal *kek-kek-kek* when alarmed; also, soft nasal *kyew*.

14a. Lightly marked type (fall): *Eye is pale gray or pale yellow; sometimes bright yellow.* Head is pale, with short, white supercilium patch. Supraorbital ridge skin is gray. Head feathers are lowered.

14b. Moderately/heavily marked type (fall): Eye is pale gray or pale yellow; sometimes bright yellow. Head is tawny; may have short, pale supercilium. *Partially raised hackles create square-headed look.*

14c. Moderately/heavily marked type (midwinter–spring): *Eye is bright yellow or orange-yellow,* but can be orange at this season on southern-born birds, especially males. Hackles are raised, creating square-headed look.

14d. Tail, western variant (fall, ventral): *Each feather tip is rounded. Tail is very rounded, with stepping-stone pattern, each outer feather set much shorter than inner set. White terminal band is wide.* Undertail coverts have thin, brown streaks. *Note:* Fairly common variation west of Rocky Mts.

14e. Tail (fall, ventral): As on *14d*, but undertail coverts are unmarked (all regions).

14f. Tail (fall, dorsal): 3 or 4 equal-width black bands; broad white terminal band. Uppertail coverts have broad white tips.

14g. Tail (spring, dorsal): As on *14f* but faded. White terminal band wears thinner and often wears off on central feather set.

14h. Male, lightly marked type (fall): Pale head; hackles are lowered. Distinct tawny edges on dark brown feathers of upperparts. Underparts have thin brown streaks, including flanks and leg feathers. Tail is very long, with rounded tip.

14i. Male, moderately marked type (fall): Tawny head; hackles are lowered. Upperparts are tawny-edged. Underparts have moderately thin brown streaks, including flanks. Leg feathers may be partially barred. Breast often has tawny wash and may appear as tawny bib. Tail as on *14h*.

14j. Female, heavily marked type (spring): Dark head; hackles partially raised. Eye can be bright yellow or orange by spring. Faded, medium-brown upperparts with faded, pale feather edges. Broad streaking on breast and belly; partially barred flank and leg feathers. Tail is very long (tail tip as on *14g*).

14k. Female, all types (fall): Tawny-edged dark brown upperparts, including wing coverts. *Dark uppertail coverts have broad, pale tips. Very long tail has 3 or 4 neat, black bands and broad, white terminal band.*

14l. Male, lightly marked type: Pale head. White underparts are thinly streaked on breast and belly; lower belly often unmarked. *Trailing edge of wing has black band.* Tail is very long, with rounded tip and wide, white terminal band.

14m. Tail (ventral): Undertail coverts are flared and retracted during some courtship flights.

14n. Male, moderately marked type: *Tawny head.* Breast and belly have moderately thin streaking; often less on lower belly. *Often has tawny wash on breast and/or belly.*

14o. Female, heavily marked type: Tawny head. Underparts are thickly streaked, including lower belly. *Wings held at perpendicular angle to body when soaring. Trailing edge of wing has wide black band. Very long tail is very rounded when fanned, with wide, white terminal band.*

14p. Secondary flight feather (ventral): *Black band on rear edge* (separates from most Sharp-shinned Hawks; see *12t*).

a
b
c
d
e
f
g
h
i
j
k
l
m
n
o
p

Plate 15. COOPER'S HAWK (*Accipiter cooperii*) 1-Year-Old and Adult

Ages: 1-year-old and adult. Sexes have dimorphic plumages. *Males are gray-blue and females are gray or gray-brown; can have bluish tinge. Western type*: Many birds w. of Great Plains are more richly colored, with more dense rufous barring on ventral areas, including underwing coverts. *Often raises hackles on nape when alarmed or cold, creating square-headed look. Supraorbital ridge skin is gray. Crown (cap) of head is always darker than pale gray or rufous nape and upperparts, and nape is paler than upperparts. Underwing coverts markings are rufous.* Dorsal tail has 3, sometimes 4 equal-width black bands. *Tip of each tail feather is rounded. Tail shape is very rounded, with outer feather sets progressively shorter than inner sets. 1-year-old.*—Plumage similar to respective sex of adult, including head patterns as noted below for adults. On average, auriculars and nape tend to be more rufous than on adults; uncommon to be all-gray. A few juvenile feathers *may* be retained on upperwing coverts and rump; p10, some secondaries in early fall.. Eye varies orange-yellow (as *14c*) to orange; sometimes yellow (not shown). Plumage attained in 2nd fall (all adult flight feathers are also attained). *Adult.*—Plumage attained in fall of 3rd year. *Males have gray auriculars and nape, but can have rufous on both areas. Females can have similar all-gray auriculars and nape, but often have partial rufous or all rufous on these regions.* Eye orange-yellow, even on older birds, to medium red; often darker on males. **Subspecies:** Monotypic. **Color morphs:** None. **Size:** Male L: 14–16" (36–41cm), W: 28–30" (71–76cm). Female L: 16–19" (41–48cm), W: 31–34" (79–86cm). Sexually dimorphic and separable in the field. **Habits:** Moderately tame to tame. A secretive species, but obvious in some areas, especially during migration. Perches on poles, posts, and branches; also concealed branches. **Food:** Aerial and perch hunter of songbirds up to small upland game birds, rodents, and lizards. **Flight:** Regular series of fairly quick wingbeats interspersed with short glides; soars. **Voice:** Nasal *kek-kek-kek* when alarmed; also, soft, nasal *kyew*.

15a. Male, full-capped type: *All-black crown. Auriculars are rufous, and nape is pale gray.* Eye is orange. Head feathers are lowered.

15b. 1-year-old male, partial-capped type: *Rufous on forehead and over eye reduces size of black cap. Auriculars and nape are rufous.* Head feathers are lowered. *Note: All-rufous head pattern* is very uncommon on adult.

15c. Female, partial-capped type: *Rufous forehead and supercilium line create black eye line. Auriculars and nape are rufous.* Eye is orange-yellow to medium red. Forehead feathers are raised.

15d. Adult male: *Full black cap. Auriculars and nape are pale gray (sometimes tinged with pale rufous).* Eye is medium or dark red. *Nape feathers are raised, making head square-shaped.*

15e. Female, partial-capped type: *Forehead and supercilium are pale gray, creating a thin black eye line and narrow black cap. Auriculars and nape pale gray.* Eye is orange-yellow to medium red.

15f. Female, black-capped type: *Auriculars and sides of nape are pale rufous-gray; rear nape is pale gray. Crown is black.* Eye is orange-yellow to medium red. Nape feathers are raised.

15g. Female, brown type (spring): *Cap is dark brown and upperparts are medium brown. Auriculars and sides of nape are pale rufous; rear nape is pale gray.* Eye is orange-yellow to medium red. Nape feathers are partially raised.

15h. Tail (ventral): *Each feather tip is rounded. Tail tip is rounded. White terminal band is wide.*

15i. Tail (dorsal): Bluish gray on males and gray on females. On both, 3 equal-width black bands and broad white tip.

15j. Male: *Black cap; rufous auriculars. Underwing coverts are rufous-barred.* Flight feathers fully barred. *Rounded tip on tail.*

15k. Tail (ventral): Undertail coverts are flared and retracted during some courtship flights.

15l. 1-year-old male: *Black cap darker than grayish-blue upperparts.* Upperwing coverts and rump *may* have brown juvenile feathering. Juvenile p10, some secondaries may show in early fall.

15m. Adult male: *Black cap, gray auriculars. Wings held perpendicular to body. Black barring on outer primaries; inner primaries, secondaries are lightly barred or lack barring. Tail is rounded, with broad, white terminal band.*

15n. Median underwing covert: *Rufous-barred. Note:* Adult Sharp-shinned has brown or black barring (*9y*).

15o. 1-year-old male (Western type): *Black cap and rufous auriculars and nape. Underparts,* especially breast and legs, are more densely marked than *15p. Note:* Rufous head pattern very uncommon on adult.

15p. Adult male: Head is as on *15d* with black cap and pale gray auriculars and nape. *Upperparts are gray-blue.* Underparts are finely barred with rufous. *Note:* All-gray head pattern is uncommon on 1-year-old.

15q. Female (late summer–early fall): *Head as 15e,* but with yellow-orange eye. *Upperparts are medium gray.* Underparts are coarsely barred. Tail molt, with partially grown feathers with white tips on inner part (r2, r5 sets; fully grown by late fall).

15r. Female, brown type (spring): Head as on *15g.* Upperparts are medium or pale brown. Underparts are coarsely barred with rufous.

a
b
c
d
e
f
g
h
i
l
m
n
j
k
o
p
q
r

COOPER'S HAWK (*Accipiter cooperii*)

HABITAT: *Summer.*—Lowland and upland wooded areas to moderate montane elevations with coniferous and mixed woodlands of mid-aged trees, usually near quiet water source and forest openings. In Southwest and on Great Plains, inhabits semi-open and open areas with riparian stretches especially of cottonwood. Also uses deciduous shelterbelt plantations. In se. areas of West, it favors coniferous plantations. Readily adapts to rural and suburban locales and nests in residential areas, campuses, cemeteries, business parks, and city parks that have taller trees with dense canopies. *Winter/migration.*—Inhabits areas similar to breeding habitat; also, semi-open and open lowland areas with scattered, low vegetation that have fence posts and utility poles for perches. **STATUS:** Uncommon to fairly common. Birds are secretive during breeding season. Population is stable and increasing. **NESTING:** Mar.–Jul.; timing depends on latitude and elevation. Typically breeds when 2 years old, but 1-year-old females often mate with older males; less commonly, 1-year-olds mate. Nests are placed in upper part of densely foliaged tree. Conifers are preferred in suburban areas to aid in secrecy. Male does most building of the moderately large stick nest, which is decorated with greenery. Once hatched, male delivers prey to female away from the nest. Adults can be aggressive toward humans intruding at nest sites. **MOVEMENTS:** Sedentary to moderate-distance migrant, moving as far south as n. Costa Rica. *Spring.*—Adults move earlier than returning juveniles. Males precede females. Adults move mid-Mar.–Apr., peak early–mid-Apr. Juveniles move latter part of Apr.–May, peak late Apr. *Fall.*—Juveniles, females earliest, move before adult ages, starting in mid-Aug., and peak in mid–late Sep. Adults, with females moving before males, peak in late Sep.–early Oct. **COMPARISON:** ***Cooper's Hawk male.*** **Sharp-shinned Hawk (female).**—*All ages:* About 3" (7.6cm) shorter than male Cooper's Hawk. High-pitched, soft *kee-kee-kee*, *vs.* loud, nasal *kek-kek-kek*. Sharp-shins soar with wings angled forward of perpendicular line, *vs.* perpendicular. Females with very rounded/wedge-shaped tails are similar but lack Cooper's wide white terminal band. Often have dark ring encircling eyes or short dark eye line *vs.* area around eyes pale. *Juveniles:* Eyes are medium to bright yellow in fall and orange by winter; some Cooper's Hawks similar, but eyes usually pale gray or pale yellow in fall, bright yellow in winter. Only heavily marked Sharp-shins have black band on trailing edge of underwing, and these birds are much more heavily streaked than any Cooper's. Rufous/tawny-tipped uppertail coverts, *vs.* tawny/white-tipped coverts. *Adults:* Crown *and* nape are medium gray, *vs.* dark crown and pale gray or rufous nape. Auriculars are rufous and nape is medium gray, *vs.* rufous auriculars *and* rufous side of nape or all of nape. Upperparts are gray (females), *vs.* gray-blue (of male Cooper's). ***Cooper's Hawk juvenile.*** **Northern Goshawk (juvenile).**—Eyes usually brighter yellow, except gray when recently fledged. Voice is harsh, repetitive *ack, ack*, *vs.* nasal *kek-kek*. Upperwing greater coverts have pale, tawny bar, *vs.* all-dark feathers. Undertail coverts are similar if have thin dark streak or unmarked. Thin white linear edges along black dorsal tail bands, *vs.* lack of white edges. **Red-shouldered Hawk (juvenile).**—Eyes are medium brown, *vs.* pale gray or yellow. Throat often has dark patch or is all-dark, *vs.* thinly streaked or unmarked. Primaries have crescent-shaped tawny/white window/panel, *vs.* lack of pale marks. Wide, pale gray bars on dorsal secondaries, *vs.* brown with dark bars. Undertail coverts can be barred, vs. streaked. **Broad-winged Hawk (juvenile).**—Eyes are medium brown, *vs.* pale gray or yellow. When perched, wingtips reach nearly to tip of tail, *vs.* much shorter. Ventral flight feathers have minimal barring, *vs.* full, thick barring.

Figure 1. Duncan, Vancouver Island, B.C. (Aug.) Year-round habitat in mid-aged conifer groves.

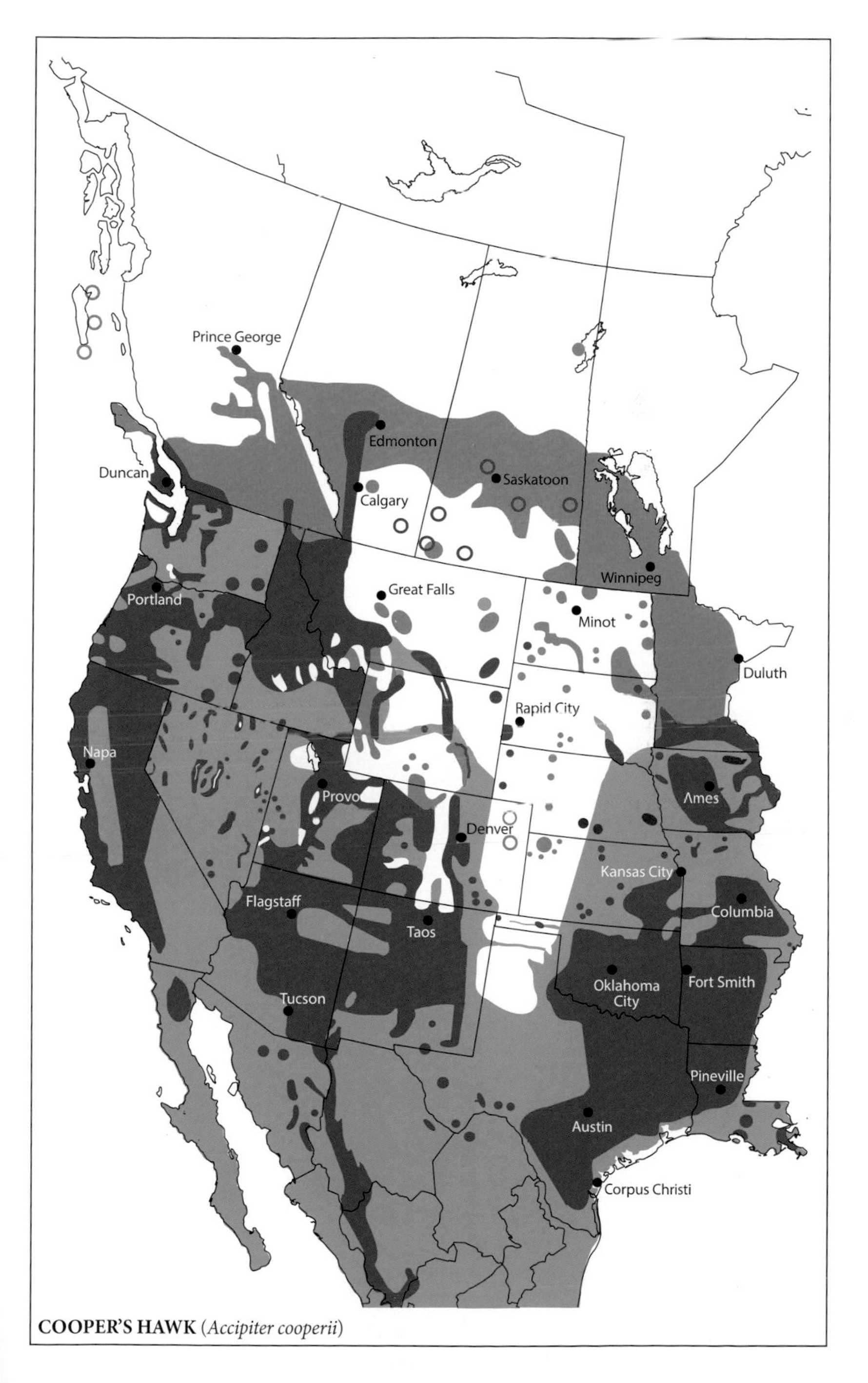

COOPER'S HAWK (*Accipiter cooperii*)

Plate 16. NORTHERN GOSHAWK (*Accipiter gentilis*) Juveniles
"Northern" (*A. g. atricapillus*) and "Queen Charlotte" (*A. g. laingi*)

Ages: Juvenile plumage is retained for 1st year; identical for both sexes. *Tail feathers have pointed tips. Tail is long with 4, sometimes 3 equal-width black bands that are bordered by thin white lines; tip is wedge-shaped.* Underside of flight feathers often has wavy dark bars. Wings are broader and tail is longer than on older ages. *"Northern"*—Typically has a distinct white supercilium over eyes. Malar mark is fairly distinct. *There is a pale rim around rear of auriculars. 1 or 2 pale bars adorn greater median upperwing coverts.* Undertail coverts are often fluffy; marked with *variable-sized heart-shapes* or streaks; regularly unmarked on either sex. *"Queen Charlotte"*—Often much darker upperparts; pale supercilium is reduced or nearly absent. Pale bar on upperwing covert is absent. *Undertail coverts have large dark marks.* **Subspecies:** "Northern" is in continental West (and all of East); "Queen Charlotte" is on Haida Gwaii (Queen Charlotte Islands) and Vancouver Island, B.C., and on islands and coastal mainland to se. Alaska. Eurasia has 6 subspecies. **Color morphs:** None. **Size:** Male L: 18–20" (46–51cm), W: 38–41" (97–104cm). Female L: 21–24" (53–61cm), W: 41–45" (104–114cm). Sexually dimorphic but difficult to tell in the field. "Queen Charlotte" is smaller. **Habits:** Fairly wary. Nesting 1-year-old females in mainly juvenile plumage are very defensive in nest territories. A secretive species, but becomes more obvious in migration and winter. Perches on large concealed branches; does *not* perch on utility poles or posts. **Food:** Aerial and perch hunter of Snowshoe Hare (*Lepus americanus*) and Ruffed Grouse (*Bonasa umbellus*); also squirrels, large passerines and woodpeckers, and waterfowl. "Queen Charlotte" feeds on Sooty Grouse (*Dendragapus fuliginosus*) and Red Squirrel (*Tamiasciurus hudsonicus*); also large passerines and woodpeckers, and species of Alcidae. **Flight:** Series of moderate-speed wingbeats interspersed with glides; soars. **Voice:** Silent, except around nest, when emits loud, harsh *cack-cack-cack. Note:* "Queen Charlotte" figures based on examples from Masset, Graham Island, Haida Gwaii (Queen Charlotte Islands), B.C, in AMNH and RBCM.

16a. Lightly marked type (recently fledged): Eye is pale gray but changes to color of *16d* by late summer–early fall. White supercilium is broad. *Pale rim borders rear of rufous auriculars.* Dark malar mark is obvious.

16b. Moderately marked type, brown-eyed variant (fall): Eye is medium brown. *Note:* Very uncommon variation.

16c. Heavily marked type (spring): Eye is orange on males and orange-yellow on most females. Plumage darker than on *16a* and *16b*; ill-defined supercilium. Underparts are broadly streaked.

16d. "Queen Charlotte" (fall): Dark head may show a faint pale supercilium and often lacks pale rim around auriculars. Eye is classic orange-yellow. Dark tawny underparts are moderately to heavily streaked. *Note:* Darkest type.

16e. "Northern" tail (dorsal): *4 irregular black bands edged with thin white lines. Pale-barred uppertail coverts. Note:* "Queen Charlotte" may have faint pale edges along inner dark bands as on *16i*.

16f. "Northern" tail, lightly marked coverts type (ventral): *Wedge-shaped tip, with each feather pointed.* Coverts lack dark marks. *Note:* Common variant on *either* sex of lightly marked birds.

16g. Tail, moderately/heavily marked coverts type (ventral): *Moderate-size to large, heart- or diamond-shaped marks on white coverts. Note:* On moderately and heavily marked types of "Northern" and all "Queen Charlotte."

16h. Lightly marked type: Thinly streaked underparts; leg feathers lightly streaked. Upperparts are pale-blotched. Undertail coverts as on *16f*. *Greater and median upperwing coverts have pale bars.*

16i. "Queen Charlotte": Dark brown head and upperparts; faint pale supercilium. Dark tawny underparts broadly streaked, including leg feathers. Dark brown wing coverts lack pale bar. *Faint, pale thin lines along black tail bands. Undertail coverts have large dark marks. Note:* Darkest type.

16j. Lightly marked type: Lightly streaked underparts with less streaking on lower belly, or lower belly is unmarked. *Undertail coverts have thin streaks (can be unmarked).* Trailing edge of wing has black band (or can be gray). Wedge-shaped tail tip. *Note:* As pale as juvenile Cooper's Hawk (*14l*).

16k. Tail, lightly marked coverts type (ventral): Coverts are flared during courtship.

16l. Moderately marked type: Moderately wide ventral streaking. Some have gray band on trailing edge of wing (black band on juvenile Cooper's Hawk; see *14n*). Undertail coverts as on *16g*.

16m. Moderately marked type: Pale bar on upperwing greater coverts. Tail is marked as on *16e*.

16n. Heavily marked type: Broad ventral streaking. *Wings are held slightly forward of perpendicular angle to body.* Rear edge of wing bows outward; usually has black band. *Undertail coverts are always marked on heavily marked birds.*

a
b
c
d
e
f
g
h
i
j
k
l
m
n

Plate 17. NORTHERN GOSHAWK (*Accipiter gentilis*) 1-Year-Olds and Adults, "Northern" (*A. g. atricapillus*) and "Queen Charlotte" (*A. g. laingi*) Perching

Ages: 1-year-old and adult. Sexes are somewhat dimorphic in both age classes. 1-year-olds and adults have narrower wings and shorter tails than juveniles. White, unmarked undertail coverts are often quite fluffy. *"Northern."—Head has a black cap, a long white supercilium extending from forehead onto nape, and a distinct black auricular patch. "Queen Charlotte."—Short, white supercilium extends from over or behind eye to side of nape.* Head and upperparts darker but underparts are identical to "Northern." *1-year-olds.*—Adult-like, but retain a few or many brown juvenile feathers on lesser upperwing coverts and rump, often on a few secondaries, and sometimes tail; primaries typically molt into adult feathering (Pyle 2005b; pers. obs.). Plumage attained by late fall of 2nd year. Males can be similar to either adult sex. Females are darker, more coarsely marked. *Adults.*—Plumage acquired in 3rd fall. Females are usually more grayish and more barred than males; some share dorsal color of males. **Subspecies:** "Northern" in continental West (and East); "Queen Charlotte" is on Haida Gwaii (Queen Charlotte Islands), B.C. and islands and coastal mainland from Vancouver Island to se. Alaska. Eurasia has 6 subspecies. **Color morphs:** None. **Size:** Sexually dimorphic but difficult to see in field. Male L: 18–20" (46–51cm). Female L: 21–24" (53–61cm). "Queen Charlotte" is smaller. **Habits:** Fairly wary. Birds are aggressive in nest territories. Secretive; are more obvious in migration and winter. Sits on large concealed branches but does *not* perch on utility poles. **Food:** Aerial and perch hunter of Snowshoe Hare and Ruffed Grouse; also tree and ground squirrels, large passerines and woodpeckers, and waterfowl. "Queen Charlotte" feeds on Sooty Grouse and Red Squirrel; also on passerines and woodpeckers, and species of Alcidae. **Voice:** Silent, except at nest sites, where it emits loud, harsh *cack-cack-cack. Note:* Figures are "Northern" unless noted; 1-year-olds from Duluth, Minn., in Oct. "Queen Charlotte" from Haida Gwaii and Vancouver Island, B.C.; also Kuiu and Kupreanof Islands, Alaska, in Jun. (in-hand birds; C. Flatten).

17a. 1-year-old male/adult female: Eye is orange to orange-red. White supercilium wraps around nape. *Gray underparts are coarsely barred and have black feather-shaft streaks. Note*: Adult females have darker eyes, whiter supercilium.

17b. 1-year-old female: Eye is yellow-orange to orange. White supercilium is heavily streaked and wraps around nape. *Gray underparts very coarsely barred; distinct black feather-shaft streaks.* Scattered juvenile feathers are visible.

17c. "Queen Charlotte" adult female: Eye is pale red. White supercilium is short, streaked. *Note:* Kupreanof Island.

17d. "Queen Charlotte" 1-year-old female/1-year-old male: Eye color as on *17b*. Whitish supercilium is very small. *Underparts very coarsely barred, with thick black feather-shaft streaks. Note:* Masset, Haida Gwaii (Feb.).

17e. Adult male: Eye is medium red or dark red. Long, white supercilium wraps around nape. Pale gray underparts are finely marked.

17f. "Queen Charlotte" adult female: Eye can be reddish brown. Whitish supercilium is very small. *Note:* Uncommon eye color on either subspecies. Comox, Vancouver Island (Jun.).

17g. 1-year-old male, bibbed type: *Heavily marked breast forms bib; coarsely barred underparts. Upperparts are medium gray. A few brown juvenile feathers retained on wing* (s8 visible on secondaries). Tail is fully banded.

17h. 1-year-old male (adult-male-like): *Finely marked underparts and pale bluish-gray upperparts are as 17l.* Only a few upperwing coverts have brown juvenile feathering. Tail partially banded; can be as on *17l*.

17i. 1-year-old female: *Underparts are very coarsely barred, with wide black streaking. Upperparts are dark gray. Dorsal surface has scattered brown juvenile feathers* (s7–9 are retained juvenile secondaries). Gray tail has wide black bands and 2 sets (r2, r5) of old juvenile feathers.

17j. "Queen Charlotte" 1-year-old male: *Coarsely barred underparts and dark gray upperparts, as on 17o.* Thin, white supercilium is over the eye but not on nape. *Note:* Comox, Vancouver Island (Nov.).

17k. "Queen Charlotte" 1-year-old female: Very short, thin, white supercilium over auriculars. *Underparts very coarsely barred. Upperparts are very dark gray. Note:* Haida Gwaii (Feb.).

17l. Adult male: *Pale gray, finely marked underparts.* Tail can be as on *17h*; rarely as on *17g*.

17m. Adult female: *Coarsely barred underparts. Upperparts are medium gray. Blue-gray upperparts. Tail is always banded.*

17n. "Queen Charlotte" adult male: *Short white supercilium. Dark blue-gray upperparts and tail. Tail can be as on 17l. Note:* Kuiu Island.

17o. "Queen Charlotte" adult female: *Head as on 17f. Upperparts are dark gray. Underparts as on 17j, 17m. Note:* Comox, Vancouver Island (Jun.).

a
b
c
d
e
f
g
h
i
j
k
l
m
n
o

Plate 18. NORTHERN GOSHAWK (*Accipiter gentilis*) 1-Year-Olds and Adults, "Northern" (*A. g. atricapillus*) and "Queen Charlotte" (*A. g. laingi*) Flying

Ages: 1-year-old and adult. Sexes are somewhat dimorphic in both age classes. 1-year-olds and adults have narrower wings and shorter tails than juveniles. White, unmarked undertail coverts are often quite fluffy. *"Northern."—Head has a black cap, a long white supercilium extending from forehead onto nape, and a distinct black auricular patch (visible even at great distances in flight). Flight feathers are darker than rest of upperparts. Gray-marked underparts on flying birds appear solid gray on all ages and sexes, except some 1-year-old females (not shown). "Queen Charlotte."—Short, white supercilium extends from over or behind eye to side of nape.* Head and upperparts darker but underparts are identical to "Northern." *1-year-olds.*—Adult-like, but retain a few or many brown juvenile feathers on lesser upperwing coverts and rump, often on a few secondaries, and sometimes tail; primaries typically molt into adult feathering (Pyle 2005b, pers. obs.). Plumage attained in late fall of 2nd year. Males can be similar to either adult sex. Females are darker, more coarsely marked. *Adults.*—Plumage acquired in 3rd fall. Females are usually more grayish and more barred than males; some share dorsal color of males. **Subspecies:** "Northern" in continental West (and East); "Queen Charlotte" is on Haida Gwaii (Queen Charlotte Islands), B.C. and islands and coastal mainland from Vancouver Island to se. Alaska. Eurasia has 6 subspecies. **Color morphs:** None. **Size:** Sexually dimorphic but difficult to see in field. Male L: 18–20" (46–51cm). Female L: 21–24" (53–61cm). "Queen Charlotte" is smaller. **Habits:** Fairly wary. Birds are aggressive in nest territories. Secretive; are more obvious in migration and winter. Sits on large concealed branches but does *not* perch on utility poles. **Food:** Aerial and perch hunter of Snowshoe Hare and Ruffed Grouse; also tree and ground squirrels, large passerines and woodpeckers, and waterfowl. "Queen Charlotte" feeds on Sooty Grouse and Red Squirrel; also on passerines and woodpeckers, and species of Alcidae. **Voice:** Silent, except at nest sites, where it emits loud, harsh *cack-cack-cack. Note:* Figures are "Northern" unless noted; 1-year-olds from Duluth, Minn., in Oct. "Queen Charlotte" from Vancouver Island, B.C. (Jun.).

18a. 1-year-old male, bibbed type: *Black auricular patch. Uniformly pale gray underwings have barred flight feathers and a few retained pale, longer juvenile secondaries (s4, s8). Unique to this age and sex, breast can be more heavily marked than rest of gray-barred underparts and form a bib.* Tail is banded adult female pattern (as on *18i*).

18b. 1-year-old male (adult-male-like): *Black auricular patch. Pale blue-gray upperparts are as on adult male (18d). Flight feathers are darker than coverts.* A few brown juvenile feathers adorn upperwing coverts; a large patch on rump. 5 old juvenile flight feathers (s3, s4, s7–9) are on secondaries. Tail is adult-male-like: gray with minimal banding; retains juvenile feathers in r2, r5 locations. *Note:* Slow molt stage.

18c. 1-year-old male (adult-female-like): *Black auricular patch.* Gray upperparts are similar to those of adult female (*18e*). *Flight feathers are darker than coverts; all are molted.* A few brown juvenile feathers on scapulars, lower back, upperwing coverts, and 1 feather on rump. Tail is banded as on most adult females. *Note:* Advanced molt stage.

18d. Adult male: *Black auricular patch. Pale blue-gray upperparts contrast with darker flight feathers. Tail has only black subterminal band* (may have partial bands on central [deck] feather set; sometimes on more feathers or, rarely, all feathers).

18e. Adult female: *Black auricular patch. Upperparts are medium gray and contrast with darker flight feathers. Tail is banded.*

18f. "Queen Charlotte" adult female: *Black head with small white supercilium, which makes black auricular patch less prominent. Upperparts are dark brownish gray, showing little or no contrast with dark flight feathers.* Tail has 3 dark bands. Distal scapulars, rump, and uppertail coverts are paler bluish.

18g. Adult male: *Black auricular patch. Underparts are very finely marked and appear solid gray at a distance. Uniformly pale gray underwings barred only on primaries.*

18h. Adult male tail (ventral): White undertail coverts are flared during courtship flights.

18i. Adult female: *Black auricular patch. Underparts are coarsely barred. Underwing is uniformly pale gray, with fully barred primaries and partially barred secondaries; secondaries sometimes unmarked as on male (18g). Tail is banded.*

18j. Flight silhouettes: *Gliding (top):* Wings bowed slightly downward or are flat. *Soaring (bottom):* Wings held flat.

a
b
c
d
e
f
g
h
i
j

NORTHERN GOSHAWK (*Accipiter gentilis*)
"Northern" (*A. g. atricapillus*) and "Queen Charlotte" (*A. g. laingi*)

HABITAT: "Northern." *Summer.*—Inhabits moderately large to large tracts of older 2nd-growth and old-growth coniferous, deciduous, or mixed woodlands. Forested montane elevations are inhabited in all areas, especially in U.S. Found in low-elevation boreal forest in Alaska and Canada. Woodlands may be interspersed with younger 2nd growth and small or large meadows; also small or moderate-size clear-cut regions, as long as large enough forested areas of several acres remain intact. On mountainsides, commonly nests in narrow "stringers" (clonal groups in thin strips) of tall, mature Quaking Aspens (*Populus tremuloides*) that are surrounded by open areas, especially sagebrush. Most areas have quiet water source, such as stream, pond, lake, or ground seepage. *Winter/migration.*—Habitat is similar to summer, especially for older males. All other ages/sexes may access more semi-open low-elevation regions without regard for tree size or density, including rural and suburban areas. **"Queen Charlotte."** *All seasons.*—Found within larger expanses of low-elevation rain forests of old-growth and mature 2nd-growth Sitka Spruce (*Picea sitchensis*), Western Hemlock (*Tsuga heterophylla*), and cedars. All areas are adjacent to forest openings, including clear-cut expanses. Younger birds may move into rural locations (and become more visible during winter). **STATUS: "Northern."** Very uncommon to uncommon and secretive. Birds become more noticeable in fall and winter, when some move out of secluded woodlands. Until early 2000s, boreal forest population of Canada and Alaska was affected by dramatic 8- to 11-year irruptive cycles of Snowshoe Hare (*Lepus americanus*) and, to lesser extent, Ruffed Grouse (*Bonasa umbellus*). These irruptions force most goshawks—including typically sedentary adult males—to move south to find ample food supply. Irruptions had occurred regularly since late 19th century, with last massive one in 1971–73; lesser (but still large) irruptions occurred in 1981–82 and 1993; a small irruption occurred in 2001–2 (with large percentage of adult birds, which is typical); no irruptions have occurred since 2002. (Snowshoe Hares are found at montane elevations in w. U.S. but these populations do not produce population fluctuations as boreal forest populations did.) Local parts of West are affected by regional status of Douglas's Squirrel (*Tamiasciurus douglasii*) and Belding's Ground Squirrel (*Urocitellus beldingi*), both of which undergo periodic population fluctuations. **"Queen Charlotte."** Canada listed this subspecies "Threatened" in 2002 under the Species at Risk Act. Alaska labeled it a "Species of Concern" in 1998. Massive habitat destruction due to clear-cut logging on especially island regions of Canada is main cause of reduced numbers. Logging techniques have since been altered in many areas to light-cut or patchwork systems to accommodate habitat requirements of goshawks. USFWS (2012) says habitat alteration will continue due to logging and will affect Canadian range more than Alaskan range. *Distribution.*—Populations on Vancouver Island and islands and mainland British Columbia are considered a "stable hybrid zone" because of mixing with "Northern" subspecies (USFWS 2012). This is also the case in se. Alaska, based on in-hand images (captured for telemetry) by C. Flatten: Breeding season adults and juveniles showed mainly "Queen Charlotte" traits of black head with much-reduced white supercilium on adults and pale mottled dorsal areas and pale dorsal wing bar on juveniles from Kuiu, Kupreanof, Mitkof, and Prince of Wales islands. One juvenile on Kupreanof Island exhibited a classic "Northern" plumage. The small number of both ages from Juneau and nearby Douglas Island possessed more, if not all, "Northern"-like traits. These data suggest that *all* of this subspecies range is a "hybrid zone," except the miniscule population on isolated Haida Gwaii (Queen Charlotte Islands), B.C. "Queen Charlotte" phenotype is still more prevalent, however, with overall darker birds and smaller size (smallest in southern range areas; USFWS 2012). *Haida Gwaii (Queen Charlotte Islands, B.C.)*—Currently, only a "handful" of pairs nest on Graham and Morseby islands, which historically had a large population of these darkly colored birds (Doyle 2016). Breeding success has been low for several years and their population is not sustainable at this rate (Doyle 2008, 2016). The massive amount of clear-cut logging, pressure from the burgeoning population of introduced Black-tailed Deer (*Odocoileus hemionus columbianus*), and 100-plus (protective) chicken farmers, this subspecies is battling for survival (Doyle 2016). Sooty Grouse (*Debdragapus fuliginosus*) has historically been the *primary* prey of Haida Gwaii goshawks—as it is in much of this subspecies range—and its population has also been reduced on these islands for many years due to food competition with large number of deer (Doyle 2016). Spanning 1882–1925, deer were introduced to the islands (Golumbia 2000). Red Squirrel (*Tamiasciurus hudsonicus*) is also a major prey item—along with the grouse—in this subspecies' range (except on Prince of Wales Island, Alaska, where both species are absent; USFWS 2012). The squirrel was not native to Haida Gwaii, but was introduced in 1950 and has proliferated (Sealy 2012). Apparently, it is not filling the goshawk diet-void from low grouse numbers. *Vancouver Island and Coastal Mainland British Columbia.*—Data based on USFWS (2012): An estimated 165 pairs are on Vancouver Island. Formulated on "modeling" techniques, 177–191 pairs inhabit the coastal mainland. *Southeast Alaska Islands and Coastal Mainland.*—This large region has an unknown number of pairs. Most are in the 17 million acre (4.5 million ha) Tongass N.F. **NESTING: Both subspecies.** Mar.–Sep. Males typically remain on territory all year. Large stick nest is placed within canopy of large live tree, and typically in major fork of deciduous tree or next to trunk of coniferous tree; "Queen Charlotte" mainly nest in hemlock and spruce; also sometimes dead trees. Nest trees are near clear-cuts, logging trails, or lightly used roads; also have fallen logs nearby for "plucking posts." Male hunts, female tends nest. 1-year-old females in mainly juvenile plumage often mate with older males. **MOVEMENTS: "Northern."** May be sedentary (especially older males), disperse short distances, or migrate short distances. Some 1-year-olds and many juveniles move each fall. *Spring.*—Seen mostly at ridgetop hawk watches in n. Rocky Mts., Feb.–May. Adults peak in mid-Mar., returning juveniles in mid–late Apr. *Fall.*—Largest numbers in N. America during any migration season are seen at Duluth, Minn. (Hawk Ridge Nature Reserve), Sep.–Dec. Juveniles move the entire period but peak early–mid-Oct., and older birds peak in mid-Nov. In previous irruption years, adults showed up in large numbers beginning in mid-Oct., with peak in late Oct. and another peak in mid-Nov. Adults far outnumbered juveniles during *all* irruption years. **"Queen Charlotte."** Mainly sedentary. Younger birds may move to nearby islands or along the mainland coast. Haida Gwaii (Queen Charlotte Islands), B.C. population remain on those islands. **COMPARISON:** ***Northern Goshawk juvenile.*** **Northern Harrier (adult female).**—Similar facial disk on auriculars, yellow eye, and streaked underparts. White uppertail coverts, wings held in dihedral, and dark axillaries, *vs.* marked uppertail coverts, flat-winged flight, and barred axillaries. Undertail coverts can have similar dark marks. **Cooper's Hawk (juvenile female).**—Voice is nasal *kek-kek-kek*, *vs.* harsh *cack-cack-cack*. Upperwing coverts are plain brown, *vs.* pale bar on greater coverts. Lacks pale edges along dorsal black tail bands. **Red-shouldered Hawk (juvenile).**—Similar pale supercilium and darker malar mark. In flight, tawny crescent panel/window on base of primaries, *vs.* lack of panel/window. Has 5 primary fingers, *vs.* 6 fingers. Voice is *kee-air*, *vs.* harsh *cack-cack-cack*. **Broad-winged Hawk (juvenile).**—Upperwing coverts are solid brown, *vs.* pale bar on greater coverts. When perched, wingtips are near tail tip, *vs.* much shorter than tail tip. Lacks thin white edges along black tail bands. Thinly barred ventral flight feathers, *vs.* strongly barred feathers. Wingtips have 4 fingers, *vs.* 6 fingers. ***Northern Goshawk, all ages.*** **Gyrfalcon (gray morph, all ages).**—Eye is dark brown and encircled by blue or yellow orbital skin, *vs.* yellow, orange, or red eye and white lores. Head is uniformly colored and lacks black crown and ear patch as on adult goshawk. In flight, underside of primaries is pale, *vs.* distinct black banding as on all goshawks. No finger separation noted on wingtips, *vs.* 6 defined fingers.

Figure 1. Larimer Co., Colo. (Jun.)
"Northern" year-round habitat (nesting territory) in Quaking Aspen (*Populus tremuloides*) and Douglas-fir (*Pseudotsuga menziesii*); nests are often placed in tall aspen near woodland edges.

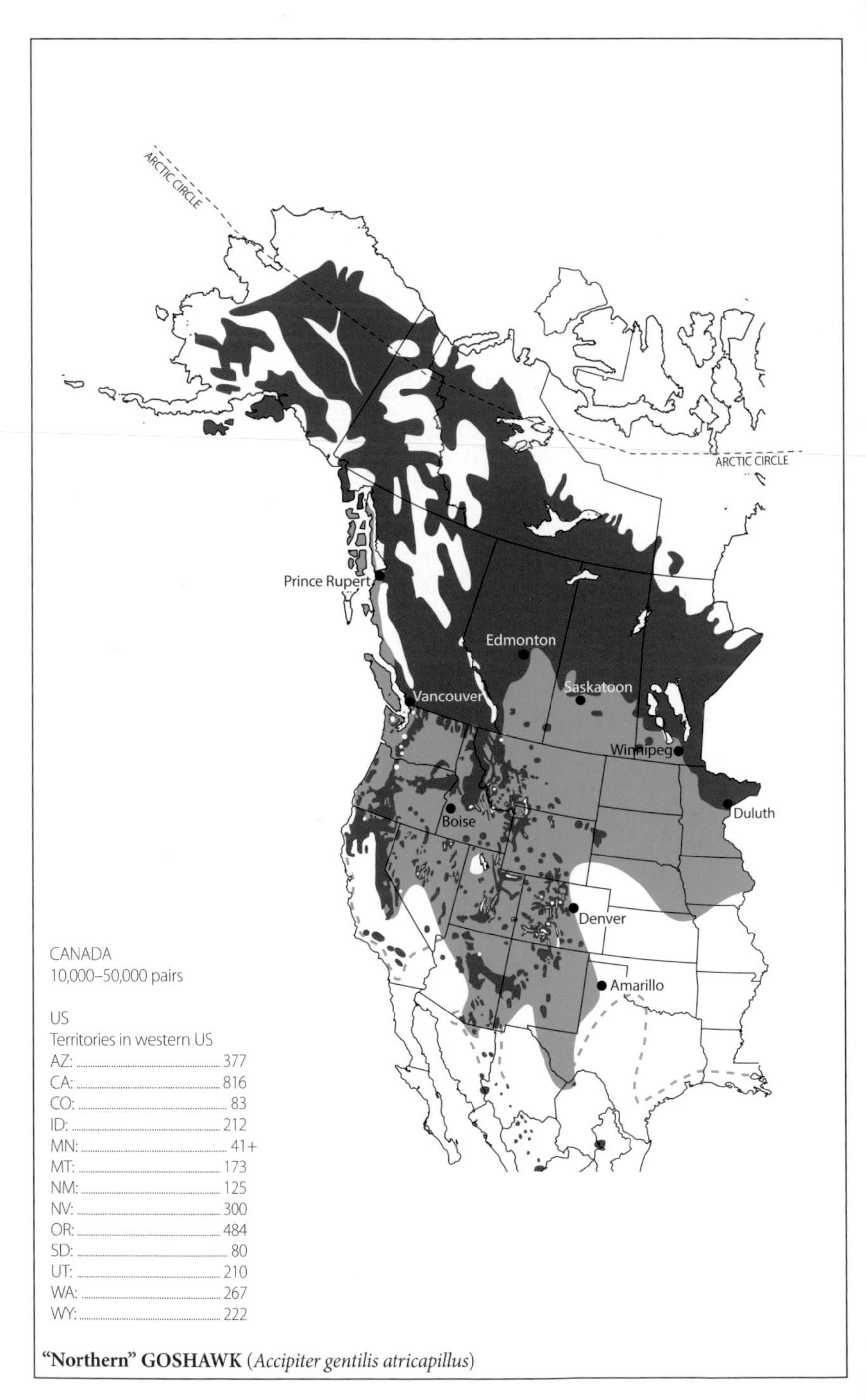

"Northern" GOSHAWK (*Accipiter gentilis atricapillus*)

Figure 2. Rocky Mt. Nat'l. Pk., Larimer Co., Colo. (Jun.)
"Northern" year-round habitat; nests near woodland openings and may forage in openings.

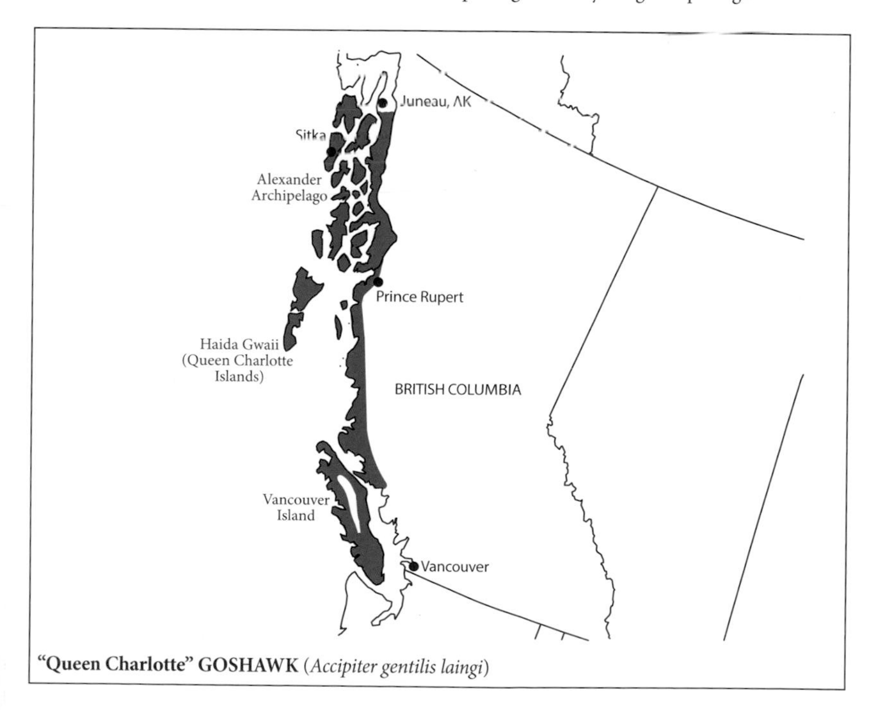

"Queen Charlotte" GOSHAWK (*Accipiter gentilis laingi*)

Plate 19. RED-SHOULDERED HAWK (*Buteo lineatus*) "Eastern" (*B. l. lineatus*) and "Southern" (*B. l. alleni*) — Juveniles

Ages: Juvenile plumage is held for 1st year. Eyes are pale or medium gray or brown. Fairly small, black bill has small to large, pale blue base under or in front of yellow cere. Head feathers are often raised in cool weather, making head appear large and bushy. Dark malar mark is broad. Supercilium is short, wide patch. Throat is white with dark center streak, or *all-dark with white edges; rarely all-white. Legs are long. 4 tail patterns; each has alternate variation that has rufous basal bands. Secondaries have white spots on outer edge forming white/pale-gray-checkered bars. In flight, in both subspecies, dorsal wing has crescent-shaped tawny panel on base of outer primaries; ventral side shows crescent-shaped tawny window in translucent light.* Juvenile's tail is longer than adult's. Minor differences in ventral markings in these 2 subspecies: "Eastern" has streaked or blob-type pattern of diamond or heart shapes; "Southern" is typically partially or fully barred. Both subspecies share variably barred leg feathers, but "Eastern" may have other marks or be unmarked. Undertail coverts can be barred. **Subspecies:** 1 other subspecies in West and 1 more in East. **Color morphs:** None. **Size:** L: 15–19" (38–48cm), W: 37–42" (94–107cm). Wing chord: "Eastern" male 309–346mm, female 315–353; "Southern" male 284–330, female 281–340. Females are larger. "Eastern" is largest subspecies. **Habits:** Wary to fairly tame; "Southern" birds are tamer. Both subspecies have adapted well to urbanization. Use concealed and exposed objects for perches. **Food:** Mainly a perch hunter. Eats small mammals, amphibians, reptiles, and invertebrates; also lesser numbers of small birds, fish, and insects. Also eats carrion. **Flight:** Soars with wings held on flat plane; glides with wings bowed downward. **Voice:** Highly vocal. Mainly emits loud, drawn-out *kee-yair* or short, squeaky *kee-aah.*

19a. Pale-headed type: Thin, dark stripe on mid-throat. Eye is pale gray. Cheek is pale. Underparts streaked.

19b. Dark-headed type: Dark throat with thin, white edge. Eye is tan. Cheek is dark. Underparts are streaked. Head feathers are raised, usually with longer spike on nape.

19c. "Eastern" breast feathers: *Left:* streaked; *middle:* diamond-shaped; *right:* heart-shaped.

19d. "Southern" breast feathers: *Left:* diamond-shaped; *middle:* streaked and diamond-shaped; *right:* heart-shaped. All have an inner crossbar that may be small and indistinct or large and distinct.

19e. Secondary flight feather: *White spots on outer (left) web form white bars when feathers are aligned.*

19f. Thin-banded tail, with rufous (dorsal): *5 or 6 narrow, pale gray bands with rufous on inner 2 or 3 bands.*

19g. Thin-banded tail (dorsal): *6 narrow, pale gray bands.*

19h. Wide-banded tail, with rufous (dorsal): 4 or 5 equal-width bands; *rufous on inner 1 or 2 bands. Note:* Uncommon type.

19i. Very thin-banded tail (dorsal): *5 or 6 very narrow, often V-shaped, pale gray bands.*

19j. "Eastern," lightly marked type: Underparts are thinly streaked; unmarked leg feathers. *Secondaries are barred.*

19k. "Eastern," moderately marked type: Underparts have diamond- or heart-shaped markings. *Secondaries are barred or checkered.* Leg feathers are spotted.

19l. "Eastern," heavily marked type: Underparts more heavily marked than on *19k*. Leg feathers are barred. Secondaries are dark, lacking pale barred/checkered pattern (uncommon trait on "Eastern," rare on "Southern").

19m. "Southern," lightly marked type: Similar to *19j* but partially barred on breast; thinly barred leg feathers.

19n. "Southern," heavily marked type: Underparts are always distinctly barred; heavily barred leg feathers.

19o. "Eastern," lightly marked type: Dark mid-throat streak. *Crescent-shaped window on base of outer primaries.* Forward part of underparts is lightly streaked. *Wings angle forward of perpendicular line when soaring.*

19p. "Eastern," moderately/heavily marked type: Dark throat. *Base of outer primaries has crescent-shaped window.* Underparts have heavy blob pattern. *Wings angle forward of perpendicular line.*

19q. "Eastern"/"Southern": *Crescent-shaped, tawny wing panel. Secondaries are checkered with white.*

19r. "Southern," heavily marked type: Dark throat. Heavily barred. Tawny windows do not show in direct light.

19s. Flight silhouettes: *Soaring (top):* Soars with wings on flat plane. *Gliding (bottom):* Glides with wings bowed downward.

a
b
c
d
e
f
g
h
i
j
k
l
m
n
o
p
q
r
s

Plate 20. RED-SHOULDERED HAWK (*Buteo lineatus*) 1-Year-Olds and Adults
"Eastern" (*B. l. lineatus*) and "Southern" (*B. l. alleni*)

Ages: 1-year-old and adult. Fairly small, black bill has small to large, pale blue base under or in front of yellow cere. Head feathers are often raised in cool weather, making head appear large and bushy. Throat is all-white, streaked, or *all-dark.* Males tend to have grayish heads and scapulars, females brownish heads and scapulars; however, many males are similar to females. Belly and flanks are finely barred with rufous; breast can be barred or solid rufous. Head, underparts fade paler by spring. *Rufous upperwing lesser coverts (shoulder). Secondaries and greater secondary coverts have large white spots on outer edge that form black-and-white-checkered pattern. In flight, dorsal side of wing has crescent-shaped, white-checkered panel on base of outer primaries; on ventral side it shows as white, crescent-shaped window in translucent light. Black tail has 3 thin, white bands on dorsal surface, 1 thin, white band on ventral surface. Legs are long.* "Eastern" may have *thin or wide dark streaks or blobs on each ventral feather shaft, especially on breast.* "Southern" lacks dark shaft marks. *1-year-olds.*—Adult-like plumage; ages are separable only at close range. Eyes are tan to medium brown, as on juvenile (based on known-age *B. l. elegans*, S. Moore, unpubl. data). May retain s4 secondary (W. S. Clark, unpubl. data). Rest of wing and all of body and tail molt into adult plumage. *Adults.*—Eyes dark brown. **Subspecies:** 1 other subspecies in West and 1 more in East. **Color morphs:** None. **Size:** L: 15–19" (38–48cm), W: 37–42" (94–107cm). Wing chord: "Eastern" male 309–346mm, female 315–353; "Southern" male 284–330, female 281–340. Females are larger. "Eastern" is largest subspecies. **Habits:** Wary to fairly tame; "Southern" birds are tamer. Both subspecies have adapted well to urbanization. Use concealed and exposed objects for perches. **Food:** Mainly a perch hunter. Eats mainly small mammals, amphibians, reptiles, and invertebrates; also lesser numbers of small birds, fish, and insects. Also eats carrion. **Flight:** Soars with wings held on flat plane; glides with wings bowed downward. **Voice:** Highly vocal. Mainly emits loud, drawn-out *kee-yair* or short, squeaky *kee-aah.*

20a. "Eastern" 1-year-old male (fall–winter): Dark grayish head. Throat is white. Eye is medium brown.

20b. "Eastern" male (spring): *Pale grayish head. White throat is streaked. Eye is dark brown.*

20c. "Eastern" female (fall–winter): *Dark brown head. Throat is black. Eye is dark brown.* Feather shafts form dark, thick streaks on breast. *Note:* Spike on top of nape.

20d. "Eastern" female/some males (spring): *Faded, pale brownish head. Throat is partially dark. Eye is dark brown.* Thin, dark feather-shaft streaks on breast.

20e. Secondary flight feather: *Outer web (left side) has distinct white bars on black feather.*

20f. Tail (dorsal): *3 thin, white inner bands; very rarely 2 thin white bands.*

20g. "Eastern" female (fall–winter): Solid rufous breast forms a bib; underparts have dark streaks. **20gg. Breast feather:** *Wide, black shaft streak.*

20h. "Eastern" male (fall–winter): *All-barred underparts have wide, black shaft streaks (as on 20gg).* Head and scapulars are grayish.

20i. "Eastern" female (spring): Rufous underparts are all-barred. Upperparts and especially rufous underparts fade considerably by spring. **20ii. Breast feather:** *Thin, black shaft streak.*

20j. "Eastern"/"Southern" (fall–winter): Underparts are all-barred with dark rufous in fresh plumage; lacks dark streaking.

20k. "Eastern"/"Southern" (spring): Faded rufous underparts lack dark streaks; solid rufous bib on breast. **20kk. Breast feather:** *Lacks dark shaft streak.*

20l. "Eastern" female, aberrant type: Broad, dark ventral streaking with minimal rufous crossbarring.

20m. "Eastern" (fall–winter): *White-checkered, crescent-shaped window on primaries. Dark shaft streaking is wide (as on 20gg).* Throat is dark; breast is a solid rufous bib; streaked underparts. *When soaring, wings angle forward of perpendicular line.*

20n. "Eastern" (fall–winter): *White-checkered, crescent-shaped window. Dark shaft streaking is thin (as on 20ii).* Throat is white; underparts are all-barred. *Wings angle forward of perpendicular line.*

20o. "Eastern"/"Southern (fall–winter): *White-checkered crescent on outer primaries; white checkering on secondaries. Upperwing coverts form rufous shoulder patch. 3 thin, white tail bands.*

20p. "Eastern"/"Southern" (spring): White throat. Faded rufous underparts lack streaks. Flight feathers are fully barred. *Thin, white tail bands. Note:* White window crescent does not show when lit by direct light.

a
c
d
f
e
b
j
i
g
h
l
kk
gg
ll
k
m
n
o
p

Plate 21. RED-SHOULDERED HAWK (*Buteo lineatus*) "California" (*B. l. elegans*)

Ages: Juvenile and adult. Juvenile plumage is held for 1st year. Adult plumage acquired as 1-year-old. Fairly small, black bill has small to large, pale blue base under or in front of yellow cere. Head feathers are raised in cool weather, making head appear large and bushy. *In flight, base of outer primaries has white crescent-shaped panel on dorsal side and white crescent-shaped window on ventral side when seen in translucent light. Secondaries and greater upperwing coverts have white spots that form white bars and checkered pattern. Legs are long. Upperwing coverts are rufous. Juvenile.*—Eyes are tan or medium brown. 2 main types of breast patterns; either pattern can be rufous or dark brown. Leg feathers are barred. *Dorsal side of tail has 3–5 thin, gray or white bands; 3-banded type is most common.* Juvenile's tail is longer than adult's. Juvenile can be quite adult-like, except breast is not solid. *Adult.*—Eyes are tan on younger adults, dark brown on older birds (based on known-age *B. l. elegans*, S. Moore, unpubl. data). Throat is all-white or streaked. Males have grayish heads and scapulars, and females have brownish heads and scapulars. Breast is usually solid rufous. By late spring–early summer, rich, rufous underparts fade and heads of many males bleach to whitish (S. Moore, unpubl. data). *Black tail has 2 or 3 fairly thin, white bands on dorsal surface.* **Subspecies:** 2 other subspecies in East and West, another in Fla. only. **Color morphs:** None. **Size:** L: 15–19" (38–48cm), W: 37–42" (94–107cm). Wing chord: male 288–305mm, female 298–312. Females average larger. This is 2nd smallest of 4 subspecies. **Habits:** Tame and has adapted well to urbanization. Uses concealed and exposed objects for perches. **Food:** Mainly a perch hunter. Feeds on small mammals, amphibians, reptiles, and invertebrates; also lesser numbers of small birds, fish, and insects. Eats carrion. **Flight:** Soars with wings held on flat plane; glides with wings bowed downward. *Wings angle forward of perpendicular line in soaring flight.* **Voice:** Highly vocal. Mainly emits loud, drawn-out *kee-yair* or short, squeaky *kee-aah.*

21a. Juvenile, streak-breasted type: Dark brown head. Throat is streaked. Breast is streaked.

21b. Juvenile, bar-breasted type: Dark brown head. Throat is dark. Breast is barred. Head feathers are raised.

21c. Juvenile breast feathers: *Left:* streaked type; *right:* barred type (heart-shaped and inner bars).

21d. Juvenile tail (dorsal): *3-banded type, with dusky, slightly wider outermost band.* Pattern is adult-like.

21e. Juvenile secondary feather: *White spots on outer (left) web form bars when feathers are aligned.*

21f. Juvenile, streak-breasted type (recently fledged): *Streaked breast; barred flanks and belly. Rufous wing coverts form patch on shoulder. Tail is 3-banded type. Note:* Markings are brown or rufous. Base color of underparts is often buffy when recently fledged, then fades to white.

21g. Juvenile, bar-breasted type: Breast and underparts are barred. *Wing coverts are rufous. Tail is 4-banded type.* Head feathers are raised. *Note:* Markings can be brown or rufous.

21h. Juvenile: *White, crescent-shaped primary panel. Secondaries are checkered. Tail is 5-banded type.*

21i. Juvenile, bar-breasted type: *White, crescent-shaped windows on primaries. Wing coverts are rufous.* Underparts are barred. *When soaring, wings angle forward of perpendicular line.*

21j. Young adult male (fall–winter): Pale brown eye. *Crown and auriculars are grayish.* Throat is white.

21k. Adult male (spring): Whitish crown and auriculars. Breast is faded rufous. Eye is dark brown.

21l. Adult female (fall–winter): Brown head, heavily streaked throat. Eye is dark brown. Head feathers are raised.

21m. Adult male (fall–winter): *Grayish head. Rufous coverts. Checkered wings. Tail is 2-banded type.*

21n. Adult male (spring): *Whitish head. Underside is faded rufous. Checkered wings. Tail is 3-banded type.*

21o. Adult female (fall–winter): *Similar to male (21m), except head and scapulars are more brownish.* Head feathers are raised.

21p. Adult tail (dorsal): *Usually 3, fairly thin, white bands; sometimes only 2 white bands (as on 21m).*

21q. Adult: *Rufous shoulder patch. Checkered wing; white-checkered wing panel. Tail is 3-banded type.*

21r. Adult: *White, checkered, crescent-shaped windows. Wings angle forward of perpendicular line.*

21s. Adult secondary feather: *White bars on outer (left) web align as checkered bars.*

a
b
c
d
e
f
g
h
i
l
j
k
o
p
s
r
m
n
q

RED-SHOULDERED HAWK (*Buteo lineatus*)
"Eastern" (*B. l. lineatus*), "Southern" (*B. l. alleni*), "California" (*B. l. elegans*)

HABITAT: "Eastern." *Summer.*—Inhabits lowland riparian areas with streams, rivers, ponds, swamps, and marshes adjacent to mature, mainly deciduous woodlands. Regularly occupies wooded suburban locales that are adjacent to riparian habitat. *Winter.*—Inhabits areas similar to breeding habitat but is also found in semi-open moist and dry habitat along wooded edges of meadows, woodlots, and shoulders of highways. It also accesses dry yards with suet feeders in rural and suburban areas. **"Southern."** *Summer.*—Breeds in lowland riparian areas with mature deciduous woodlands, but also mixed woodlands; a few upland areas may have primarily coniferous trees. Commonly found in moist rural and suburban locales with mature trees and ample tree density. *Winter.*—Utilizes similar areas as in nesting season but may be found in even more open areas along canals and irrigation ditches, as well as along wet or dry, tree- and pole-edged highway shoulders with minimal tree size and density. **"California."** *Year-round.*—Sea level up to 3,000' (900m) in Coast Ranges and Sierra Nevada and up to 4,500' (1370m) in montane areas of s. Calif. Uses a variety of habitats, from mainly dry areas in core range in Calif. to more humid, wet areas in w. Oreg. and w. Wash. Found in natural areas within riparian corridors with tall, mainly deciduous trees in flat, hilly, or canyon terrain; also at moderate montane elevations. This species has adapted to densely populated suburban areas in Calif. that have tall eucalyptus trees. Suburban areas may or may not be near riparian habitat. **STATUS:** A highly adaptable species, especially "California" subspecies. "Eastern" and "Southern" subspecies have been affected by wetland drainage and deforestation. **"Eastern."** Uncommon. It is local in e. Okla., e. Iowa, and Minn., due to limited riparian habitat and extensive agricultural areas adjacent to riparian stretches. **"Southern."** Common. This subspecies is widespread in much of its range. **"California."** Common in its core Calif. range, but uncommon and local northward. It is regular nester in Oreg. since early 1970s. Isolated nesting occurs in Nev. and Ariz. **NESTING:** Highly vocal during nesting period (and year-round). Builds well-made, compact, fairly large stick nests high up in live, tall, mature tree, in main crotch of deciduous or next to trunk of coniferous. Brood is 2 or 3. Male mainly hunts but performs some incubation duties. Mainly does not breed until 2 years old, but 1-year-old females may mate with adult males. **"Eastern."** Begins in Feb. in s. part of range and Mar.–Apr. in n. part; ends in Jun.–Jul. **"Southern."** Begins in Jan.–Feb.; ends in May–Jun. **"California."** Begins nesting in Nov. in s. Calif. and Jan.–Feb. in n. Calif. and northward. In its vast usage of suburban habitat, non-native eucalyptus trees are vital for nest sites. **MOVEMENTS: "Eastern."** Moderate-distance migrant in n. part of range and short- or very short-distance migrant or sedentary in s. part of range. *Spring.*—Adults move before returning juveniles. Adults move 1st, starting in early Feb. in s. areas and may peak in early–mid-Mar. Returning juveniles peak in mid-Apr. *Fall.*—Juveniles move prior to adults. Juveniles may begin moving in early Sep. and peak in mid-Oct. Adults may move in mid-Oct.–mid-Nov., with some still heading south into Dec. *Extralimital movements.*—1 record exists of apparent adult of this subspecies from cen. Calif. in Sep. There are several records for N.Dak., including 1 winter record, and many mainly spring records for s. Man. (along Red and Assiniboine Rivers). Multiple records exist for Jan.–Nov. in e. Colo. of "Eastern" and "Southern" subspecies, mainly of juveniles. **"Southern."** Sedentary or very short-distance migrant. Southward-bound juveniles show up at Tex. coastal hawk watches beginning in late Aug.; steady movement of both ages occurs in very low numbers through Nov. Migrating adults heading north have been seen in inland s. Tex. in early Feb. This subspecies has nested with Gray Hawk at Big Bend Nat'l. Pk., Tex. **"California."** Sedentary but engages in dispersal movements east and north of core Calif. range. It regularly occurs in fall and winter in Wash. *Extralimital movements.*—Moves east to e. Nev., w. Utah, and cen. Ariz.; rare in Idaho and w. Mont. **COMPARISON:** ***Red-shouldered Hawk juvenile.*** **Northern Goshawk (juvenile).**—Similar gray eye color when recently fledged, but eye becomes yellowish before fall. Streaked leg feathers, *vs.* unmarked or lightly barred on some "Southern" and all "California." Streaked or diamond-shaped undertail coverts on many, *vs.* unmarked or barred coverts on some "Southern" and all "California." Tawny bar on greater upperwing coverts and black barring on gray/brown secondaries, *vs.* no tawny bar, and checkered gray ("Eastern," "Southern") or white ("California") secondaries. **Broad-winged Hawk (juvenile).**—Heads can be identical in birds of both species that have unmarked throat or thin dark throat streak, but Red-shouldered can also have large dark throat patch or all-dark throat. Streaking on underparts is often sparser or absent on breast, *vs.* uniformly streaked. Similar to "Eastern" if leg feathers are unmarked, *vs.* thinly barred leg feathers on most "Southern" and all "California." Undertail coverts are *always* unmarked, *vs.* can be barred on some "Eastern," many "Southern," and all "California"; or unmarked on some "Eastern" and "Southern." **Red-tailed Hawk (juvenile).**—Similar eye color if tan or pale gray. Cere is greenish or

sometimes bright yellow, *vs.* cere always bright yellow. Numerous thin, dark tail bands, *vs.* dark tail with numerous thin, pale bands. Dorsal secondaries are barred with black, *vs.* checkered with gray or white. In flight, whitish rectangular window on ventral primaries or tan rectangular panel on dorsal primaries and primary greater coverts, *vs.* tawny, crescent-shaped wing window/panel on primaries on "Eastern" and "Southern" juveniles or white crescent-shaped window/panel on "California." ***Red-shouldered adult.***
Broad-winged Hawk (adult).—Can have similar rufous barred underparts, but they are coarsely barred, *vs.* finely barred. Upperparts are mainly solid brownish, including wings, *vs.* black-and-white-checkered wings. Eye is pale orange-brown, *vs.* dark brown. Voices readily separate them: high-pitched, piercing *pee-heeee*, *vs.* drawn-out *kee-yair*.

Figure 1. Sonoma Co., Calif. (May)
Eucalyptus grove, now a classic habitat for "California." Photo by Stan Moore

Figure 2. Nueces Co., Tex. (Sep.)
Year-round habitat for "Southern" in riparian growth along Nueces River.

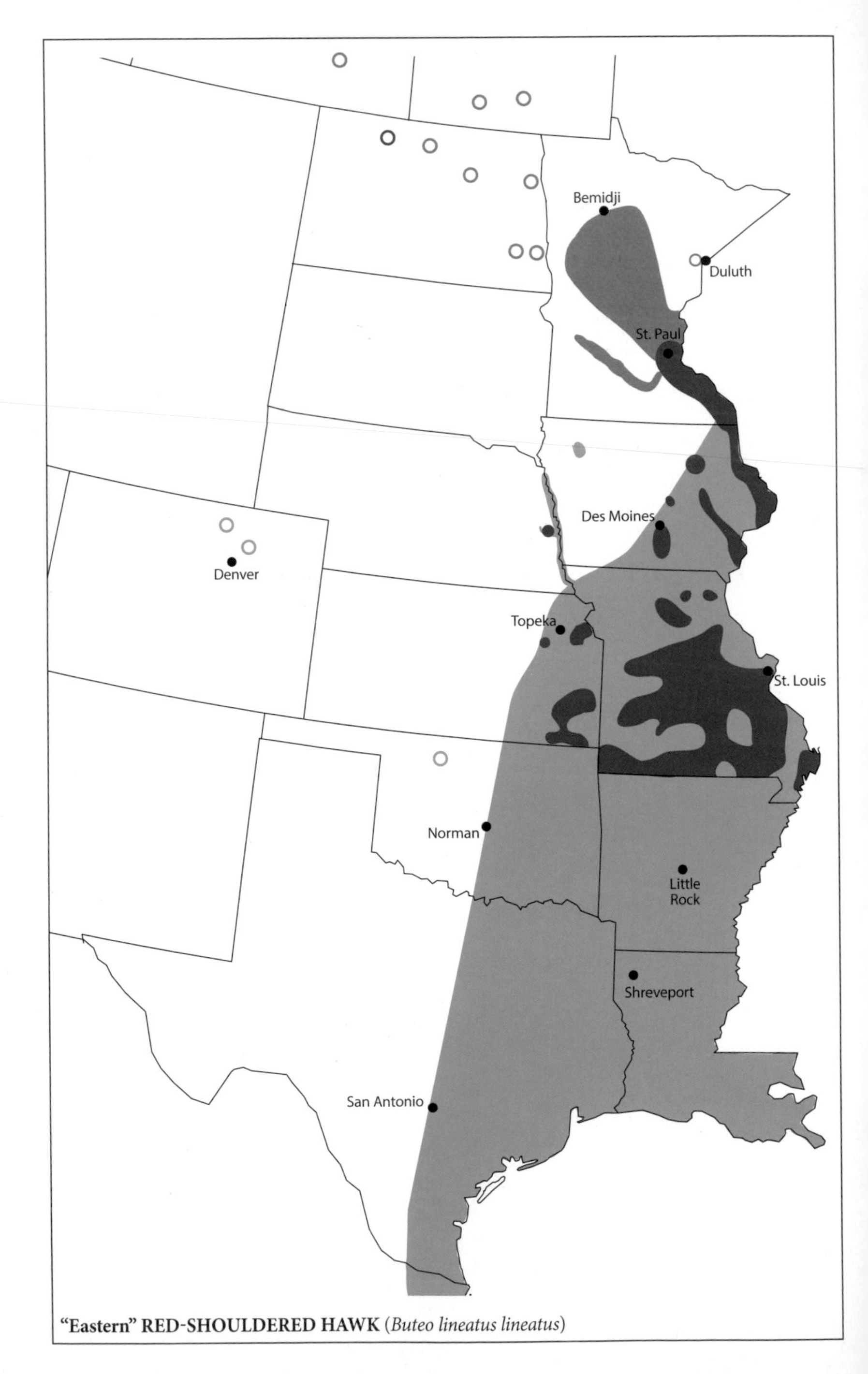

"Eastern" RED-SHOULDERED HAWK (*Buteo lineatus lineatus*)

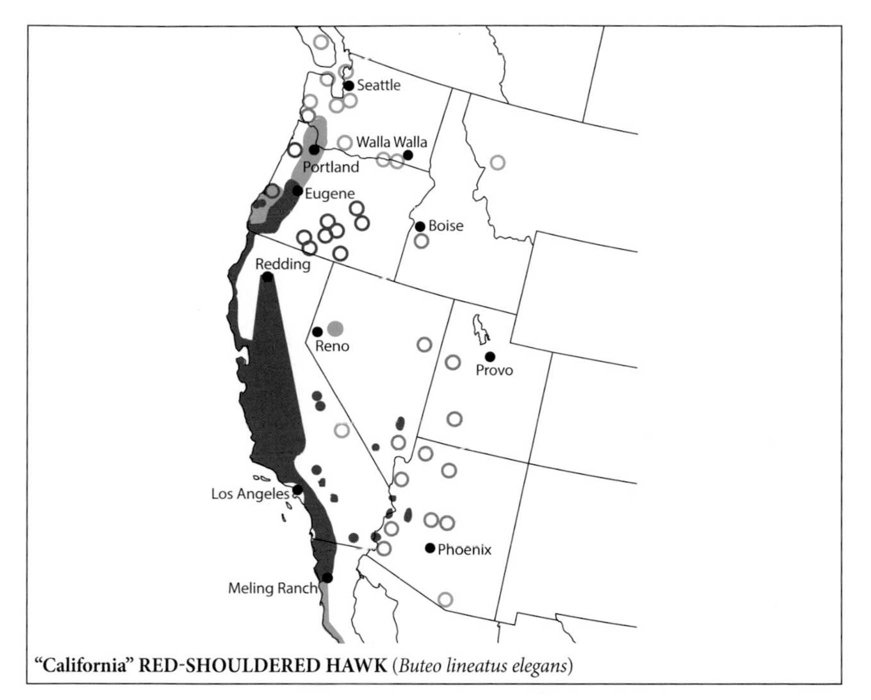

"California" RED-SHOULDERED HAWK (*Buteo lineatus elegans*)

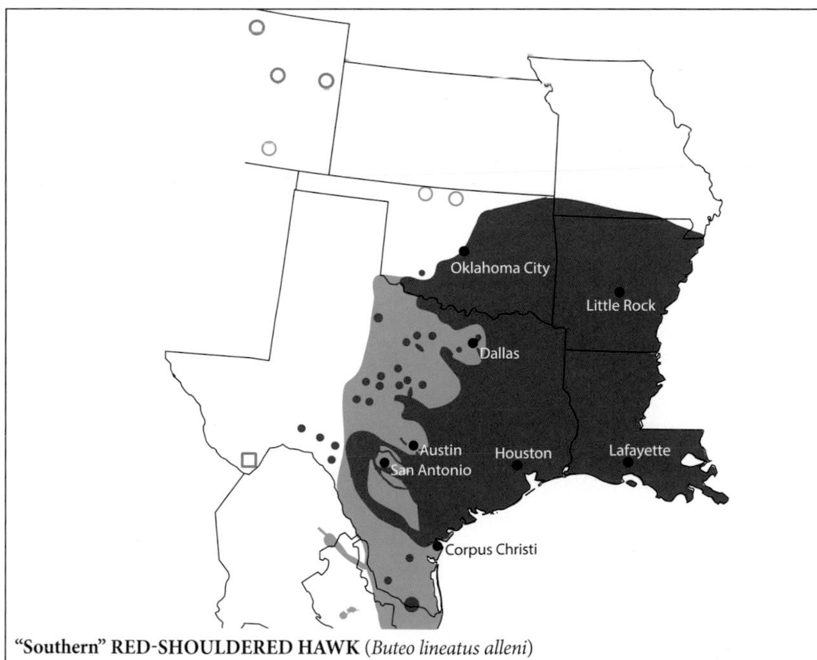

"Southern" RED-SHOULDERED HAWK (*Buteo lineatus alleni*)

Plate 22. BROAD-WINGED HAWK (*Buteo platypterus*) Juvenile

Ages: Juvenile plumage is held for 1st year. Bill has small, pale blue spot on lower base of upper mandible and base of lower mandible. Cere is yellow. Eyes are medium brown. *Dorsal surface of wing is uniformly dark, or primaries are paler and show as lighter panel in flight; secondaries are solid dark or faintly barred.* Pale rectangular panel often shows on underside of primaries in translucent light. Trailing edge of underwing has narrow gray band. Outer primaries can be thinly barred or solid dark. Wingtips are fairly pointed when soaring and very pointed when gliding. When perched, wingtips are moderately shorter than tail tip. There are 4 tail patterns, shared by all color types. *Light morph.*—Small white patch on forehead. Head has short, white or tawny supercilium, and tawny cheek. Dark malar mark connects to lower and front eye region. *Pale throat often has thin, dark center streak or is unmarked.* There are 3 main variations of markings on underparts, including on leg feathers. Tawny base color fades to white by spring. White tips on uppertail coverts wear off by spring. Undertail coverts are *never* marked (barred on many juvenile Red-shouldered Hawks). *Dark morph.*—Head is dark except all-gray lores. Juvenile has variations in ventral markings. **Subspecies:** *B. p. platypterus* in N. America; 5 others in Caribbean. **Color morphs:** Light and dark, with no cline between color morphs. **Size:** L: 13–17" (33–43cm), W: 32–36" (81–91cm). Females average larger. **Habits:** Tame. Perches on exposed and concealed objects, including utility poles and wires. **Food:** Perch hunter. Eats small amphibians, insects, reptiles, and rodents. Prey is caught on the ground or on outer branches in direct flight, rarely by diving. **Flight:** Soars with wings held on flat plane; glides with wings held flat or bowed slightly. Often migrates in large flocks (kettles). **Voice:** Vocal. Drawn-out, piercing, whistled *pee-heeeee.*

22a. Light morph (fall): May have a thin dark mid-throat streak. Small white patch on forehead, short whitish supercilium, tawny auriculars, and broad dark malar mark. Lacks a dark eye line.

22b. Light morph (spring): May have an all-white throat. Auriculars fade to white by spring. Lacks a dark eye line.

22c. Dark morph, streaked type: *Dark head, including forehead. Lores are gray.* Underparts are streaked.

22d. 6-banded tail (dorsal): *Wide black subterminal band, 5 thin inner bands; distal inner band is thinnest.*

22e. 5-banded tail (dorsal): *Wide black subterminal band, 4 thin inner bands; distal inner band is thinnest.*

22f. 4-banded tail (dorsal): *Wide black subterminal band, 3 thin inner bands; distal inner band is thinnest.*

22g. 3-banded tail (dorsal): *3 nearly equal-width, wide dark bands; middle band may be thinner.*

22h. Secondary flight feather: Dark brown and faintly barred or solid brown.

22i. Light morph, light marked type (fall): Dark mid-throat streak. Side of breast and flanks are lightly streaked. Belly and leg feathers are unmarked.

22j. Light morph, moderately marked type (fall): Unmarked throat. Side of breast and belly are streaked and flanks are barred. Central breast is unmarked. Leg feathers have spots.

22k. Light morph, heavily marked type (fall): Dark mid-throat streak. Breast and belly are heavily blotched or streaked; flanks are barred. Leg feathers are heavily barred.

22l. Light morph, heavily marked type (spring): As on *22k*, but all areas are faded.

22m. Dark morph, streaked type: Underparts are variably streaked rufous or tawny. Upperparts are dark brown.

22n. Dark morph, all-dark type: All of body is uniformly dark brown.

22o. Light morph, lightly marked type (fall): Belly and flanks are lightly streaked. Primaries have rectangular-shaped window.

22p. Light morph, moderately marked type (fall): Unmarked throat. Belly and flanks are moderately streaked. Primaries have rectangular-shaped window.

22q. 3-banded tail (ventral): Nonaligned bands on outer feather set; 2 wide black bands.

22r. Light morph, heavily marked type (May): Underparts are heavily streaked. 2 inner primaries are molting and dropped, creating a gap. Primaries have rectangular-shaped window. 2 mid-tail feathers are dropped due to molt.

22s. Light morph, all types (fall): *Pale rectangular panel on primaries.* White tips on dark uppertail coverts.

22t. Dark morph, streaked type: Rufous-streaked brown underparts. Flight feathers are uniform white in direct light.

22u. Dark morph, all-dark type: Uniformly dark brown underparts. Primaries have rectangular-shaped window.

a
b
c
d
e
f
g
n
i
j
k
l
m
h
q
s
r
p
o
t
u

Plate 23. BROAD-WINGED HAWK (*Buteo platypterus*) 1-Year-Old and Adult

Ages: 1-year-old and adult. Bill has small pale blue spot on lower base of upper mandible and base of lower mandible. Cere is yellow. *Eyes are pale or medium orange-brown to brown; also tan. Flight feathers are white on underside, with lightly barred secondaries* and wide black band along trailing edge. Wingtips are fairly pointed when soaring and very pointed when gliding. When perched, wingtips are moderately shorter than tail tip. *1-year-old.*—May retain a few faded, brown juvenile flight feathers, visible in flight during fall and winter. Up to 3 juvenile outer primaries (p8–10) and some secondaries (s3, s4, s8, s9) are regularly retained; also some feathers on upperwing coverts and rump. 1-year-olds share below-noted adult plumage and tail variations. *Adult.*—There are 2 dorsal and 3 ventral tail patterns; shared by both color morphs. Dorsal surface of secondaries is uniformly dark gray/ brown, with black band on trailing edge. Undertail coverts are unmarked or barred. *Light morph.*—Small white forehead. Malar mark is dark. *Pale throat has thin, dark center streak or is fully streaked.* Underparts are variably barred with rufous or dark brown; some patterns have a bibbed or streaked pattern on the breast. Leg feathers are always barred. Broad, white tips on uppertail coverts wear off by spring. *Dark morph.*—Head is dark, except all-gray lores. No variation in ventral markings in adults. **Subspecies:** *B. p. platypterus* in N. America; 5 others in Caribbean. **Color morphs:** Light and dark, with no cline in between. **Size:** L: 13–17" (33–43cm), W: 32–36" (81–91cm). Females average larger. **Habits:** Tame. Perches on exposed and concealed objects, including utility wires. **Food:** Perch hunter. Eats small amphibians, insects, reptiles, and rodents. Prey is caught on the ground or on outer branches in direct flight, rarely by diving. **Flight:** Soars with wings held on flat plane; glides with wings held flat or bowed slightly. Often migrates in large flocks (kettles). **Voice:** Quite vocal, especially during nesting season. Drawn-out, piercing, whistled *pee-heeee*.

23a. Light morph: *Pale to medium orange-brown eye. Head is rufous-brown with white forehead and dark malar mark; lacks pale supercilium.*

23b. Dark morph: *Eye is as on 23a.* Head, including forehead, is uniformly dark brown; lores are all-gray.

23c. 1-banded tail (dorsal): *1 broad pale gray band. Note:* Mainly on males.

23d. 2-banded tail (dorsal): *1 wide and 1 thin pale gray band. Note:* On either sex.

23e. 1-banded tail (ventral): *1 complete wide white band; 1 partial thin white band on outer set.*

23f. 2-banded tail (ventral): *1 complete wide white band; 2 partial thin white bands on outer set.*

23g. 2-banded tail, alternate (ventral): *1 complete wide white band; many narrow bands on outer 1 or 2 sets. Note:* Mainly on females.

23h. Secondary flight feather, either color morph: A wide black trailing edge band and 2 thin, dark bands on gray surface.

23i. Light morph, streak-breasted, lightly marked type: *Breast is streaked; rest of underparts lightly barred brown.*

23j. 1-year-old light morph, lightly marked, bibbed type: Rufous bib; rest of underparts lightly barred rufous. *Note:* Remnant faded brown juvenile feathers on wing; otherwise as adult.

23k. Light morph, heavily marked, bibbed type: Dark brown bib; rest of underparts heavily barred with brown.

23l. Light morph, heavily marked, all-barred type: Underparts are uniformly barred with rufous.

23m. Dark morph: Uniformly dark brown, including undertail coverts. Tail as on light morph.

23n. Light morph, streak-breasted, lightly marked type: Body as on *23i.* Tail as on *23e.*

23o. Light morph, lightly marked, bibbed type: Body as on *23j* but brown markings. Tail has 1 wide white band; except on 23g type. Wings are very pointed when gliding.

23p. 1-year-old light morph, heavily marked bibbed type: Body as on *23k* but rufous markings. Tail as on *23g.* Juvenile feathers are on primaries (p9, p10) and secondaries (s4, s9; lack black band).

23q. Light morph, heavily marked, all-barred type: Body as on *23l* but brown markings. Tail has 1 wide white bandand wings are very pointed when gliding.

23r. Light morph, all types: Uniform brownish, with white-tipped uppertail coverts. Tail as on *23d.*

23s. Dark morph: Uniform dark brown, including uppertail coverts. Tail as on *23c.*

23t. Dark morph: Uniform dark brown body and wing coverts. Tail as on *23f.*

23u. Flight silhouettes: *Top:* Soars and often glides on flat wings. *Bottom:* May glide with bowed wings.

a
b
c
d
e
f
g
h
i
j
k
l
m
n
o
p
q
r
s
t
u

BROAD-WINGED HAWK (*Buteo platypterus*)

HABITAT: *Summer.*—In n. Minn. and Canada, breeds in s. boreal forest biome in mixed woodlands of Quaking Aspen, Balsam Poplar (*Populus balsamifera*), and White Spruce (*Picea glauca*) with small forest openings and some water source. South of this region, breeds in low- to moderate-elevation mid-aged and mature deciduous and deciduous-coniferous woodlands of at least 100 ac. (40ha) in size that have small openings, trails, roads, and some source of water. Mainly found in remote woodlands but may occupy wooded rural areas with minimal human activity. Habitat in Great Plains region is isolated wooded locales, often at higher elevations but also in low-elevation wooded tracts or, less commonly, in suburban areas. *Winter/migration.*—Found in semi-open subtropical woodlands and along wooded openings. During migration it passes over a variety of land types, including vast expanses of mesquite in s. Tex. **STATUS:** Fall migration counts near Veracruz, Mexico, tally 1.6–2 million birds. Dark morph is less than 1% of population. Dark morphs nest mainly in Alta. but also in e. B.C. and sw. N.W.T.; isolated records for sw. Man. Large numbers are seen in fall migration along Lake Superior (Duluth, Minn.) and largest numbers in North America are tallied along Gulf of Mexico (near Corpus Christi, Tex.). Small numbers of Broad-wings winter in s. Fla., s. La., and s. Tex. Bulk of population winters from sw. Mexico south to n. South America. **NESTING:** Begins late Apr. in s. regions and early Jun. in n. areas. 1-year-old females in still mainly juvenile plumage may mate with adult males. Nesting is completed by late Jul.–Aug. Medium-size, poorly made stick nest is placed in 1st major crotch, relatively high in deciduous tree or, less commonly, next to trunk in conifer. Nest tree is near an opening. **MOVEMENTS:** Highly migratory, with punctual timing. *Spring.*—Mid-Mar.–early Jul. Adults move before returning juveniles. Adults begin to arrive in mid-Mar. and peak late Mar.–early Apr. in s. Tex., in mid-Apr. on Great Plains, and in early May along Lake Superior. Northern breeders may not arrive until late May–early Jun. Returning juveniles are less punctual: Peak numbers occur in s. Tex. in late Apr. and along Lake Superior in late May; movement continues into early Jul. *Fall.*—Both ages move together in mid-Aug.–mid-Oct. They peak in mid-Sep. along Lake Superior and late Sep.–early Oct. in s. Tex. **COMPARISON:** *Broad-winged Hawk juvenile.* **Red-shouldered Hawk (juvenile).**—Both species have similar dark mid-throat streak, but only Red-shouldered can have all-dark throat. Dorsal surface of secondaries has pale bars, *vs.* all-dark or faint dark bars. In flight, pale tawny crescent window/panel on base of primaries, *vs.* rectangular whitish window. Pale tail bands are thinner than dark bands, *vs.* equal-width or wider pale bands. **Red-tailed Hawk (juvenile).**—Cere is pale greenish, *vs.* yellow. Pale window on ventral side of wing is similar, but Red-tail has dark patagial mark on underwing, *vs.* lack of dark mark. Upperwing primary coverts are pale, *vs.* dark upperwing coverts in Broad-wing. **Cooper's Hawk (juvenile).**—Gray eyes of recently fledged birds are similar. Uniformly colored head, *vs.* head with dark, distinct malar. When perched, wingtips are much shorter than tail tip, *vs.* wingtips nearly reaching tail tip. Underside of wings is distinctly barred, *vs.* little barring. **Northern Goshawk (juvenile).**—Pale gray eyes on recently fledged birds are similar, but goshawk's eyes quickly turn yellow (by late summer). Upperwing greater coverts have tawny bar, and secondaries are barred, *vs.* all-dark dorsal wing surface. *Broad-winged adult.* **Red-shouldered Hawk (adult).**—Black-and-white checkering and crescent-shaped panel on dorsal wing surface, *vs.* solid dark dorsal wing. Dorsal side of tail has 3 thin white bands, *vs.* 1–3 wide, pale gray bands.

Figure 1. Hazel Bazemore Co. Pk., Corpus Christi, Tex. (Sep.)
Bulk of U.S. population passes here each fall; large migrant flock.

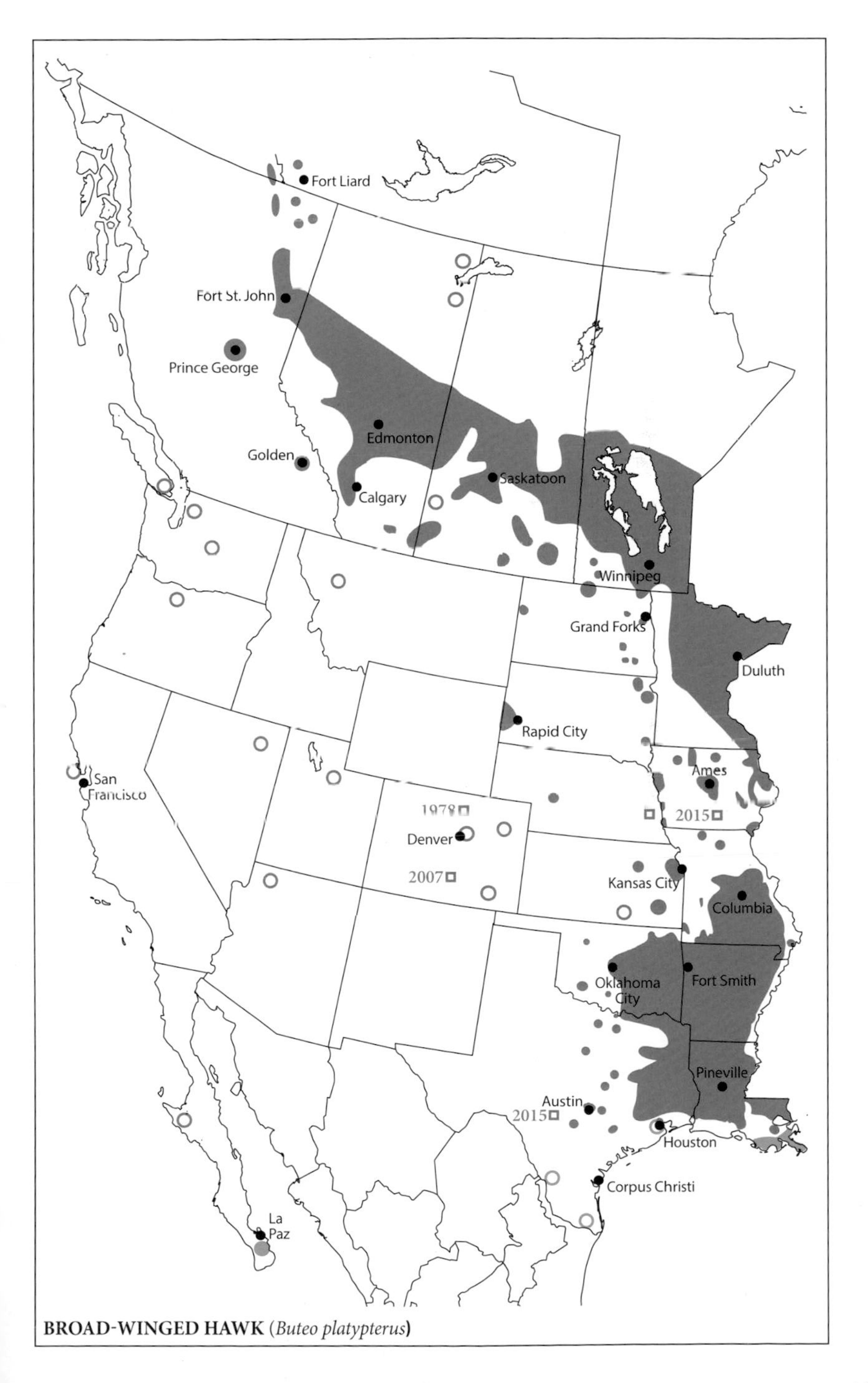

BROAD-WINGED HAWK (*Buteo platypterus*)

Plate 24. SWAINSON'S HAWK (*Buteo swainsoni*) Juvenile and 1-Year-Old, Perching

Ages: Juvenile and 1-year-old have small, pale blue spot on lower base of upper mandible. Eyes are medium brown. *Inner ½ of lores is dark. Thin dark eye line. Dark malar connects to dark patch on sides of neck and, on lighter morphs, forms partial bib or necklace.* Wings are long. *Juvenile.*—Plumage held for nearly 1 year. Cere is mainly *yellow* but can be pale green. *Throat may have dark center streak or patch* (separates it from most large buteos). Pale supercilium connects to pale forehead; pale auriculars. Lighter morphs have pale scapular patches. Tail is slightly longer than on older ages. *When perched, wingtips are barely shorter than tail tip.* Head and underparts are mainly tawny, though can be white (white type) on lighter morphs, but all fade to white by spring. *1-year-old.*—Aug.–Oct. span is shown. Plumage is worn in 2nd year. This age retains some faded brown juvenile flight feathers, with outermost primaries and some secondaries most visible. *Distinct white patch adorns ventral side of retained faded juvenile outer primaries.* A few juvenile tail feathers may be retained into fall. Similar to *all* juvenile plumage types, except head is darker, and wings and tail are mainly adult-like, with wide subterminal band on tail; some acquire adult-like body traits. Cere is yellow. *Wingtips equal tail tip.* **Subspecies:** Monotypic. **Color morphs:** Cline from light to dark; intermediate is median color morph. **Size:** L: 17–21" (43–53cm). Females average larger. **Habits:** Tame. Uses exposed perches and perches on ground. **Food:** Perch and aerial hunter of open-area small rodents and large insects, especially grasshoppers; some small birds and reptiles. Juveniles eat mainly insects after fledging. **Voice:** Juvenile emits food-begging *skree*. 1-year-old emits adult's *keeyah*.

24a. Juvenile light morph, lightly marked, white type: Variant with pale green cere. *Inner lores are dark.* Head is very pale. *Thin, dark eye line. Dark malar and side-of-neck patch.*

24b. Juvenile light to dark intermediate morph: *Dark inner lores. Cere is yellow. Dark malar and side-of-neck patch.*

24c. 1-year-old head (spring, Apr.–Jun.): Head often fades and wears to white by spring. *Thin dark eye line. Distinct malar and side-of-neck patch. Note:* Darker morphs similar.

24d. Juvenile dark morph: Similar to most paler morphs but darker on nape and hindneck.

24e. 1-year-old light morph: Pale forehead; short supercilium begins behind eye. Auriculars are dark. *Note:* Darker than juvenile head.

24f. Juvenile tail feather: *Narrow, equal-width black bars*; very rarely as on *24u*. Tip is pointed.

24g. Secondary flight feather: Solid dark. *Note:* Barred on other large species of *Buteo*.

24h. Juvenile light morph, lightly marked type: Head as on *24a*, but tawny, including underparts. *Dark malar and side-of-neck mark form partial dark bib.* Flanks are lightly marked; leg feathers unmarked.

24i. Juvenile light morph, heavily marked, white type: Darker head, more heavily marked on flanks than *24h*. Leg feathers are unmarked.

24j. Juvenile light intermediate morph: More heavily marked flanks than *24i*; lightly spotted leg feathers.

24k. Juvenile intermediate morph: More heavily marked underparts than *24j*; moderately barred leg feathers.

24l. Juvenile light/intermediate morph (spring): Faded plumage, whitish head as on *24c*. *Dark malar and side-of-neck patch.* White mid-scapular patch (similar to Red-tailed Hawk, Figs. *22c, f, h*); may not show on some birds. Wingtips equal tail tip.

24m. Juvenile dark intermediate morph, streaked type: Darker than *24k*. Leg feathers are thickly barred.

24n. Juvenile dark intermediate morph, dark-bellied type: As on *24m* but streaked only on breast. Leg feathers are thickly barred.

24o. Juvenile dark morph, dark-bellied type: More heavily marked than *24n*. Breast has small amount of tawny markings. Dark legs have tawny speckling. *Undertail coverts are pale and barred.*

24p. Juvenile dark morph (spring): Head may fade to whitish by spring as on *24c*, even on dark morphs. Ventral body as on *24o*. Upperparts are dark. *Whitish undertail coverts.*

24q. 1-year-old light morph (adult-like): Head as on *24e*. *Partial dark bib and barred flanks.*

24r. 1-year-old intermediate morph (adult-like): Head as on *24e*. *Underparts are tawny-rufous, barred, including leg feathers.*

24s. 1-year-old dark intermediate morph (adult-like): Darker than *24r*. Some pale brown juvenile feathers are retained on wing and tail.

24t. 1-year-old dark morph (adult-like): Dark with some tawny spotting. Some pale brown juvenile feathers are retained on wing and tail. Pale, barred undertail coverts.

24u. 1-year-old/adult tail feather: Narrow black inner bands and wide subterminal band. Tip is rounded.

a
b
c
d
e
f
g
h
i
j
k
l
m
n
o
p
q
r
s
t
u

Plate 25. SWAINSON'S HAWK (*Buteo swainsoni*) Juvenile and 1-Year-Old, Flying

Ages: Juvenile and 1-year-old have *dark malar mark that connects to dark patch on side of neck and breast, which often wraps around breast as partial bib or necklace on all but dark morphs.* Long wings have pointed tips. *Outer 3 primaries are notched. Juvenile.*—Plumage is held for nearly 1 year. Pale supercilium extends over eye and connects to pale forehead; pale auriculars. Distinct pale scapular patches on all but dark morph. *Throat may have dark center streak or patch. Underside of flight feathers is medium gray, with whitish spot on base of outer primary and narrow dark gray band on trailing edge.* Outer 5 primaries have large black tips. Wings are narrower and tails longer than on older ages. Head and underparts are mainly tawny, though can be white (white type) on lighter morphs, but all fade to white by spring. *1-year-old.*—Aug.–Oct. shown. Similar to *all* juvenile plumage types, except head is darker and tail is mainly adult-like; some acquire adult-like body traits. Plumage is worn in 2nd year; wing molt is not completed. *Outer 2–4 primaries are worn retained juvenile feathers with abrupt large white patch on basal area.* A few or many faded brown and shorter juvenile secondaries are retained (figures are shown in sequence of molt). *Underside of new flight feathers is adult-like dark gray with broad black band on trailing edge.* With highly visible retained juvenile flight feathers, this age class is easily aged in flight. Juvenile tail feathers are retained until early–late fall. New adult-like tail feathers have wide, black subterminal band. **Subspecies:** Monotypic. **Color morphs:** Cline from light to dark; intermediate is median color morph. All color types are seen during migration. **Size:** L: 17–21" (43–53cm), W: 47–54" (119–137cm). Females average larger. **Habits:** Tame. **Food:** Perch and aerial hunter of open-area small rodents and large insects, especially grasshoppers, which may be captured and eaten while in flight; some small birds and reptiles. Juveniles eat mainly insects after fledging. **Flight:** Soars with wings in high or low dihedral. When gliding, wings are held in low or modified dihedral. Hovers and kites. Wingbeats are of moderate speed. **Voice:** Juvenile emits food-begging *skree.* 1-year-old emits adult's *keeyah.*

25a. Juvenile light morph, lightly marked type: *Dark malar and side-of-neck patch. Long, pointed wings have medium-gray flight feathers with distinct large white patch on outer 3 primaries.* Dark tail bands are equal width.

25b. Light intermediate morph, 1st prebasic molt (May–Jun.): *White head with dark malar and side-of-neck patch.* White underparts lightly marked. Wingtips are very pointed when gliding. *Note:* 1st stage of molt into 1-year-old, with dropped inner 2 primaries creating mid-wing gap. Plumage is still juvenile. *Very distinct white patch on outer 2–4 primaries.*

25c. Juvenile light/intermediate morph: *Upperwing's uniformly dark flight feathers and greater coverts are darker than lesser and median coverts.* Uppertail coverts are pale. Tail bands of equal width. Pale scapular patches.

25d. Juvenile intermediate morph: Moderately marked underparts and underwing coverts. *Medium-gray flight feathers with dark wingtips.* Tail bands of equal width.

25e. Juvenile dark intermediate morph: As on *25d* but more heavily marked. *Undertail coverts are pale.*

25f. Juvenile dark morph, dark-bellied type: *Medium-gray flight feathers, as on all juvenile plumages.* Body and underwing coverts are dark with some tawny mottling. *Tawny undertail coverts are barred.*

25g. Juvenile tail (dorsal): Gray, with black subterminal band that is slightly wider or equal in width to inner bands.

25h. 1-year-old light morph: *White patch on outer 4 juvenile primaries.* Also retains juvenile s3, s4, and s7–10. Tail is molting; retains juvenile r2 and r5 sets.

25i. 1-year-old light intermediate/intermediate morph (juvenile-like): Plumage similar to *25d. 4 retained juvenile primaries with large white patch.* Also retains juvenile s3, s4, and s8–10. Tail is mix of juvenile and adult feathers.

25j. 1-year-old intermediate morph (adult-like): Dark bib on breast. Tawny/rufous-barred underparts. Molt more advanced than on *25i*; retains juvenile p8–10, s3, s4, s8, s9.

25k. 1-year-old tail (dorsal): Retains long juvenile feathers at r2, r5 locations. New adult-like feathers have wide, black subterminal band.

25l. 1-year-old dark morph, streaked type (juvenile-like): Advanced molt, with retained juvenile outer 2 primaries (p9, p10) and 2 secondaries (s4, s9). *Undertail coverts are pale. Note:* Body plumage is similar to that of juvenile.

25m. 1-year-old dark morph (adult-like): Body mainly dark; undertail coverts pale. Wing molt is as on *25l.*

a
b
c
d
e
f
g
h
i
j
k
l
m

Plate 26. SWAINSON'S HAWK (*Buteo swainsoni*) — Adult, Perching

Ages: Acquires adult plumage as 2-year-old but may not appear fully adult until 3-year-old. *Lores are white on outer ½ and dark on inner ½ on all color morphs. However, all color morphs but dark morph have white forehead that merges with white lores and whitish throat to create white face mask.* Eyes are dark brown. Crown and auriculars are gray on males, brown on females. There is much sexual and individual variation within each color morph. Light through intermediate morph males have more distinctly pale-edged dorsal feathers than females. Light through dark intermediate males with rufous breast may also have rufous nape. *When perched, wingtips equal tail tip.* Gray tail has wide black subterminal band and several narrow inner bands. **Subspecies:** Monotypic. **Color morphs:** Cline from light to dark; intermediate (rufous) morph is median color morph, and light intermediate (light rufous) and dark intermediate (dark rufous) morphs are paler and darker, respectively. All types may be seen in all parts of West during migration and winter. **Size:** L: 17–21" (43–53cm). Females average larger. **Habits:** Tame. Uses exposed perches. **Food:** Perch and aerial hunter of open-area small rodents and large insects, especially grasshoppers; some small birds and reptiles. Eats vertebrates mainly during nesting season. **Voice:** *Keeyah*, given when agitated.

26a. Light morph male, pale-headed type: *Gray crown and partially gray auriculars. Supercilium and cheeks are white. Note:* It is unknown if this is an aberrant adult pattern or 2-year-old.

26b. Light morph male: *Gray crown and auriculars. Breast is rufous. Note:* Typical bird shown.

26c. Light/intermediate morph female: *Uniformly brown head and breast. Note:* Typical bird shown.

26d. Light/intermediate morph female, pale-headed type: Tawny supercilium and pale auriculars. *Breast is rufous-brown. Note:* This is the most rufous that bibs get on females.

26e. Dark intermediate morph male: Streaked throat. Head is grayish. *Note:* Female shown in *26s.*

26f. Dark morph male: *White outer lore spot.* Head is grayish. *Note:* Female shown in *26t.*

26g. Light morph male (possible 2-year-old), pale-headed type: Head as on *26a*. Rufous bib is partial, which can occur on adults.

26h. Light morph male: Gray head. Rufous bib is incomplete on center breast. White underparts are unmarked.

26i. Light morph male, tawny type: Gray head. Solid rufous bib. Tawny underparts are lightly barred on flanks and legs. This is the heaviest markings for a light morph. *Note:* Tawny underparts can be unmarked.

26j. Light morph female: *Brown head and bib.* Underparts are white, unmarked (uncommon on females).

26k. Light morph female (probable 2-year-old): White partial supercilium and cheeks. Brown bib has broken pattern. Underparts are very lightly marked. *Note:* Separable from 1-year-old only by lack of faded, worn juvenile outer primaries (visible only in flight).

26l. Light morph female, pale-headed type: Head and breast as on *26d.* White underparts are lightly barred on flanks and legs. *Note:* This is heaviest amount of ventral markings for light morph.

26m. Light morph female, scallop-breasted, tawny type: *Brown bib with pale-edged (scalloped) feathers is common variation on females.* Tawny underparts are barred on flanks and leg feathers.

26n. Light intermediate morph male, rufous-barred type: *Rufous bib; rufous barred belly, flanks, and leg feathers. Note:* Male-only plumage.

26o. Light intermediate morph male, brown-barred type: *Dark brown bib; brown barred belly, flanks, and leg feathers. Note:* Palest color morph where males can have dark brown bibs. Female would have dark brown head.

26p. Intermediate morph male, all-rufous type: Rufous underparts; white undertail coverts (not shown). *Note:* Male-only plumage.

26q. Intermediate morph male, brown-bibbed type: *Brown breast; rufous flanks, belly, and leg feathers are barred with brown. White undertail coverts. Note:* Female would have dark brown head.

26r. Dark intermediate morph male, dark-bellied type: Rufous breast and legs; brown bellyband. *Note:* Female can be similar but has dark brown head.

26s. Dark intermediate morph female: Brown. Leg feathers and lower belly are rufous tinged or rufous barred. Undertail coverts are pale.

26t. Dark morph female: All-brown head and body. Head can be paler, as on *26s. Undertail coverts are pale, with barring.*

a
b
d
c
e
f
m
g
i
k
l
j
h
o
n
p
q
r
s
t

Plate 27. SWAINSON'S HAWK (*Buteo swainsoni*) Adult, Flying

Ages: Acquires adult plumage as 2-year-old but may not appear fully adult until 3-year-old. *Lores are white on outer ½ and dark on inner ½ on all color morphs. However, all color morphs but dark morph have white forehead that merges with white lores and whitish throat to create white face mask.* Crown and auriculars are gray on males and brown on females; however, at typical field-viewing distances of flying birds, this sex character is not visible. *Underside of flight feathers is dark gray, with black band on trailing edge and whitish area on outer 1 or 2 primaries.* Upperside of wing has black flight feathers and paler brown coverts. *Wings are long-pointed and become very pointed when gliding. Outer 3 primaries are notched. Tail is gray* with wide black subterminal band and several narrow inner bands. There is much sexual and individual variation within each color morph. Light through intermediate morph males are more distinctly pale-edged on dorsal feathers than females. Males of light through dark intermediate morphs that have rufous breast (bib) may also have rufous hindneck. **Subspecies:** Monotypic. **Color morphs:** Cline from light to dark; intermediate (rufous) morph is median color morph, and light intermediate (light rufous) and dark intermediate (dark rufous) morphs are paler and darker, respectively. **Size:** L: 17–21" (43–53cm), W: 47–54" (119–137cm). Females average larger. **Habits:** Tame. Uses exposed perches. **Food:** Perch and aerial hunter of open-area small rodents and large insects, especially grasshoppers; some small birds and reptiles; large insects often captured and eaten while in flight. Vertebrates are eaten mainly when nesting. **Flight:** Soars with wings in high or low dihedral. Glides with wings held in low or modified dihedral or on flat plane. Hovers and kites. Wingbeats are of moderate speed. **Voice:** *Keeyah*, given when agitated.

27a. Light morph male: *Gray head. Bib is all-rufous.* Underparts and underwing coverts are white. *Flight feathers are dark gray and contrast with white underwing coverts.* Flanks are lightly barred with rufous. *Note:* Pale underparts can be tawny instead of white, as on tawny type (*26i*).

27b. Light morph male, pale-headed type: Pale head and cheeks. *Rufous bib has broken pattern.* Rest of body is similar to *27a. Wingtips are very pointed when gliding. Note:* Uncommon aberration of adult or 2-year-old; common to have partial bib.

27c. Light morph female: Brown head and bib. Rest of underparts as on *27a*. Wings are very pointed when gliding. *Note*: Classic female plumage.

27d. Light intermediate morph male, rufous-barred type: *Gray head. Rufous bib and flanks; rufous-barred belly and legs. White undertail coverts. Dark gray flight feathers contrast with white/tawny coverts. Note*: Male-only plumage.

27e. Light intermediate morph female, brown-barred type: *Brown bib and flanks; brown-barred belly and legs.* Pale underwing coverts can be marked (as shown) or unmarked. *Tawny undertail coverts may be barred. Note:* Male of this color morph can be similar but has gray head.

27f. Light intermediate morph female, unbarred type: Brown head and brown bib (male has gray head, brown bib). Dark tawny, unbarred underparts often have dark blotches. *Note:* Unbarred type is less common than types with barred bellies.

27g. Tail (dorsal): *Gray, with thin black bands and wide black subterminal band.* Whitish uppertail coverts have U-shaped white edge.

27h. Flight silhouette: *Soaring.* Wings held in high dihedral. Wings are held more flat during high-speed glides.

27i. Light/intermediate morph female: Black flight feathers contrast with paler brown coverts and rest of upperparts. Uppertail coverts have thin U-shaped white edge. *Note:* Male has paler-edged upperparts and grayer head.

27j. Dark intermediate/dark morph: Dark brown upperparts, including uppertail coverts.

27k. Intermediate morph female: Brown bib and rufous belly. Undertail coverts are pale and often barred. *Underwing coverts are rufous.*

27l. Intermediate morph male, all-rufous type: Uniformly rufous underparts; *pale undertail coverts. Note*: Male-only plumage.

27m. Dark morph: Brown body. *Undertail coverts are white or tawny and often barred. Underwing coverts are rufous. Note:* Sexes are difficult to separate on flying dark morphs.

a
b
c
d
e
f
g
h
i
j
k
l
m

SWAINSON'S HAWK (*Buteo swainsoni*)

HABITAT: *Summer.*—Open and semi-open prairies and mixed light agricultural areas that have a single bush or tree (at least), scattered bushes or trees, a wooded riparian corridor, or shelterbelt plantations. Found in semi-open and open, high-elevation montane valleys in natural vegetation areas or especially locales with hayfields. In s. areas, also found in semi-open mesquite tracts. In very arid regions, occupies irrigated light agricultural areas that have some tall trees. Found in rural and suburban and even urban locales. *Winter/migration.*—In winter, found in coastal grasslands and agricultural areas of s. Tex. (and sw. Mexico) and in pasture and mixed light agricultural areas of n.-cen. Calif. During migration roosts and feeds in vast wheat fields of Great Plains and in mesquite expanses of s. Tex. **STATUS:** Common and adaptable. Population may be 1 million. Lighter morphs account for 65% and darker morphs 35% of population (dark morph at 10%); 10–15% of Calif. birds are lighter types. Darker morphs nest only west of *dashed line* on range map. All colors mix on migration. Nested in Eagle Plains region of Y.T., Canada, in 2005 (C. Eckert pers. comm.). Irregular, rare in interior Alaska in summer, with no confirmed breeding. **NESTING:** Begins in Apr.–May and ends Jul.–mid-Sep., depending on latitude and elevation. Nests are placed in small bushes or small trees in undisturbed areas and in tops of tall trees in rural areas. Moderate-size stick nest is built in live or dead tree. Brood size is 2 or 3. 1st breeds when 3 years old. **MOVEMENTS:** Highly migratory. Northern populations engage in 7,000-mi. (11,300km) fall and spring journeys to and from main winter quarters in n. Argentina; a few spend winter in n.-cen. Calif., s. Tex., and sw. Mexico. Birds of cen. Calif. may be sedentary, move into Mexico, or journey to Argentina. *Spring.*—Adults leave Argentina in early–mid-Feb., peak in s. Tex. mid–late Mar., and arrive on Great Plains in early Apr., peaking mid-Apr. Pairs often form prior to arrival on breeding territories (pers. obs.). Calif. birds may arrive in early Mar. 1- and 2-year-olds do not return to natal areas; arrive May–June and are nomadic. *Fall.*—Begins late Aug.–late Sep., but not until Oct.–early Nov. in cen. Calif. population. All ages move together, but some non- or failed-breeding adults move 1st. Peak occurs in e. Colo. in late Sep., in s. Tex. in mid-Oct. Has been recorded at Barrow, Alaska, in late summer–fall. **COMPARISON:** ***Swainson's Hawk juvenile.*** **"Eastern" Red-tailed Hawk (juvenile).**—Eye is tan, pale gray, or pale yellow, *vs.* medium brown. Both species may share similar throat patterns of dark patch, rarely thin dark center-streak. When perched, wingtips are far shorter than tail tip, *vs.* nearly equal tail tip. Dorsal secondaries are pale brown with dark bars, *vs.* all-dark brown. Dark patagial mark on underside of wing, *vs.* uniform underwing coverts. Underside of flight feathers is whitish, *vs.* gray. In flight, pale dorsal wing panel and ventral window, *vs.* lack of wing panel/window. **"Western" Red-tailed Hawk (juvenile light morph).**—As "Eastern" but often has streaking on throat, or throat is all-dark. **"Krider's" Red-tailed Hawk (juvenile).**—Similar white head as on returning springtime juvenile/1-year-old Swainson's Hawk. Wingtip-to-tail-tip data as noted above for "Eastern." Wing panel/window as noted for "Eastern" but more whitish. **Broad-winged Hawk (juvenile).**—Can have identical dark mid-throat streak and dark eye. When perched, wingtips shorter than tail tip, *vs.* equal to tail tip. Similarly pointed wingtips in flight, but flight feathers are white on ventral side, *vs.* gray. ***Swainson's adult.*** **Northern Harrier.**—White on uppertail coverts is large patch, *vs.* thin, U-shaped white crescent. **Rough-legged Hawk (light morph adult male).**—Dark bib on breast can rival Swainson's dark bib. Wingtip-to-tail-tip ratio similar. White flight feathers on underwing, *vs.* gray flight feathers.

Figure 1. Weld Co., Colo. (Jul.) Typical nest site in small tree in open grassland.

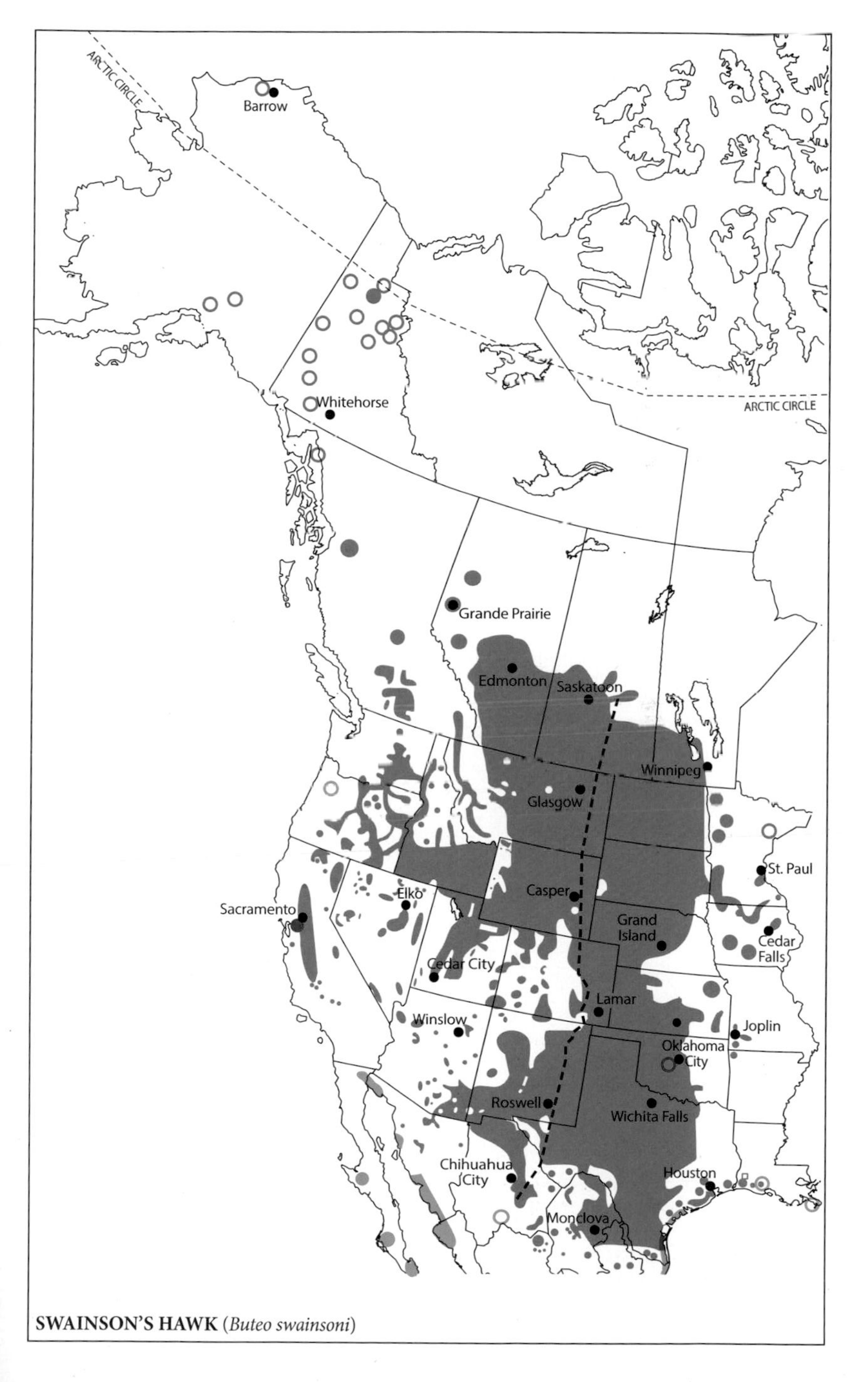

SWAINSON'S HAWK (*Buteo swainsoni*)

Plate 28. RED-TAILED HAWK (*Buteo jamaicensis*) "Eastern" (*B. j. borealis*) and "Krider's" (*B. j. kriderii*) — Juveniles, Perching

Ages: Juvenile plumage is held for 1st year. Eye color can be pale yellow, tan, or pale gray. Cere is pale green. When perched, wingtips are far short of tail tip (by 2–3"/5–7.5cm). *"Eastern."—Tawny-edged hindneck feathers.* Throat is white, even on heavily marked type; can have dark collar and/or thin streaking. Secondaries are barred; greater coverts may be solid dark or barred. Tail has 4 main color variations: brown, gray, partially rufous, or rufous. Tail has 3 main banding patterns: 7–10 equal-width, thin, dark bands; 7–10 dark bands with subterminal band widest; or partially banded, with lack of banding on basal area. White undertail coverts are typically unmarked but can be barred (as on many "Western" Red-tails; Plates 32–35). Heavily marked type is found in *any* breeding region. *"Krider's."*—Hindneck feathers are streaked. May have thin, dark eye line. *On dorsal side, greater coverts and secondaries are barred. Tail has mainly equal-width, thin bands on distal ½, in 3 main patterns: all-white; tan on distal part; or rufous on distal part. Note:* Juvenile light morph "Harlan's" (Plates 36, 37) can be *identical*, but tail is fully banded on basal area. **Subspecies:** 7 subspecies in U.S. and Canada (6 in West); 6 more in Mexico, Cen. America, and Caribbean. **Color morphs:** None in these 2 subspecies. **Size:** L: 17–22" (43–56cm). Females average larger. Juveniles are longer than adults because of typically longer tails. **Habits:** Tame to wary; tamest on winter grounds. **Food:** Perch and aerial hunter of small and medium-size amphibians, birds, mammals, and reptiles. **Voice:** Clear, high-pitched, drawn-out *skee-ah*; high-pitched *klee* when begging.

28a. "Eastern," heavily marked type: Pale yellow eye. White throat has dark collar.

28b. "Eastern," moderately/lightly marked type: Tan eye. Throat is white. Supercilium joins to white forehead. Hackles are raised.

28c. "Eastern," all types: *2 white-mottled scapular patches. Equal-width bands* as on *28k.*

28d. "Eastern" scapular feather: Large, dark distal arrowhead shape; dark basal crossbar.

28e. "Eastern," heavily marked type (recently fledged): Head as on *28a.* Breast is tawny; fades to white by late summer–fall. Dark bellyband can be quite dark on some birds. Leg feathers barred.

28f. "Eastern," moderately marked type: Moderately dark bellyband; spotted legs. Tail is gray type, with bands of equal width.

28g. "Eastern"/"Krider's" secondary flight feather: *Brown with thin, dark barring. Note:* Compare to juvenile Swainson's Hawk (*18g*).

28h. "Eastern," lightly marked type: Lightly marked bellyband; unmarked legs. Tail is brown type, with wide subterminal band.

28i. "Eastern," partial-banded gray tail with wide subterminal band (dorsal): Lacks dark banding or has partial banding on basal part. Basal region may be mottled.

28j. "Eastern," brown tail with wide subterminal band (dorsal): Brown and gray are most common color types.

28k. "Eastern," partially rufous tail with equal-width bands (dorsal): Rufous color is on a portion of most feathers.

28l. "Eastern" all-rufous tail with equal-width tail bands (dorsal): Rufous on all feathers (can be as red as on adults).

28m. "Eastern" tail feather: Wide subterminal band type (as on *28h, 28i, 28j*). Tip is pointed.

28n. "Krider's" tail feather: Equal-width, thin bands on up to ½ of distal part of each feather. Type with tan on distal portion. Tip is pointed.

28o. "Krider's," darker type: Pale gray eye. Head is whitish, with dark marks on crown and moderately dark malar mark.

28p. "Krider's," palest type: Tan eye. White; streaked on nape. Lacks a dark malar mark.

28q. "Krider's" tail, darker (dorsal): *Distal ½ of tail can be tan with equal-width, thin dark bands.*

28r. "Krider's" tail (dorsal): *Tan, with whitish base and equal-width bands; bands absent on base.*

28s. "Krider's" × "Eastern" intergrade: Head as on *28h* but with thin, dark eye line. Body is similar to *28h*, but upperwing greater coverts are paler, barred. Tail is tan or rufous with white base; distal part has equal-width, thin bands.

28t. "Krider's": White head with some streaking on crown; partial dark malar (as on *28o*). Belly and flanks are lightly marked. *Scapulars and median and lesser wing coverts are mottled with white.* Pale brown greater wing coverts are barred. *Distal ½ of tail is tan with equal-width, thin bands.*

28u. "Krider's," palest type: White head (as on *28p*), but has thin, dark eye line (as on younger light morphs of Swainson's Hawk, *24a, 24c*). *Scapulars and median and lesser wing coverts are extensively mottled with white. Greater coverts are white with dark barring. Tail is all-white, with up to ½ of distal part thinly banded.*

28v. "Krider's" scapular feather: Arrowhead-shaped with thin crossbar (marking can also occur on juvenile light morph "Harlan's").

a
b
c
e
f
d
h
g
p
i
j
k
l
o
v
m
n
u
t
q
s
r

Plate 29. RED-TAILED HAWK (*Buteo jamaicensis*) "Eastern" (*B. j. borealis*) and "Krider's" (*B. j. kriderii*)

Juveniles, Flying

Ages: Juvenile plumage is held for 1st year. *Large pale panel on dorsal surface of primaries and greater primary coverts creates dark inner ½ and light outer ½ of wing.* In shadowed light, this creates white translucent window on underwing (shared by other species of *Buteo*). *White wrist "headlight" is visible when viewed head on. Dark patagial mark on leading edge of underwing from body to wrist.* Wingtips are rounded, with 5 fingers. Dorsal secondaries are barred on both subspecies. Juveniles have narrower wings and typically longer tails than adults. *"Eastern."*—Throat is white, rarely lightly streaked; often has dark collar; *never* all-dark (as on many juvenile "Western"; Plates 32, 33). Primary tips can be solid gray with dark tips; gray with thin, dark barring; or all-dark. Tail has 4 main color variations: brown, gray, partially rufous, or rufous. Tail has 3 main banding patterns: 7–10 equal-width, thin, dark bands; 7–10 dark bands with subterminal band widest; or partially banded, with lack of banding on basal portion. White undertail coverts rarely have barring on moderately/heavily marked types. Heavily marked type found in all regions but *may* be more common in boreal forest. *"Krider's."*—Primary tips are mainly gray with thin dark bars. *Tail has mainly equal-width, thin bands on distal ½, in 3 main patterns: all-white, tan on distal part, or rufous on distal part.* **Subspecies:** 7 subspecies in U.S. and Canada (6 in West); 6 more in Mexico, Cen. America, and Caribbean. **Color morphs:** None in these 2 subspecies. **Size:** L: 17–22" (43–56cm), W: 43–56" (109–142cm). Females average larger. Juveniles are longer than adults because of their longer tails. **Habits:** Tame to wary; tamest on winter grounds. **Food:** Perch and aerial hunter of small and medium-size amphibians, birds, mammals, and reptiles. **Flight:** Often soars. Wings are held on flat plane or in low dihedral. Glides with wings held flat or bowed slightly downward. Powered flight is with moderately slow wingbeats in variable cadence. May hover and kite when hunting. **Voice:** Clear, high-pitched, drawn-out *skee-ah*; high-pitched *klee* or *kree* when begging.

29a. "Eastern," heavily marked type: White throat with dark collar. Heavily marked bellyband. *Black patagial mark is distinct.* Inner primaries show white, square-shaped window in translucent light. Wingtips can be partially or fully barred or all-dark. *Tail is brown type, with wide, dark subterminal band and narrow inner bands.*

29b. "Eastern," moderately marked type: All-white throat. Moderately marked bellyband. *Dark patagial mark is fairly distinct.* Inner primaries show white, square-shaped window in translucent light. Wingtips are thinly barred . *Tail orange/rufous, with equal-width dark bands. Note*: Undertail coverts have crossbars, which is very uncommon.

29c. "Eastern," lightly marked type: All-white throat. Lightly marked bellyband. *Faint dark patagial mark.* Inner primaries show white, square-shaped window in translucent light. Wingtips are partially barred. *Tail pale brown, with wide, dark subterminal band and narrow inner bands.*

29d. "Eastern," heavily marked type (recently fledged): White throat with dark collar. Breast is tawny; fades to white by late summer–fall. *Black patagial mark.* Underwing is uniformly pale when lit in direct light. *Tail bands are equal width.*

29e. "Eastern": Large, mottled white patch on each scapular feather. *Pale brown primaries and primary coverts contrast with darker inner ½ of wing. Brown tail, with equal-width, narrow bands.*

29f. "Eastern": Large, mottled white patch on each scapular feather. *Pale brown primaries and primary coverts contrast with darker inner ½ of wing. Orange/rufous tail with equal-width, narrow bands. Note:* Orange or rufous dorsal tail surface is common on "Eastern" juveniles and can be nearly as rufous as on adults.

29g. "Krider's": May have partial dark malar. Belly and flanks are lightly speckled. *Dark patagial mark is faint but visible. Wingtips are thinly barred. White or tan tail has equal-width, thin bands on distal ½.*

29h. "Krider's" darker type or "Krider's" × "Eastern" intergrade): Pale head with white or dark crown and malar mark. *Primaries and greater primary coverts are very pale brown and much paler than inner ½ of wing. Greater secondary coverts are medium to pale brown and barred. Tail is pale brown on distal ¾; white, lacking narrow bands on basal ¼.*

29i. "Krider's": All-white head. Scapulars and lesser and median coverts are extensively mottled with white; greater secondary coverts are white and thinly barred. *Primaries and greater primary coverts are white and contrast with darker (but barred) secondaries. Tail is all-white or very pale brownish and narrowly banded on distal ½.*

29j. Flight silhouette: *Soaring; both subspecies show white "headlight" patch on wrist of wing.*

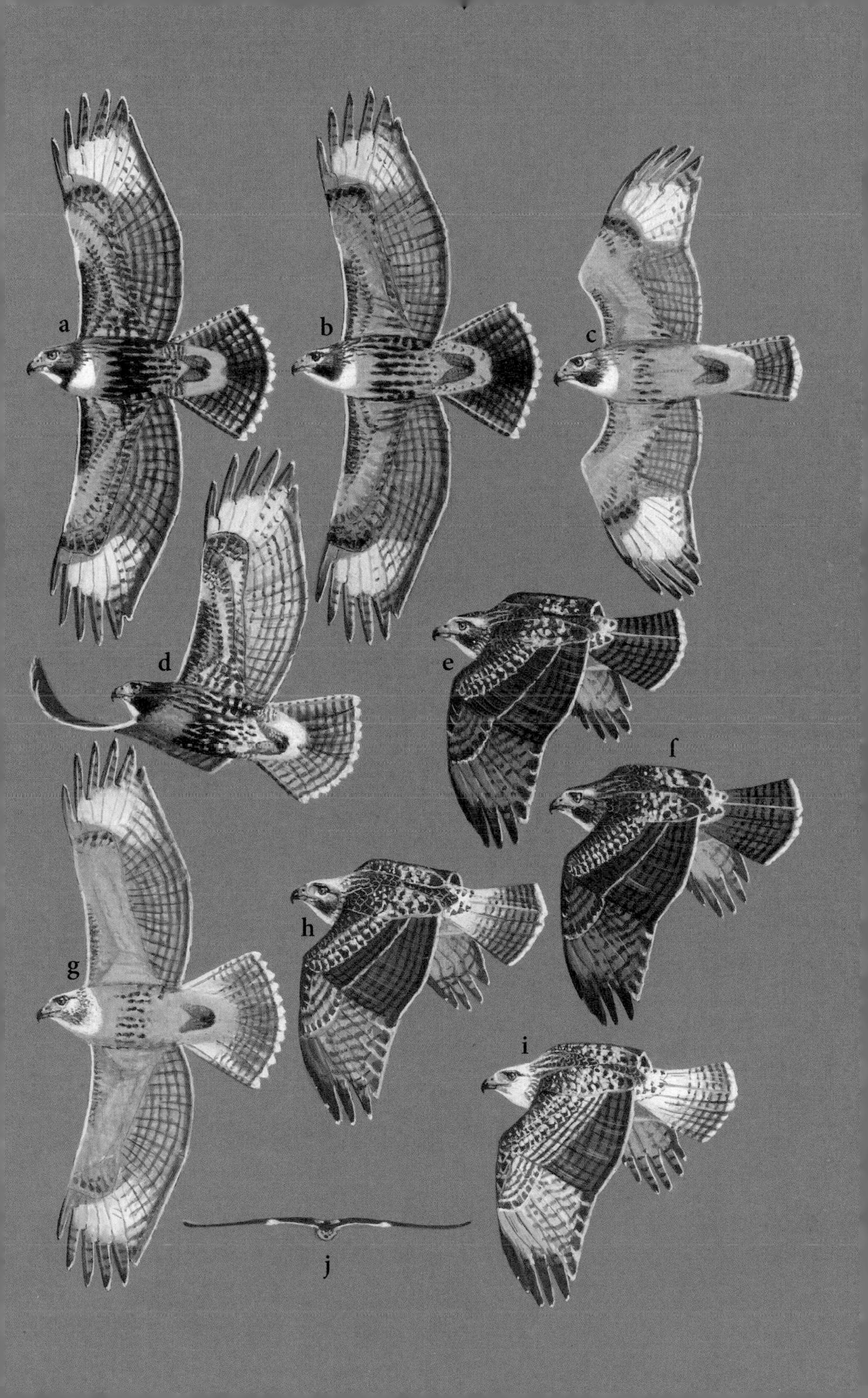
a
b
c
d
e
f
h
g
i
j

Plate 30. RED-TAILED HAWK (*Buteo jamaicensis*) 1-Year-Olds and Adults, Perching "Eastern" (*B. j. borealis*) and "Krider's" (*B. j. kriderii*)

Ages: 1-year-olds and adults. Cere is green or yellow. *Rufous patch adorns sides of neck.* Ventral parts are pale tawny or white on both subspecies. Undertail coverts are *always* unmarked. When perched, wingtips are *shorter* than tail tip. Tail usually shorter than on juvenile. *1-year-olds.*—Plumage is adult-like, attained in fall of 2nd year. Faded brownish juvenile feathers retained on wing (1–4 outer primaries, mid-wing secondaries, and their coverts) and sometimes tail. Eye is pale yellowish on most 1-year-olds; usually darkens with age. *Adults.*—Plumage attained when 2 years old in fall of 3rd year. *"Eastern."*—Throat is white or streaked, often with dark collar; can be all-dark (in *any* part of breeding range). Scapulars have moderate to large white patches. Birds with tawny and/or streaked or blob-patterned breasts are uncommon. Bellies and flanks (bellyband) vary from unmarked to heavily marked; flanks often are barred. Moderately marked and lightly marked are most common types in all areas; heavily marked type adults are in mid- and northerly latitudes. Uppertail coverts are mainly white but can be rufous with white tips (as on many "Western"; Plates 32–35). *Tail is rufous and fades by spring.* Partial or full tail banding is very common in w. Canada, but uncommon to very uncommon in East. Barring on leg (thigh) feathers is common in w. Canadian range but uncommon in East (trait on all light "Western"). *"Krider's."*—Most lack ventral markings. *Very large, white scapular patches. Most have rufous/tawny edges on feathers of upperwing coverts and forward upperparts. Tail is white on basal ½ and has thin, black subterminal band.* **Subspecies:** 7 subspecies in U.S. and Canada (6 in West); 6 more in Mexico, Cen. America, and Caribbean. **Color morphs:** None in these 2 subspecies. **Size:** L: 17–22" (43–56cm). Females average larger. Juveniles are longer than adults because of typically longer tails. **Habits:** Tame to wary; tamest on winter grounds. **Food:** Perch and aerial hunter of small and medium-size amphibians, birds, mammals, and reptiles. **Voice:** Raspy, hoarse *skee-ya*, *skee-yer*, or *squee-ah*. *Note:* For some figures, location information for specific birds is supplied for plumage verification; see also "Plumage Variation in 'Eastern' Red-tailed Hawk," at end of Red-tailed Hawk species account.

30a. "Eastern" 1-year-old, dark-throated/streak-breasted type: Head is dark. Breast is streaked. Head can be all-dark with unmarked breast. *Note:* Lake Athabasca, Alta. (Jun.); N.W.T. (Apr.); "Bobby" and "Rosie," N.Y.C. (residents, 2012).

30b. "Eastern," dark-collared type: Forehead is white. Throat is white with dark collar.

30c. "Eastern," white-throated type: Forehead is white; short, whitish supercilium. *Note:* Head feathers are raised.

30d. "Eastern," heavily marked type: All-dark head. Breast can be streaked. Bellyband is dark, but is not solid black. Thigh feathers unmarked or barred. *Note:* Lake Athabasca, Atla. (Jun.); N.W.T. (Apr.); "Lima" N.Y.C. (resident, 2011).
30dd. Undertail covert: *Unmarked coverts.*

30e. "Eastern," moderately marked type: Dark-collared type. Tawny may fade to white by winter. Belly and flanks are moderately marked.

30f. "Eastern," lightly marked type: White-throated type. White-mottled scapulars. White underparts are typically barred on flanks but can be unmarked; belly is unmarked. *Tail is rufous.*

30g. "Eastern," with banded tail (spring): Pale rufous tail is banded. *Note:* Summer in Alta, B.C., N.W.T., Sask ; also residents of Ga., N.C., N.Y., Va.

30h. "Krider's" 1-year-old: Whitish head with dark malar mark. *Rufous neck patch.* Cere is green.

30i. "Krider's": Mainly white head; faint malar mark. Cere is yellow; eye is dark brown. *Pale rufous neck patch.*

30j. "Eastern" × "Krider's" 1-year-old: Lightly marked underparts. Wing retains brown juvenile feathers. *Tail is white on base and has banding on distal part* [N.Dak., Jul.]).

30k. "Krider's": Pale head (as on *30h*). *Large white scapular patches.* Underparts can be pale tawny or white. *Tail is rufous on distal ½.*

30l. "Krider's," palest type: White head. *Large white patch on scapulars. Rufous side-of-neck patch. Tail is white, except pale rufous on very distal part; thin, neat subterminal band.*

30m. "Eastern," thin-banded tail (dorsal): *All-rufous, with thin subterminal band.* White uppertail coverts. *Note:* Classic pattern.

30n. "Eastern," partial-banded tail (dorsal): Partial bands on center feathers. White coverts.

30o. "Eastern," banded tail (dorsal): Partial to full bands on much of tail. Uppertail coverts rufous with white tips (uncommon variation).

30p. "Eastern," unbanded tail, spring (dorsal): Lacks subterminal band, which is fairly common. Rufous color fades by spring on all types.

30q. "Eastern" × "Krider's" tail (dorsal): *Rufous on more than ½ of distal part; white base (base is never barred).*

30r. "Krider's," all-white tail (dorsal): *White, with thin subterminal band.*

a
b
c
e
f
g
d
dd
h
i
m
n
o
p
q
r
j
k
l

Plate 31. RED-TAILED HAWK (*Buteo jamaicensis*) 1-Year-Olds and Adults, Flying "Eastern" (*B. j. borealis*) and "Krider's" (*B. j. kriderii*)

Ages: 1-year-olds and adults. *Rufous patch adorns sides of neck.* Ventral parts are pale tawny or white. *Dark patagial mark extends from body to wrist. White "headlight" adorns wrist when viewed head on.* Wingtips are rounded, with 5 fingers. Wings are much wider than on juveniles and have wide black band on trailing edge. Undertail coverts are *always* unmarked. Tail is usually shorter than on juvenile. *1-year-olds.*—Plumage is adult-like, attained in fall of 2nd year. Faded, brownish juvenile feathers retained on wing (1–4 outer primaries, mid-wing secondaries, and their coverts) and sometimes tail. *Adults.*—Plumage attained as 2-year-old in fall of 3rd year. *"Eastern."—Patagial mark is brown, rufous, or black; can be fairly large on any plumage type, especially on heavily marked type.* Underwing coverts mainly unmarked or lightly marked; more heavily marked birds may have dark marks. Axillaries are unmarked or streaked; can be barred with brown/black on some U.S. and many Canadian birds. Throat is white or streaked, often with dark collar; sometimes all-black (in *any* part of breeding range). Scapulars have small to large white patches. Tawny and/or streaked or blob-patterned breasts are uncommon in all of range. Belly and flanks (bellyband) vary from unmarked to heavily marked. Barring on thigh feathers is common in w. Canadian range, uncommon in e. U.S. and e. Canada. Uppertail coverts are mainly white; can be rufous with white tips. *Tail is rufous; color fades by spring.* Partial or full tail banding is very common in w. Canada but uncommon in East. Moderately marked and lightly marked are most common types in all areas; heavily marked type adults are in mid- and northerly latitudes. *"Krider's."—Patagial mark is rufous or brown.* Most lack other ventral markings. *Very large, white scapular patches. Most have rufous/tawny edges on upperwing coverts and feathers of forward upperparts.* White uppertail coverts. *Tail is white on basal ½ and has thin, black subterminal band.* **Subspecies:** 7 subspecies in U.S. and Canada (6 in West); 6 more in Mexico, Cen. America, and Caribbean. **Color morphs:** None in these 2 subspecies. **Size:** L: 17–22" (43–56cm), W: 43–56" (109–142cm). Females average larger. Juveniles are longer than adults because of typically longer tails. **Habits:** Tame to wary; tamest on winter grounds. **Food:** Perch and aerial hunter of small and medium-size amphibians, birds, mammals, and reptiles. **Flight:** Often soars; wings held flat or in low dihedral. Glides with wings flat or bowed slightly. Moderately slow wingbeats. May hover, and kite. **Voice:** Raspy, hoarse *skee-ya*, *skee-yer*, or *squee-ah*. *Note:* For some figures, location information for specific birds is supplied for plumage verification; see also "Plumage Variation in 'Eastern' Red-tailed Hawk," at end of Red-tailed Hawk species account.

31a. Flight silhouette: *Soaring; white "headlights" on wrist of wing.*

31b. "Eastern" axillary feathers: 1. Rufous streak with arrowhead. **2.** Black-barred. **3.** Brown-barred. *Note:* Axillary may be unmarked as on *31e.*

31c. "Eastern" 1-year-old, heavily marked type: Outer 3 primaries and middle and inner secondaries (s4, s8, s9) are retained juvenile feathers. Throat is dark. *Black patagial mark.* Bellyband is heavy. Axillaries are barred. *Rufous tail can be banded* (especially basal area). *Note:* Lake Athabasca, Alta. (Jun.); N.W.T. (Apr.); "Lima," N.Y.C. (resident, 2011).

31d. "Eastern," moderately marked type: White throat with dark collar. *Black patagial mark.* Belly and flanks are moderately marked; flanks barred. *Rufous tail has thin subterminal band.*

31e. "Eastern," lightly marked type: White throat. *Brown patagial mark.* Belly and flanks are sparsely marked. Axillaries are unmarked.

31f. "Eastern" 1-year-old, all types: Retains faded brown juvenile outer 4 primaries (p7–10), secondaries (s3, s4, s8, s9), and tail feathers (r5 set). Rufous uppertail coverts have white tips (uncommon type). *Note:* A slow molt stage.

31g. 2-year-old upperwing, all types/subspecies: Middle primaries (p5–7) and respective greater primary coverts are faded 1-year-old adult-type feathers; rest of flight feathers are newer (and darker) 2nd-generation adult feathers.

31h. "Eastern," all types: *White scapular patches.* White uppertail coverts. *Rufous tail.*

31i. "Krider's": White head with dark malar. Underparts are pale tawny, unmarked. Often has rufous side-of-neck patch. Wingtips are barred. *Patagial mark is rufous or brown. Tail is mainly white, with neat, thin subterminal band.*

31j. "Eastern" × "Krider's": Whitish head with dark cap and malar mark. *Rufous tail has white on inner ¼.*

31k. "Krider's": All-white head. Large white scapular patches. Feathers of upperparts are rufous-edged. *Tail is ½ white , ½ pale rufous; thin band.*

31l. "Krider's": White head with dark malar mark. Feathers of upperparts are rufous-edged. *Tail is mainly white, with thin, neat subterminal band.*

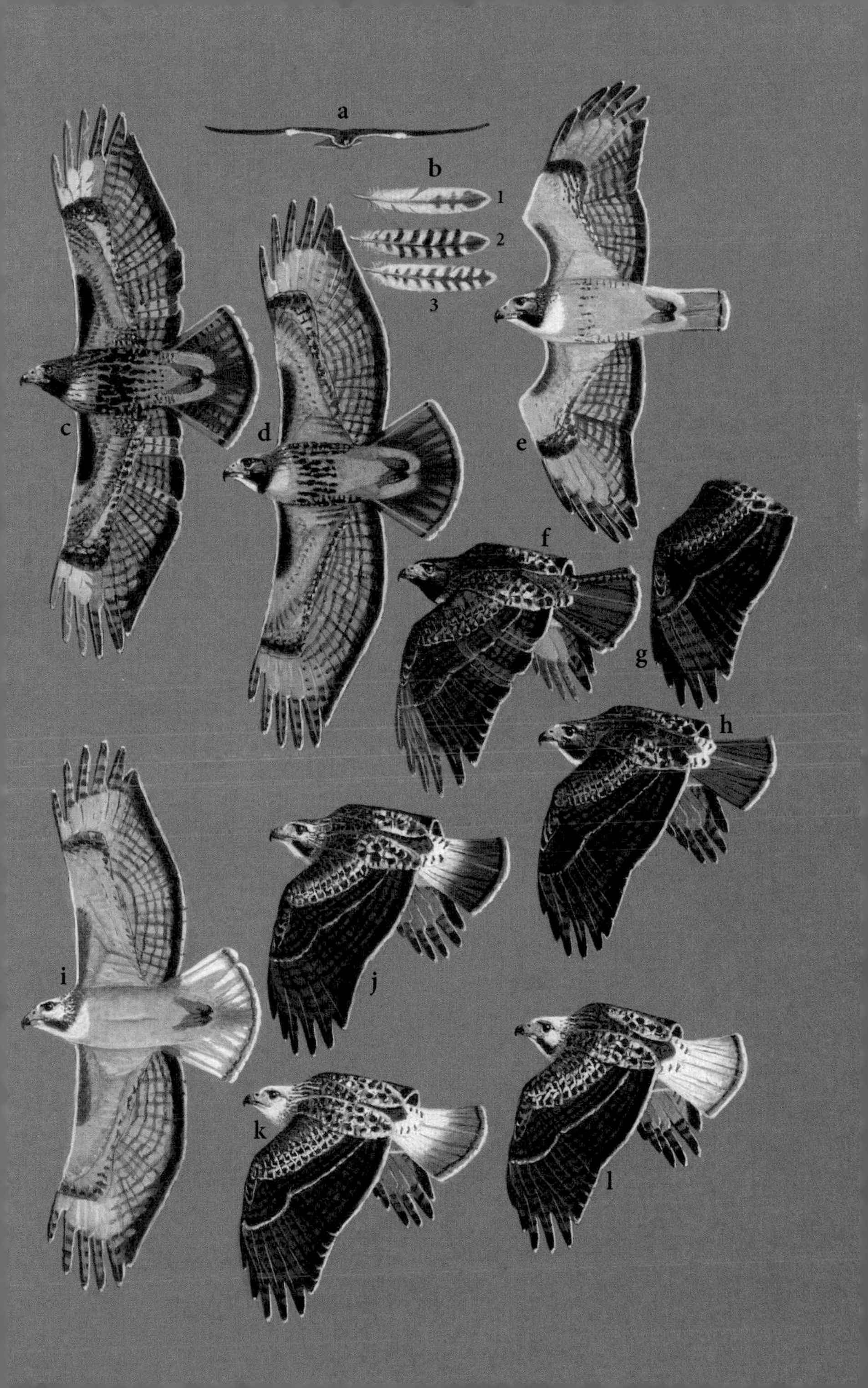
a
b
1
2
3
c
d
e
f
g
h
i
j
k
l

Plate 32. RED-TAILED HAWK (*Buteo jamaicensis*) "Western" (*B. j. calurus*)

Juvenile, Perching

Ages: Juvenile plumage is retained for 1st year. Eye color is pale yellow, tan, or pale gray. Bill has moderate-size, pale blue base with pale green cere. Forehead is *always* dark. Throat is white with dark collar, white with streaking, or commonly all-dark (never all-dark on juvenile "Eastern" Red-tail; Plates 28, 29). *Hindneck feathers on all but dark morph are edged with tawny.* Wingtips are somewhat shorter than tail tip (1–2"/2.5–5cm) when bird is perching. As with "Eastern," tail has 4 main color variations—brown, gray, partially rufous, or rufous—and 4 main banding patterns: 7–9 equal-width, thin, dark bands; 7–9 dark bands with subterminal band widest; 6 or 7 wide, nearly equal-width bands; or partially banded, with banding on distal part (see *28i–m*). Tail bands are equal in width or wider than on "Eastern" and often wider on darker morphs. Undertail coverts of light morph usually have a crossbar on each feather (shared by a few "Eastern" juveniles). **Subspecies:** 7 subspecies in U.S. and Canada (6 in West); 6 more in Mexico, Cen. America, and Caribbean. **Color morphs:** Cline from light to dark; intermediate is median color morph. **Size:** L: 17–22" (43–56cm). Females are larger. Juveniles are longer than adults because of their longer tails. **Habits:** Tame to wary; tamest on winter grounds. **Food:** Perch and aerial hunter of small and medium-size amphibians, birds, mammals, and reptiles. **Voice:** Clear, high-pitched, drawn-out *skee-ah*; high-pitched *klee* when begging.

32a. Light morph, streak-throated type: Throat can be white and streaked, with dark collar. Forehead is dark. Tawny supercilium does not reach forehead. Eye can be pale yellow. Breast is unmarked, but side of neck is streaked.

32b. Intermediate morph: Dark head has paler, dark tawny, short supercilium. Throat is typically all-dark. Eye can be tan. Breast is heavily streaked.

32c. Dark morph: Dark brown head, including forehead. Lores are all-gray. Hackles are raised in cool weather and may show white patch (as shown). Eye can be pale gray.

32d. Light morph, moderately marked type (recently fledged): White throat with dark collar is uncommon in all of subspecies' range. Rich tawny breast fades to white by late summer or fall. Thigh feathers are *always* barred (except on dark morph). Belly and flanks are moderately marked.

32e. Light morph, heavily marked type: Dark throat. Breast is white and unmarked. Belly and flanks are heavily marked; thigh feathers are *always* barred. *Mid-scapulars have moderate white patch.* Undertail coverts are barred. Gray tail has 7 equal-width bands.

32f. Light morph, all types: Throat is dark. Brown tail has 7 equal-width bands.

32g. Gray tail with equal-width bands (dorsal): 9 equal-width dark bands on gray surface.

32h. Brown tail with wide subterminal band (dorsal): Wide subterminal band, with 7 thinner bands on brown surface.

32i. Partial rufous tail with equal-width bands (dorsal): 7 equal-width bands, with rufous on a portion of most feathers.

32j. Light intermediate morph: Breast is lightly streaked. Thigh feathers are heavily barred. Throat is streaked or dark.

32k. Intermediate morph: Breast is heavily streaked. Thigh feathers are heavily barred. Tail is partial rufous type. *Note:* A rare type (not depicted) has dark breast and pale tawny spotting on rest of underparts.

32l. Scapular feather, lighter morphs: Large, dark arrowhead shape, wide streak, and basal crossbar. *Note:* Larger amount of dark area than juvenile "Eastern" (*28d*); compare to juvenile "Harlan's" (*36e, 36n*).

32m. Tail feather, gray with equal-width bands: Gray with 8 equal-width bands. White, pointed tip. *Note:* Compare to juvenile "Harlan's" (*36i–k*).

32n. Secondary flight feather (dorsal): Barred, with white tip. *Note:* Compare to juvenile "Harlan's" (*36h*). Other large buteos of any color morph lack barring on this feather tract.

32o. Dark intermediate morph: Breast is dark with tawny/rufous streaking. Belly and flanks are speckled. Dark thigh feathers are speckled. *2 rows of white spots on wing coverts. Tail is rufous type,* with 6 dark, equal-width bands.

32p. Dark morph: Dark brown. *Wing coverts have 2 rows of white spots.* Tail is brown, with 6 wide, equal-width bands.

32q. Dark morph tail (dorsal): *Brown, with 6 wide, equal-width bands.* White terminal band is thin on darker morphs.

32r. Dark morph tail feather (dorsal): Brown, with 6 or 7 wide bands and wide subterminal band. White terminal band is thin. *Note:* Compare juvenile "Harlan's" tails (*36s–u*).

a
b
c
d
e
f
g
h
i
j
k
l
m
n
o
p
q
r

Plate 33. RED-TAILED HAWK (*Buteo jamaicensis*) "Western" (*B. j. calurus*) — Juvenile, Flying

Ages: Juvenile plumage is held for 1st year. *Large pale panel on dorsal surface of primaries and greater primary coverts creates dark inner ½ and light outer ½ of wing.* In shadowed light, creates white, square-shaped, translucent window on underwing. Wingtips are rounded, with 5 fingers (fingers are longer than on "Eastern"). Wingtips are usually solid black or gray with dark tips; on darker morphs *rarely* gray with dark barring and identical to juvenile "Harlan's" (J. Liguori, unpubl. data). Juveniles have narrower wings and longer tails than adults. As with "Eastern," tail has 4 main color variations—brown, gray, partially rufous, or rufous—and 4 main banding patterns: 7–9 equal-width, thin, dark bands; 7–9 dark bands with subterminal widest; 6 or 7 wide, nearly equal-width bands; or partially banded, with banding on distal part (see *28i–m*). Tail bands are equal in width or wider than on "Eastern," and often wider on darker morphs. Secondaries are barred dorsally. *Lighter morphs.—Dark patagial mark on leading edge of wing from body to wrist. Pale "headlight" patch on forward edge of wrist is visible in head-on view.* White undertail coverts have dark bar or heart-shaped mark (shared by a few juvenile "Eastern"; see *29b*). *Darker morphs.*—Outer primaries are typically all-dark; if barred, separation from darker morphs of juvenile "Harlan's" is difficult (see Plate 37). Dark areas of plumage generally are warm dark brown, with any pale markings tawny (*vs.* blackish with white markings on similarly plumaged "Harlan's"). **Subspecies:** 7 subspecies in U.S. and Canada (6 in West); 6 more in Mexico, Cen. America, and Caribbean. **Color morphs:** Cline from light to dark; intermediate is median color morph. **Size:** L: 17–22" (43–56cm), W: 43–56" (109–142cm). Females average larger. Juveniles are longer than adults because of their longer tails. **Habits:** Tame to wary; tamest on winter grounds. **Food:** Perch and aerial hunter of small and medium-size amphibians, birds, mammals, and reptiles. **Flight:** Often soars. Wings are held on flat plane or in low dihedral. Glides with wings held flat or bowed slightly downward. Powered flight is with moderately slow wingbeats in variable cadence. May hover and kite when hunting. **Voice:** Clear, high-pitched, drawn-out *skee-ah*; high-pitched *klee* or *kree* when begging.

33a. Light morph, moderately marked type: *Black patagial mark on leading edge of underwing.* Throat is white with collar. Belly and flanks are moderately marked. Tail is brown type, with wide subterminal band. *Note*: White-throated type not separable from "Eastern" (29a, 29b), except patagial mark is more distinct and "fingers" are longer.

33b. Light morph, all types: *Pale brown panel on outer ½ of wing.* Mid-scapulars have white-mottled patch. Throat is all-dark. Tail is brown type with equal-width bands.

33c. Light morph, heavily marked type (recently fledged): *Black patagial mark.* Throat is all-dark. Belly and flanks are heavily marked with black. Tail has equal-width dark bands. *Note:* Breast is tawny but fades to white by late summer or fall.

33d. Intermediate morph: *Black patagial mark.* Breast is heavily streaked. Rest of body similar to heavily marked light morph. Tail is rufous type, with wide subterminal band (can be as reddish as on adults). *Note:* Common type. Light intermediate morph (not depicted) is similar but has very thin breast streaking.

33e. Intermediate morph, streaked type: *Black patagial mark visible but somewhat masked by dark coverts.* All underparts are streaked. Tail has wider subterminal band. *Note:* Uncommon type.

33f. Lighter morph flight silhouette: *Soaring. White "headlight" patch on wrist area.* Wings held in flat plane but may bend upward at tips.

33g. Dark intermediate morph (soaring): Dark brown body and wing coverts; tawny streaking on breast and tawny speckling on belly and flanks. Patagial mark is masked by dark coverts. Tail is brown type, with equal-width bands. *Note:* Shows barred wingtips identical to those on juvenile "Harlan's" (which has white throat and body markings; *37g*).

33h. Dark morph: All-dark body. *Inner ½ of wing is dark and outer ½ is pale brown.* Upperwing coverts have 2 rows of white spots. Brown tail has equal-width bands.

33i. Dark morph: All-dark body and wing coverts. Wingtips are solid dark or sometimes barred (as on juvenile "Harlan's"). Brown tail has wide subterminal band.

a
b
c
d
e
f
g
h
i

Plate 34. RED-TAILED HAWK (*Buteo jamaicensis*) 1-Year-Old and Adult, Perching "Western" (*B. j. calurus*)

Ages: 1-year-old and adult. Cere is green or yellow. Tail is shorter than on juvenile. *Tail is rufous (fades by spring) and has single, variable-width subterminal band or is partially or fully banded; basal bands can be wider.* Wingtips equal tail tip when bird is perching. *1-year-old.*—Plumage is adult-like, gained in fall of 2nd year. Brownish juvenile feathers retained on wing (1–4 outermost primaries, a few secondaries, and respective greater coverts) and sometimes tail. Eye is pale yellowish on most 1-year-olds; usually darkens with age. *Adult.*—Plumage attained as 2-year-old in fall of 3rd year. *Lighter morphs.*—Forehead is dark. *Rufous patch adorns side of neck.* Breast *may* be streaked. Uppertail coverts are rufous or rufous with white tips, or can be white and barred. *Black or rufous crossbars on belly and lower belly, but may have black streaks or blobs and lack barring; can be solid black. Leg (thigh) feathers are barred rufous or brown.* Undertail coverts are *barred* or unmarked. Tail bands, if present, are thin. *Mid-scapulars have gray or tawny patch* (the darker the color morph, the smaller the pale patch). *Darker morphs.*—Throat and forehead are dark. Tail as on lighter morphs or thickly banded; subterminal band can be quite wide. **Subspecies:** 7 subspecies in U.S. and Canada (6 in West); 6 more in Mexico, Cen. America, and Caribbean. **Color morphs:** Cline from light to dark; intermediate (rufous) morph is median color morph, and light intermediate (light rufous) and dark intermediate (dark rufous) morphs are paler and darker, respectively. **Size:** L: 17–22" (43–56cm). Females average larger. Juveniles are longer than adults because of typically longer tails. **Habits:** Tame to wary; tamest on winter grounds. **Food:** Perch and aerial hunter of small and medium-size amphibians, birds, mammals, and reptiles. **Voice:** Raspy, hoarse *skee-ya*, *skee-yer*, or *squee-ah*.

34a. 1-year-old light morph: Eye is pale yellowish. White throat is streaked, often has dark collar. Head is overall dark.

34b. Intermediate morph: Throat and head are dark. Breast is rufous; can be unmarked.

34c. Dark morph: Head, including forehead, is dark brown; pale lores.

34d. 1-year-old light morph, lightly marked white type: Dark-throated type. Eye is pale. *White underparts are barred rufous or black on belly and flanks; small black spots. Thigh feathers are white or tawny and barred with rufous/brown.*

34e. 1-year-old light morph, lightly marked tawny type: Streak-throated type. *Tawny underparts barred; small black spots. Thigh feathers are barred.* Tail has single, thin subterminal band.

34f. Light morph, moderately marked white type: White-throated type. *White underparts barred rufous or black on belly and flanks. Thigh feathers barred.*

34g. Light morph, moderately marked tawny type: Dark-throated type. *Tawny underparts barred with rufous or black on belly and flanks. Thigh feathers barred. Tail band is moderate width.*

34h. Light morph, heavily marked white type: Dark-throated type. White underparts heavily marked on belly and flanks; may not show barred pattern. Thigh feathers are barred. *Note:* Separable from heavily marked "Eastern" by longer wings (reach tail tip).

34i. Light morph, heavily marked tawny type: Dark-throated type. Tawny underparts heavily marked on belly and flanks. Thigh feathers barred with rufous/brown. *Subterminal tail band is wide.*

34j. Light morph, all types (spring): *Small mid-scapular patches with gray or tawny mottling.* Wingtips equal tail tip. *Tail fades. Tail is banded*

34k. Light intermediate morph: *Tawny-rufous underparts with streaked/blobbed bellyband. Tail is banded and has moderate-width subterminal band.*

34l. Intermediate morph, lightly/moderately marked type: Underparts are rufous; barred bellyband. Thigh feathers are barred

34m. Intermediate morph, heavily marked type: Rufous breast; may be streaked. Bellyband is solid. Thigh feathers are usually rufous, barred. *Tail is partial-banded type.* **34mm:** Covert is *barred.*

34n. Dark intermediate morph: Dark brown body and thighs; rufous breast. *Tail is banded with moderate-width subterminal band.*

34o. Dark morph: Dark brown body. *Rufous tail is banded type; wider basal bands; subterminal band is wide.*

34p. Partial-banded tail, spring (dorsal): *Partial bands on basal part of center feathers.*

34q. Partial-banded tail, spring (dorsal): *Partial bands are wider on basal region.*

34r. Banded tail with thin subterminal band (dorsal): *Thin inner bands; thin subterminal band.* White uppertail coverts.

34s. Banded tail with moderate-width subterminal band (dorsal): *Complete thin inner bands; fairly wide subterminal band.*

34t. Banded tail with wide subterminal band (dorsal): *Fairly wide inner bands, wider on basal area; wide subterminal band.*

a
b
c
d
e
f
g
h
i
j
k
l
m
mm
n
o
p
q
r
s
t

Plate 35. RED-TAILED HAWK (*Buteo jamaicensis*) 1-Year-Old and Adult, Flying "Western" (*B. j. calurus*)

Ages: 1-year-old and adult. Wingtips are rounded, with 5 fingers. Wings are broad (wider than on juvenile), with wide black band on trailing edge. Tail is usually shorter than on juvenile. *Tail is rufous and fades to pale rufous by spring.* Underside of tail is pale rufous or gray when not lit by translucent light. Tail may have single, variable-width subterminal band or be partially or fully banded. Banding pattern is often wider on basal region of tail. *1-year-old.*—Plumage is adult-like, gained in fall of 2nd year. Faded brownish juvenile feathers are retained on wing (1–4 outermost primaries [s7–10], a few secondaries [s3, s4, s8, s9], and respective greater coverts) and sometimes tail. *Adult.*—Plumage attained as 2-year-old in fall of 3rd year. 2-year-olds are separable only in dorsal flight view. *Lighter morphs.—Leading edge of underwing has distinct, large, black patagial mark from body to wrist. Pale "headlight" adorns wrist of wing; visible when viewed head on.* Axillaries are barred rufous or black. Uppertail coverts are rufous or rufous with white tips; uncommonly, white and barred. *Side of neck has rufous patch.* If tail banding is present, pattern is of thin bands. *Mid-scapulars have small to moderate-size pale patch with gray or tawny marks* (the darker the color morph, the smaller the pale patch). *Darker morphs.*—Axillaries are barred, except on dark morph (shared trait with "Harlan's"; Plates 36–39). Tail as on lighter morphs, but if banded, inner bands are often thicker, and subterminal band can be wide. **Subspecies:** 7 subspecies in U.S. and Canada (6 in West); 6 more in Mexico, Cen. America, and Caribbean. **Color morphs:** Cline from light to dark; intermediate (rufous) morph is median color morph, and light intermediate (light rufous) and dark intermediate (dark rufous) morphs are paler and darker, respectively. **Size:** L: 17–22" (43–56cm), W: 43–56" (109–142cm). Females average larger. Juveniles are longer than adults because of typically longer tails. **Habits:** Tame to wary; tamest on winter grounds. **Food:** Perch and aerial hunter of small and medium-size amphibians, birds, mammals, and reptiles. **Flight:** Often soars. Wings are held on flat plane or in low dihedral. Glides with wings held flat or bowed slightly downward. Powered flight is of moderately slow wingbeats in variable cadence. May hover and kite when hunting. **Voice:** Raspy, hoarse *skee-ya*, *skee-yer*, or *squee-ah*.

35a. 2-year-old light morph: Small light patch on each scapular. Uppertail coverts are rufous; can have pale tips. *Rufous tail is unmarked except thin subterminal band. Note:* Aged by faded, paler patch of 1st-generation adult feathers on middle primaries and respective coverts. Compare to *35i* for older adult dorsal wing pattern.

35b. Axillary feathers: 1. Rufous-barred on many light morphs. **2.** Black-barred on many light morphs and darker morphs. **3.** Thick, black barring on darker morphs (except dark morph).

35c. Light morph, lightly marked white type: White-throated type with dark collar. White underparts have rufous smudge on belly and flanks, which are lightly speckled with black. Axillaries are rufous-barred. *Black patagial mark.* Tail is partially banded; thin subterminal band. *Note:* Common in Calif.; uncommon other regions.

35d. 1-year-old light morph, lightly marked white type: Whitish throat. Belly and flanks lightly speckled and with rufous crossbars. *Black patagial mark.* Retains juvenile primaries (p9, p10) and secondaries (s4, s8, s9). Axillaries rufous-barred. *Undertail coverts rufous-barred.* Tail is fully banded.

35e. Light morph, moderately marked tawny type: Dark-throated type. Bellyband is barred with black. *Black patagial mark.* Tail is unbanded type. Axillaries are rufous-barred. Undertail is pale rufous or gray when closed.

35f. Light morph, heavily marked tawny type: Plumage as on *35e*, but bellyband is heavier and can be nearly solid or solid black; barring may not show on some birds. *Black patagial mark.* Axillaries are black-barred. *Undertail coverts are distinctly barred.* Banded-type tail has wide subterminal band.

35g. Flight silhouette: *Lighter morphs show white or pale "headlight" patch on wrist.*

35h. Intermediate morph: Rufous breast, lower belly, and underwing coverts. *Black patagial mark.* Bellyband can be solid black, streaked with blobs, or barred. Axillaries are barred. Tail is banded type; subterminal band is thin.

35i. Dark morph: Uniformly dark chocolate-brown upperparts, including uppertail coverts. Tail has single, wide subterminal band (dark morphs, as with lighter morphs, may lack multiple tail bands).

35j. Dark intermediate morph: Breast is rufous, but rest of body and underwing coverts are dark brown. Axillaries are usually partially barred. Tail is partial-banded (on basal part) type. *Note:* Dark underwing coverts mask patagial mark.

35k. Dark morph: Uniformly dark chocolate-brown body and underwing coverts, including axillaries. Undertail coverts may be rufous-tipped. Tail is banded type; may show thicker bands on basal area; subterminal band is wide.

a
b
1
2
3
c
d
e
f
g
h
i
j
k

Plate 36. RED-TAILED HAWK (*Buteo jamaicensis*) "Harlan's" (*B. j. harlani*) — Juvenile, Perching

Ages: Juvenile plumage is held for 1st year. Pale areas of plumage are stark white and lack tawny tones of other subspecies. Eye color is pale yellow, tan, or pale gray. Bill has moderate-size pale blue base with pale green cere. Lighter morphs have white-edged hindneck feathers. *All birds have white throats, except most dark morphs.* When perched, wingtips are far short of tail tip (2–3"/5–7.5cm; similar proportions to juvenile "Eastern," Plate 28). On lighter morphs, scapular feather may have typical "Eastern"- or "Western"-like diamond-shaped mark (*28d*, *32l*, respectively) or *subspecies-unique dark streak or dark T-bar mark.* Tail has numerous banding patterns and color variants, often with white or rufous infusion. *Tail feathers often have subspecies-unique dark "spike" along feather shaft at its tip. Tails bands often form V-shaped pattern, and bands are often wider on basal area, or tails lack bands and show adult-like mottling, especially on distal areas.* On upperwing, greater coverts are barred, except on dark morph. *Secondaries and their greater coverts also often have dark spike on feather tip.* Dark plumage areas are often more blackish than on other subspecies. **Subspecies:** 7 subspecies in U.S. and Canada (6 in West); 6 more in Mexico, Cen. America, and Caribbean. **Color morphs:** Cline from light to dark. **Size:** L: 17–22" (43–56cm). Females are larger. Juveniles are longer than adults because of typically longer tails. **Habits:** Tame to wary; tamest on winter grounds. **Food:** Perch and aerial hunter of small and medium-size perched birds and mammals. **Voice:** Clear high-pitched, drawn-out *skee-ah*; high-pitched *klee* when begging.

36a. Light morph: Eye is pale yellow. Head is whitish with dark crown streaking and black malar.

36b. Intermediate morph: Eye is tan. Forehead and short supercilium are white; streaked throat. Breast is streaked.

36c. Dark intermediate morph: Dark head with short, white supercilium; white throat. Breast is speckled or dark.

36d. Light morph, lightly marked type: Mainly all-white head. Scapulars and upperwing coverts have white mottling (diamond-shaped scapular marks). Flanks and often belly are streaked. Tail is partially banded on basal area and mottled on distal area, which occurs *only* on this subspecies.

36e. Light morph scapular feather: *White with dark center streak and thin inner crossbar. Note:* Subspecies-unique pattern.

36f. Light morph, moderately marked type: Whitish on head forms spectacles around eyes. Scapulars marked as on *36e* (inner crossbar often concealed). Tail has V-shaped bands and spikes.

36g. Light intermediate morph: Head similar to *36f.* White breast is thinly streaked. *Whitish scapulars have T-bar mark.* Thigh feathers are streaked. *Tail has wide, equal-width, straight bands.*

36h. Secondary/greater covert: Dark spike extends into white tip of secondary flight feather and greater covert.

36i. Tail feather, partial V-shaped bands: Partial distal V-shaped bands; lacks dark spike.

36j. Tail feather, V-shaped bands, with spike: *V-shaped bands; thin, black strip to tip.*

36k. Tail feather, V-shaped bands, with expanded spike: Dark bands can be wider on basal area. White highlights. *Broad black tip.*

36l. Intermediate morph, streak-breasted type: White supercilium and throat. Breast is broadly streaked.

36m. Intermediate morph, streaked type: White supercilium and throat. Underparts are all-streaked. *Tail has V-shaped pattern, white highlights on some feathers, and spikes.*

36n. Scapular feather with T-bar: Subspecies-unique pattern on light intermediate and intermediate morphs.

36o. Dark intermediate morph, streak-breasted type: Dark head with white throat. Black body has white mottling on lower breast.

36p. Dark intermediate morph, spot-bellied type: Black head and breast form bib against white-spotted belly and flanks. White-spotted wing coverts. Tail has wide, V-shaped bands and dark tip.

36q. Dark morph: Black head and body; white spots on wing coverts. Tail has V-banded pattern and dark tip.

36r. Tail with thin straight bands (dorsal): Whitish rufous, with thin, fairly straight, equal-width bands. *Note:* Uncommon pattern found only on light morph (see Mindell 1985; nestling from Kuskokwim River, Alaska).

36s. Tail with wide straight bands (dorsal): Banding is complete or incomplete; rufous wash.

36t. Tail with equal-width V-shaped bands, with spikes (dorsal): White center part on many feathers. *Dark spike extends to tip.*

36u. Tail with unequal-width V-shaped bands, dark-tipped (dorsal): Dark bands wider on basal areas. *Broad, dark spike on tip*

a
b
c
d
e
f
g
h
i
j
k
l
m
n
o
p
q
r
s
t
u

Plate 37. RED-TAILED HAWK (*Buteo jamaicensis*) "Harlan's" (*B. j. harlani*) — Juvenile, Flying

Ages: Juvenile plumage is retained for 1st year. Pale areas of plumage are stark white and lack tawny tones of other subspecies. 5 outer primary fingers are typically distinctly barred (more so than on juveniles of other subspecies), but some birds lack barring on outermost couple of primaries (rarely on all outer primaries). Upperwing greater coverts are barred. *Secondary flight feathers and greater coverts often have dark spike on tip.* Tail has numerous banding patterns and color variants, often with white or rufous infusion. *Tail feathers often have subspecies-unique dark spike along feather shaft at its tip. Tail bands often form V-shaped pattern, and bands are often wider on basal area, or tails lack bands and show adult-like mottling, especially on distal part of tail (other subspecies often lack banding on basal region).* Dark plumage areas are often more blackish than on other subspecies. *Lighter morphs.*—Throat is always white. Scapular feathers have typical Red-tail diamond-shaped mark surrounded by white or *subspecies-unique dark streak or dark T-bar mark. Black patagial mark adorns leading edge of wing from body to wrist, as on lighter morphs of other subspecies.* Recently fledged birds have tawny breast, which quickly fades, as on recently fledged light morphs of other subspecies. *Darker morphs.*—Throat is white, except on some dark morphs. **Subspecies:** 7 subspecies in U.S. and Canada (6 in West); 6 more in Mexico, Cen. America, and Caribbean. **Color morphs:** Cline from light to dark. **Size:** L: 17–22" (43–56cm), W: 43–56" (109–142cm). Females are larger. Juveniles are longer than adults because of typically longer tails. **Habits:** Tame to wary; tamest on winter grounds. **Food:** Perch and aerial hunter of small and medium-size perched birds and mammals. **Flight:** Often soars. Wings are held on flat plane or in low dihedral. Glides with wings held flat or bowed slightly downward. Powered flight is of moderately slow wingbeats in variable cadence. May hover and kite when hunting. **Voice:** Clear, high-pitched, drawn-out *skee-ah*; high-pitched *klee* when begging. *Note*: The figures on this plate are in the same posture and placed in the same location on the plate as respective-plumaged figures of adult "Harlan's" on Plate 39.

37a. Light morph, moderately marked type: White head with dark malar. Underparts are white, with dark streaks forming light to moderate bellyband. *Black patagial mark. Tail has V-shaped equal-width bands and spikes on feather tips.*

37b. Light morph, lightly marked type: White head with small darker crown patch. White scapular patches are small or moderate-size. *Whitish panel on primaries and their greater coverts contrasts with darker inner ½ of wing. Tail has partially banded basal area and mottled distal area (subspecies-unique pattern); dark subterminal band.*

37c. Light intermediate morph: White spectacles formed by white supercilium and cheeks. Throat is white or can be lightly streaked. White breast is thinly streaked. Belly and flanks have dark, mottled band. *Tail has V-shaped bands and spikes on feather tips.*

37d. Intermediate morph, streak-breasted type: White or streaked throat. Breast is moderately streaked. Rest of body and wing coverts are black and mottled; patagial mark may be masked or show faintly.

37e. Intermediate morph: White supercilium and cheeks. *Whitish panel on primaries and greater coverts contrasts with darker inner ½ of wing. Tail is whitish, with V-shaped bands and wide subterminal band that lacks spikes.*

37f. Intermediate morph, streaked type: White or streaked throat. Blackish underparts are streaked with white. *Patagial mark on leading edge of wing. Tail has wide, equal-width bands and dark spikes.*

37g. Dark intermediate morph, streak-breasted type: Black head; throat may be white or streaked. Underparts are blackish, with small amount of white streaking or mottling on breast. *Tail is V-banded type, with black bands on basal part of tail becoming solid dark mass; dark tail tip has expanded spikes.*

37h. Dark intermediate morph, spot-bellied type: Black head (or paler as on *37g*), with *white throat.* Underparts are spotted with white on belly and flanks. *Tail has equal-width, V-shaped bands and expanded dark spikes on tips. Note:* Spotted belly pattern very rarely shared by juvenile dark intermediate morph "Western."

37i. Dark morph: Black head, body, and underwing coverts. *Tail has V-shaped bands and expanded dark spikes on tip*; bands are wider in basal area. Throat can be white, even on dark morph ("Harlan's"-only trait).

a
b
c
d
e
f
g
h
i

Plate 38. RED-TAILED HAWK (*Buteo jamaicensis*) 1-Year-Old and Adult, Perching "Harlan's" (*B. j. harlani*)

Ages: 1-year-old and adult. Cere is green or yellow. Tail is usually much shorter than on juveniles. When perched, wingtips are moderately shorter than tail tip (proportion as on adult "Eastern"; Plate 24). Dorsal surface of secondaries and greater coverts can be mottled or barred. Tail patterns are highly variable. They can be unmarked, mottled, banded, or a mix of these patterns in white, gray, or partially rufous or mostly rufous. Banded patterns commonly have V-shaped banding. See Clark (2009) for numerous tail pattern variations. Dark plumage areas of "Harlan's" are typically more blackish than in other subspecies but can be similar warm dark brown, especially by spring, due to fading. Upper- and undertail coverts regularly have rufous infusion. *1-year-old.*—Plumage is adult-like, gained in fall of 2nd year. Faded brownish juvenile feathers are retained on wing (1–4 outer primaries, middle secondaries, and respective greater coverts) and sometimes tail. This subspecies is likely to retain more juvenile feathering because of shorter molt season in its northerly range. Eye is tan or pale yellow on most 1-year-olds; usually darkens with age. *Adult.*—Plumage attained as 2-year-old in fall of 3rd year. 2-year-olds can be aged only in dorsal flight view (*39b*). *Lighter morphs.—Sides of neck are white* with dark streaking (rufous on all other subspecies). Mid-scapulars have small or moderate-size white- or gray-mottled patch. *Darker morphs.*—Blackish plumage has variable white markings. All have *white throat* except some dark morphs. **Subspecies:** 7 subspecies in U.S. and Canada (5 in East, Plates 22–34, 64); 6 more in Mexico, Cen. America, and Caribbean. **Color morphs:** Cline from light to dark, with much individual variation. **Size:** L: 17–22" (43–56cm). Females average larger. Juveniles are longer than adults because of typically longer tails. **Habits:** Usually wary, but occasional tame individuals are encountered. **Food:** Perch and aerial hunter of small and medium-size perched birds and rodents. **Voice:** Raspy, hoarse *skee-ya*, *skee-yer*, or *squee-ah*.

38a. 1-year-old light morph: Pale yellow eye. White spectacle surrounds eye.

38b. Intermediate morph: Forehead is white; throat is white or streaked. Breast is streaked.

38c. Dark intermediate morph: Black head, but can have streaked or white throat.

38d. Light morph, pale type: Head nearly all-white. Underparts are white, unmarked. *Plain white tail has rufous on distal part, and partial subterminal band. Note:* Very uncommon; appears as adult "Krider's" (compare subterminal tail bands).

38e. Light morph, moderately marked type: Head as on *38a*. Dark blobs on belly and flanks.

38f. 1-year-old light morph, heavily marked type: Head as on *38a*. Belly, flanks, and thighs are heavily marked. *White tail is mottled and has wide, black terminal band. Note:* Faded brown juvenile feathers are retained on wing.

38g. Light intermediate morph: Head as on (*38a*). White breast is thinly streaked. Belly, flanks, and thighs are very heavily marked. *White tail is rufous on distal ¾; wide, black subterminal band.*

38h. Intermediate morph, streak-breasted type (spring): White throat and forehead. Breast is moderately streaked. *Tail is gray and mottled.* Dark brown underparts are speckled. *Note*: Plumage fades to brownish on many birds by spring.

38i. Intermediate morph, streaked type: Head as on *38b*, but throat is white. Underparts are streaked. *Tail is gray with rufous on distal ¼; moderate-width black subterminal band.*

38j. Dark intermediate morph, streak-breasted type: Throat is white or streaked. Breast has minimal white.

38k. Dark intermediate morph, spot-bellied type: Throat can also be white or streaked. *Black belly, flanks, and thighs are covered with white speckling. Gray tail is rufous on distal ½.*

38l. Dark morph: All-black (but throat may be white). Dark gray mottled-type tail.

38m. Tail (dorsal): *Distal ¾ is rufous; white/gray basal area is mottled and barred; irregular black terminal band.*

38n. Tail: *Distal ½ is rufous; gray basal area is thickly banded; irregular black subterminal band.*

38o. Tail: *Pale gray with rufous on distal part; mix of banding and mottling; fairly neat subterminal band.*

38p. Tail: *Dark gray with rufous on distal part; thin black inner banding; neat subterminal band.*

38q. Tail: *White, with fairly thick complete or partial inner bands; neat, wide subterminal band.*

38r. Tail: *White, with many thin, V-shaped inner bands; neat, black subterminal band.*

38s. Tail: *Dark gray, with wide inner bands; neat, wide black subterminal band.*

38t. Tail: *White with rufous wash on distal ½; V-shaped bands wider on basal area.*

38u. Tail: *Dark gray with black-and-white mottling; rufous-infused throughout; wide terminal band.*

a
b
c
d
e
f
g
h
i
j
k
l
m
n
o
p
q
r
s
t
u

Plate 39. RED-TAILED HAWK (*Buteo jamaicensis*) 1-Year-Old and Adult, Flying "Harlan's" (*B. j. harlani*)

Ages: 1-year-old and adult. Adults usually have much shorter tails than juveniles. 5 fingers on outer primaries are either barred or solid black. Dorsal and ventral surfaces of secondaries can be unmarked, mottled, or barred. Trailing edge of underside of wing has wide black band. Tail patterns are highly variable. They can be unmarked, mottled, banded, or a mix of these patterns in white, gray, or partially rufous or mostly rufous. Banded patterns commonly have V-shaped banding. As with other subspcies, banding may be thicker on basal region. See Clark (2009) for numerous tail pattern variations. Dark plumage areas are typically more blackish than on other subspecies, but can be dark brown, especially when faded by spring. Upper- and undertail coverts regularly have rufous infusion. *1-year-old.*—Plumage is adult-like, gained in fall of 2nd year. Faded brownish juvenile feathers are retained on wing (1–4 outermost primaries, middle secondaries, and respective greater coverts) and sometimes tail. This subspecies is likely to retain more juvenile feathering because of shorter molt season in its northerly range. *Adult.*—Plumage attained as 2-year-old in fall of 3rd year. 2-year-olds can be aged only in dorsal flight view (*39b*). *Lighter morphs.*—Throat is white. Sides of neck are white with dark streaking (rufous on all other subspecies). *Black patagial mark from body to wrist on leading edge of underwing.* Mid-scapulars have small white- or gray-mottled patch. *Darker morphs.*—Blackish plumage usually has some white markings. Throat can black or sometimes white. **Subspecies:** 7 subspecies in U.S. and Canada (5 in East, Plates 22–34, 64); 6 more in Mexico, Cen. America, and Caribbean. **Color morphs:** Cline from light to dark, with much individual variation. **Size:** L: 17–22" (43–56cm), W: 43–56" (109–142cm). Females average larger. Juveniles are longer than adults because of typically longer tails. **Habits:** Usually wary but occasionally tame. **Food:** Perch and aerial hunter of small and medium-size perched birds and mammals. **Flight:** Often soars. Wings are held on flat plane or in low dihedral. Glides with wings held flat or bowed slightly downward. Powered flight is with moderately slow wingbeats. May hover and kite when hunting. **Voice:** Raspy, hoarse *skee-ya*, *skee-yer*, or *squee-ah*. *Note*: The figures on this plate are in the same posture and placed in the same location on the plate as respective-plumaged figures of juvenile "Harlan's" on Plate 37.

39a. 1-year-old light morph: White spectacles around eyes. Belly can be streaked; unmarked on palest type (*38d*). *Black patagial mark.* Outer 3 primaries (p8–10) and middle secondaries (s4, s8, s9) are retained juvenile feathers that lack wide black band at tip. Flight feathers are barred type. White tail has fairly neat, dark subterminal band.

39b. 2-year-old light morph: Head as on *39a*. White speckles on mid-scapulars. Middle primaries (p6–8) are faded adult feathers from previous year; p9, p10, and p1–5 are new, darker feathers. Dorsal side of flight feathers is brown and unmarked. Tail is white on basal ½ and rufous on distal ½; irregular dark subterminal band.

39c. Axillary feathers: 1. Light morph: dark streak. **2.** Light/intermediate morphs: equal white and dark bars (shared by "Western"). **3.** Darker morphs (except pure dark morph): thin white bars on black feather (also shared by darker "Western").

39d. Light intermediate morph: Head as on light morph (*39a*). White breast is thinly streaked. Belly and flanks are heavily marked. *Black patagial mark.* Flight feathers are unmarked or mottled; primary fingers are solid black. Whitish tail is mottled with rufous on very distal part.

39e. Intermediate morph, streak-breasted type: White throat. Breast is equally streaked with white and black. Rest of body and wing coverts are black; may have white speckling. Flight feathers are barred. White-/gray-mottled tail has wide, black terminal band. *Note:* Classic plumage.

39f. Intermediate morph: Uniformly blackish upperparts. Flight feathers are barred type. *Tail is gray on basal ½ and rufous on distal ½; uppertail coverts are rufous* (common trait).

39g. Intermediate morph, streaked type: White throat. All of underparts are black and streaked with white. *Flight feathers are combination of mottled and banded feathers; black outer primaries. Partially rufous tail has irregular banding and mottling*; irregular, black subterminal band.

39h. Dark intermediate morph, streak-breasted type: Mainly black head (throat can be white). Underparts are black with small amount of white markings on breast. Flight feathers are barred, with barred fingers. *Gray tail has irregular mix of mottling and banding.*

39i. Dark intermediate morph, spot-bellied type: White or dark throat. Underparts are black, with white spotting on belly, flanks, and some wing coverts. Mottled/barred flight feathers. White-and-rufous-banded tail has wide subterminal band.

39j. Dark morph: All-black body. *Flight feathers are mottled. Gray-mottled tail has wide, black terminal band.*

a
b
c
1
2
3
d
e
f
g
h
i
j

Plate 40. RED-TAILED HAWK (*Buteo jamaicensis*) "Fuertes" (*B. j. fuertesi*) and "Alaskan" (*B. j. alascensis*)

Ages: Juvenile, 1-year-old, and adult. Both subspecies have long wings (similar proportions as respective ages of "Western"; Plates 32–35). *Black patagial mark on underwing extends from body to wrist; visible when in flight.* In flight, wingtips are rounded, with 5 fingers. *Juveniles.*—Plumage is held for 1st year. Pale yellow, gray, or tan eye. Cere is green but sometimes pale yellow. Brown or gray tail has multiple black bands. Small white patch adorns mid-scapulars. Trailing edge of underwing has gray band. *1-year-olds.*—Adult-like, with a few retained faded brown juvenile feathers on wing and sometimes tail. Eyes are pale yellowish and usually turn darker with age. *Adults.*—Plumage is attained as 2-year-old in fall of 3rd year. Eyes darken to dark brown. Cere is green or yellow. *Tail is rufous.* Adults have broader wings and usually shorter tails than juveniles. Trailing edge of underwing has wide black band. Each side of neck has rufous patch. *"Alaskan."*—Not readily separable from many "Western." Examples herein are from AMNH and RBCM from Graham Island, Haida Gwaii (Queen Charlotte Islands), B.C. **Subspecies:** 7 subspecies in U.S. and Canada (6 in West); 6 more in Mexico, Cen. America, and Caribbean. **Color morphs:** None in either subspecies. **Size:** L: 17–22" (43–56cm), W: 43–56" (109–142cm). Females average larger. "Fuertes" is largest and "Alaskan" is smallest of subspecies of U.S. and Canada (size differences not shown). **Habits:** Tame to fairly tame. Uses exposed perches. **Food:** Perch and aerial hunter of amphibians and reptiles ("Fuertes") and small birds and mammals. **Flight:** Often soars. Wings are held on flat plane or in low dihedral. Glides with wings held flat or bowed slightly downward. Powered flight is with moderately slow wingbeats in variable cadence. May hover and kite when hunting. **Voice:** *Juveniles.*—Clear, high-pitched, drawn-out *skee-ah*; high-pitched *klee* or *kree* when begging. *Adults.*—Raspy, hoarse *skee-ya*, *skee-yer*, or *squee-ah*.

40a. "Fuertes" juvenile: Similar to juvenile "Eastern," but when perched, wingtips closer to tail tip, as on "Western." Underparts often have narrow, V-shaped pattern on belly rather than streaking as on other subspecies. *Mid-scapulars have white patch.* Legs are lightly spotted or unmarked. Tail has many thin dark bands.

40b. "Fuertes" adult (Tex.): Dark head with white or streaked throat. Upperwing coverts are dark. Wingtips reach tail tip. *Small or large, pale tawny scapular patches.* White underparts lightly marked. *Rufous tail is unmarked.*

40c. "Fuertes" adult (Ariz.): Dark head with streaked or dark throat. White or tawny underparts are lightly marked on flanks, rarely on belly. Leg feathers can be rufous-barred (as on "Western"). *Rufous tail lacks subterminal band. Note:* May be an intergrade with "Western"; also found in s. N.Mex.

40d. "Fuertes" adult (Tex.): Similar to "Eastern" adult but darker, with dark forehead. Throat is white or streaked.

40e. "Fuertes," unmarked tail (Tex.; dorsal): White uppertail coverts. *Subterminal band is lacking on rufous tail.* This tail pattern is most commonly seen on this subspecies but is also found on "Eastern" and, to a lesser extent, "Western."

40f. "Fuertes," single-banded tail (Tex.; dorsal): White uppertail coverts. Tail pattern shared by "Eastern."

40g. "Fuertes," unmarked tail (Ariz.; dorsal): *Rufous uppertail coverts. Lacks subterminal band on tail.*

40h. "Fuertes," partial-banded tail (Ariz.; dorsal): Partial banding on some feathers; possible intergrade.

40i. "Fuertes" adult: *Dark patagial marks.* Belly is usually unmarked and flanks may be lightly barred.

40j. "Alaskan" juvenile: Similar to light morph "Western" (*32a*) but darker, more blackish. Throat is white or streaked (not all-dark as on some light morph "Western").

40k. "Alaskan" juvenile: Similar to light morph "Western" (*32e*) but darker, more more blackish, and lacks pale supercilium. Small white on scapular patches. Leg feathers are heavily barred. Undertail coverts are barred (not shown). Tail has 6 or 7 wide, dark bands.

40l. "Alaskan" juvenile tail (dorsal): Brown, with 6 or 7 wide, dark bands or wider subterminal band.

40m. "Alaskan" juvenile tail feather: Brown, with 6 or 7 wide, equal-width bands or with wider subterminal band.

40n. "Alaskan" adult: Dark head, including throat. *Breast has unique, rufous-brown, heart-shaped markings.*

40o. "Alaskan" adult, lightly marked type: Dark head; throat can be streaked. Breast can be solid rufous, darker than rest of underparts. Bellyband is barred with black. Leg feathers are barred.

40p. "Alaskan" adult, heavily marked type: Dark upperparts have few pale areas on scapulars and wingcoverts. Breast as on *40n*. Heavy blotchy bellyband. Leg feathers and rest underparts are pale, with rufous barring. Rufous tail can be banded (shown) or unbanded as on *40o*.

a
b
c
d
e
f
g
h
i
j
k
l
m
n
o
p

Plate 41. RED-TAILED HAWK (*Buteo jamaicensis*) Adults
Albinos and Other Variants

Albino types.—Albinism is common in Red-tailed Hawks and is found in all subspecies and color morphs. Total albinism, with pink eyes and fleshy areas and all-white plumage, is rare and not depicted. Albinism may affect only portions of a feather. Birds with partial, or incomplete, albinism have normal-colored eyes and fleshy areas. *Leucistic types.*—Plumage is *not* albino; leucism is reduced pigmentation, which makes darker colors tan or pale rufous. Also called *dilute* plumage. *Other variations.—Left vertical row:* 3 variations of "Eastern" subspecies that occur in *all* of U.S. and Canadian breeding range: Dark-headed/dark-throated and streak- and/or colored-breasted birds are uncommon in *all* regions. Bar-legged type and band-tailed type are common to very common in w. Canada but uncommon to very uncommon in East. *Middle vertical row:* 3 variants of "Harlan's" possibly exhibiting traits of other Red-tailed Hawk subspecies, especially "Western." *Right vertical row:* 4 variations of "Harlan's" tail patterns that may show traits of either "Eastern" or "Western."

41a. Partial albino (any subspecies): Bill, cere, and eye color as on typical birds. *Note:* 1 "normal" feather is retained on neck; rest of plumage is all-white.

41b. "Western" partial albino: Much of plumage is white, except a few wing and tail feathers.

41c. "Eastern" partial albino: Virtually all-white, except 2 small normal feathers on neck and upperwing coverts. Bill, eye, and fleshy areas have normal coloration.

41d. "Western" leucistic: Plumage appears tan-colored. All rufous-, brown-, or black-pigmented areas of normal plumage are tan or pale rufous.

41e. "Eastern" partial albino: Albinism affects wings and dorsal body surface more commonly than ventral body surface and tail.

41f. Talon of partial albino: Talon is pale bluish or pinkish. Legs and toes are normal yellow.

41g. Partial albino (any subspecies): At field distances, palest partial albinos can appear all-white. *Note*: "Eastern" because of short fingers on primaries.

41h. "Eastern," bar-legged type: White throat with wide dark collar. Breast is lightly streaked (shown) or unmarked. Thigh feathers partially or fully barred. *Note:* Common in w. Canada east of Rocky and Mackenzie Mts. In East: Atlantic Co., N.J. (resident, 2015); "Christo," Tompkins Sq. Pk., N.Y.C. (resident, 2014); Oakland Co., Mich. (May).

41i. "Eastern," streak-breasted, heavily marked type: Dark-headed type. Rufous/tawny on breast often fades to white by late fall–winter. Breast is streaked or marked with dark blobs all year. Bellyband is heavily marked type. *Note:* Darker types are fairly common in w. Canada. In East: "Charlotte," N.Y.C. (resident 2005–08, based on Sept. 2005). Dark-headed type in all of range.

41j. "Eastern," streak-breasted and/or band-tailed type: White-throated with thin collar. Breast is tawny and streaked; often fades paler (to white) by late fall–winter. Bellyband is moderately marked type. Thighs are partially barred. Tail is banded; can be partially or fully banded. *Note:* Band-tailed type is very common in w. Canadian range. Uncommon in East: Long Island, N.Y. (May); seen on "Scarlette" of James City Co., Va.; resident of DeKalb Co., Ga.

41k. "Western" intermediate morph × "Harlan's": Dark throat and rufous breast and thighs, as on intermediate morph "Western," but tail is gray and partially mottled on base, as on "Harlan's." *Note:* Tok, Alaska (Jul.).

41l. "Harlan's" × "Western" intermediate morph: Similar to *41k*, with dark throat of "Western," but more like "Harlan's," with white supercilium, white breast markings, and gray, mottled tail. *Note:* Holt Co., Mo. (Nov.).

41m. "Harlan's" with rufous tail: Dark intermediate morph "Harlan's". Rufous tail with genetics of other Red-tailed Hawks or an intergrade. *Note:* Fairbanks, Alaska (Apr.).

41n. "Harlan's" tail: "Harlan's" mottled base, but "Eastern"/"Western" color and banding pattern on distal ⅔, with neat subterminal band. *Note:* Head and body of this bird were of dark intermediate morph "Western"; Minn. (Oct.).

41o. "Harlan's" tail: "Harlan's" black-and-white-mottled base; somewhat "Western" banding pattern and color on distal portion, but "Harlan's" incomplete subterminal band. *Note:* This bird had light morph "Harlan's" body; s. Alaska (Aug.).

41p. "Harlan's" tail: "Harlan's" influence shows in white on much of tail. Tail banding is as found on majority of "Eastern" in w. Canada. *Note:* This bird had head and body of "Eastern" (rufous neck patch) and some "Harlan's" traits; Nevis, Alta. (Jul.).

41q. "Harlan's" tail: "Harlan's"-type mottling on basal white area. Neat subterminal band is atypical for this subspecies. *Note:* Head and body similar to "Eastern"; Delta Junction, Alaska (Jul.).

a
b
c
d
e
f
g
h
k
n
i
l
o
p
j
m
q

RED-TAILED HAWK (*Buteo jamaicensis*)
"Eastern" (*B. j. borealis*), "Krider's (*B. j. kriderii*), "Western" (*B. j. calurus*), "Harlan's" (*B. j. harlani*), "Fuertes" (*B. j. fuertesi*), "Alaskan" (*B. j. alascensis*)

STATUS: "Eastern." For in-depth information on plumage variation among "Eastern" birds, status of "Eastern" birds in summer/breeding season, and mixing of subspecies among w. populations, see "Plumage Variation in 'Eastern' Red-tailed Hawk," further below. See also "Analysis of the Proposed 'Northern' Subspecies of Red-tailed Hawk (*B. j. abieticola*)." *Winter/migration data note*: All migratory and winter data in the text for all Canadian-bred populations are based on banding data in Dunn et al. (2009) and Houston (1967). No banding data exist for N.W.T., but the author surmises that this population likely also winters far south and east of breeding areas in U.S. *Summer.*—"Eastern" is a common and widespread subspecies south of boreal forest. Range extends west across Great Plains to Rocky Mts. in U.S. Revised breeding range depicted for this subspecies in w. Canada is based on studies of large numbers of breeding birds of this subspecies found in w. Canada east of Rocky and Mackenzie Mts. (see "Plumage Variation"). Breeding range is extended—from the previously published delineation in boreal forest of cen. Man. (Preston and Beane 1993, Wheeler 2003b) westward to e. edge of Rocky Mts. of Alta. and ne. B.C. and northward east of the barren Mackenzie Mts. of N.W.T. and Richardson Mts. of Y.T. (uninhabitable for Red-tails) and farther north to Mackenzie River delta. (This demarcation also aligns with w. edge of range for many e. forest-dwelling songbirds.) *Note*: Black-dotted line on "Eastern" range map in w. Canada delineates range overlap and potential intergrading with "Western" and mainly light morph "Harlan's." Magenta dashed line denotes sparsely populated region of lowland boreal forest in n. B.C. and s. Y.T. between Rocky and Mackenzie Mts. Northern range border of "Fuertes" subspecies in Tex. is vaguely known (DNA from 9 summer birds from nw. Tex. south of panhandle) but is aligned with "Eastern" (Hull et al. 2010). *Winter.*—Majority of "Eastern" subspecies winter strongly southeast of breeding areas into cen. U.S. Banding data from Alta. and Sask. birds show they move as far east as cen. Ohio and south to nw. Fla. Population from N.W.T. is probably identical to above data. In e. Colo., wintering population of "Eastern" subspecies far outnumbers those of "Western" and "Harlan's"; lightly marked and moderately marked birds are the vast majority, and heavily marked birds are a small minority (pers. obs.). Canadian banding data show that some winter due south, mainly east of the Rockies, but a very small percentage may pass through e. Great Basin and winter as far south as Ariz.—and may be regular in very small numbers there. An adult photographed in winter at Kamloops, B.C., appears to be this subspecies (R. Howie pers. comm./unpubl. photo). **"Krider's."** This pale bird is considered a subspecies by most authorities and the AOU. It is also believed by some authors to be a pale-colored variant or color morph of "Eastern," which a great number of individuals now appear to be. (The author previously considered them as such, which is why the 2 subspecies are on the same plates.) This subspecies was probably quite isolated on the n. Great Plains for thousands of years, sharing the short-grass semi-arid region with Ferruginous and Swainson's Hawks. However, since the late 19th century, massive habitat alteration on the plains, including increased tree growth and native grassland alteration, enabled the especially adaptive "Eastern" subspecies to spread westward and southward. This massive influx of "Eastern" hawks has completely inundated "Krider's" core native grassland breeding range. Historically, "Western" has mixed with "Krider's" on the latter's w. border, and light morph "Harlan's" has sparingly mixed with it on its n. border in s. Sask. Very pale pure "Krider's" birds still exist and often mate together (Sullivan and Liguori 2010). Red-tailed Hawks of n. Great Plains are generally overall pale, though a few heavily marked birds are noted in N.Dak. (see Liguori and Sullivan 2010). With time, this subspecies will be further diluted because of continual intermixing with and dominance of "Eastern" subspecies (see Liguori and Sullivan 2010). *Summer.*—The range map has been revised, and "Krider's" breeding area reduced from previous publication (in Wheeler 2003b). The subspecies breeds in n. prairie states and in very s. Sask. and very sw. Man.; core population appears to be in N.Dak. (based on data in Liguori and Sullivan 2010). There are no sight records, photographic records, museum examples, or rehabilitation birds of this subspecies from any season reported from Alta. (W. S. Clark pers. comm.). *Winter.*—"Krider's" is uncommon in winter in states shown in cyan on the range map. Most of the population moves due south of breeding area and winters on s. Great Plains south to se. Tex. and s. La. Lesser numbers of this subspecies winter southeast of breeding areas, mainly in states adjacent to Mississippi River; very rare/vagrant to East Coast and Fla. **"Western."** An even more variable subspecies than "Eastern" because of multiple color morphs as well as individual variation. Hull et al. (2010) found that DNA of cen. Calif. and Great Basin residents is different; DNA markers from darker morphs of "Western" align with light morphs of this subspecies but are quite different from those of darker "Harlan's" (but authors note that all subspecies are overall fairly similar). For detailed status of color morphs, see "Western" coverage in "Regional Status of Red-tailed Hawk Plumages" in "Plumage Variations" essay. Summer (w. Canada).—Range map is revised from Wheeler (2003b), and "Western" breeding range is reduced from previous publications (in Wheeler 2003b; Preston and Beane 1993). Map is based, in part, on an unpublished map by W. S. Clark. Magenta dashed pattern on range map throughout n. Alta. and N.W.T. is where this subspecies historically (Preble 1908) and currently exists in small numbers, but it is far outnumbered by "Eastern" (and interbreeds with "Eastern"). This dashed pattern is based *solely* on photographic and museum records of the easy-to-identify intermediate (rufous) morph. "Western" subspecies breeds from s. Rocky Mts. and westward, north throughout B.C., with a small number of mainly darker birds in extreme s. Y.T. Summertime birds on Vancouver Island align more with this subspecies than with "Alaskan" subspecies (based on limited data collected by author). Light morph makes up bulk of "Western" population. Intermediate (rufous) morph is uncommon (5% of local breeding population is deemed average in most regions). Dark morph varies from uncommon to rare in most of its breeding range; it is even absent from s. Rocky Mts. (pers. obs.) but prevalent in main (magenta-colored) Canadian range, especially in B.C. All color morphs interbreed extensively with "Eastern" along e. range border and especially in boreal forest of n. Alta. and northward east of Rocky and Mackenzie Mts. All color morphs interbreed with "Harlan's" in n. B.C. and in s. Y.T. No verifiable "Western" found east of Alta. during breeding season. *Winter.*—Bulk of B.C. "Western" population winters mainly due south of breeding grounds in s. B.C. and w. U.S. from Rocky Mts. and westward. According to Canadian banding data (that is nonspecific to subspecies), the small percentage of this subspecies that breeds on e. side of Rocky Mts. and (undoubtedly) Mackenzie Mts. of Alta., B.C., and N.W.T. winters east of Rocky Mts. to cen. U.S. This small population shares winter grounds with "Eastern" birds but is far outnumbered by them (and "Harlan's"). Adults of more southerly latitudes are resident. "Western" is rare in this season on n. Great Plains; a few records exist for s. Alta., including a photographed dark morph juvenile. This subspecies is irregular in cen. B.C. north of Kamloops to Williams Lake (Campbell et al. 1990); not depicted on map, also occurs farther north to Quesnel (R. Howie pers. comm.) and very irregularly in low snow years at Prince George (J. Bowling pers. comm.). **"Harlan's."** This is a distinctive subspecies with blackish plumage that is, in its purest form, quite diagnostic. Most notable adult plumage feature on all but darkest of dark morphs is white or whitish throat—and even some all-dark birds have white throat. Most diagnostic juvenile trait is black "spikes" along tip of feather shaft on some or all tail feathers and often on secondaries and upperwing coverts. Darker morphs (encompassing any plumage darker than light morph) are most common in core range of Y.T. and Alaska. Based on personal observation on Apr. 10–18, 2009, at Gunsight Mt. hawk watch near Glennallen, Alaska (mile marker 118), of 725 adult "Harlan's," 68 (9%) were light morph. This is similar to percentage seen in winter in e. Colo. (pers. obs.). East of core range, in area within magenta dashed linear pattern on range map, are primarily light morph types. Of 754 total Red-tailed types from Gunsight Mt., 29 (4%) were non-"Harlan's" (which indicates potential mixing of subspecies). In DNA studies, Hull et al. (2010) showed that this subspecies is "quite similar genetically" to "Western" and "Eastern" but "shows closer evolutionary affinities" to "Eastern," and "historical and contemporary gene exchange" between both subspecies. (Most identifiable intergrade or shared *jamaicensis* traits are with "Western," particularly with rufous-breasted intermediate morph.) *Summer.*—"Harlan's" core breeding range is in extreme n. B.C. and all areas with larger trees and low cliffs along rivers in Y.T. and Alaska. The magenta dashed line on the range map encompasses regular but low-density summer and/or breeding records of adults (examples follow, by region). Majority of birds seen in this region are light morph; darker morphs recently found nesting around Edmonton, Alta. Northwest Territories: 1 "Harlan's" light morph and 1 probable intergrade type south of Great Slave Lake (G. Vizniowski unpubl. photos); 1 darker morph adult in summer near Ft. Simpson (G. Matchans pers. comm./photo); 2 darker morphs in 2016 at Ft. Hays in mid-May and Ft. Smith in late Apr. (G. Vizniowski pers. comm./photos). A juvenile from the Saline River near middle part of Mackenzie River, N.W.T., appears to be primarily this subspecies (2 dark tail spikes is indicative of "Harlan's"; dorsal/ventral photos, CMN #18549). An adult darker morph was seen in 1980s near Norman Wells, N.W.T., on n. Mackenzie River (G. Court pers. comm.). A rufous morph "western" noted in Mackenzie River delta in 1955 probably has at least some "Harlan's" in it (but lacked whitish throat typical of most "Harlan's"). Preble (1908) makes no mention of this subspecies on his Mackenzie River expeditions. It is likely the few "Harlan's" that may occur along Mackenzie River are probably the difficult-to-identify light morph. An intermediate 1-year-old nonbreeder from w. side of Lake Athabasca in late Jun. 1920 is 1 of a few summer records of darker morphs in ne. Alta. (dorsal/ventral photos, USNM #28311). ***British Columbia and Alberta***: Cen. Alta. nest records date back to Jul. 1917, of a nesting light morph female with many "Harlan's" characters from Nevis (near Red Deer), shown in Sullivan and Liguori (2010). A light morph nesting female from 1925 was also seen near Nevis (image in Sullivan and Liguori 2010; CMN #19690). Recent data show that light morphs nest in low density far east and south of their core range of n. B.C., Y.T., and Alaska: into N.W.T., south and east in ne. B.C., in much of Alta. and Sask. north of Calgary and Regina, respectively, and sparingly into w. Man. and n.-cen. N.Dak. (K. Donohue pers. comm./photos, W. S. Clark pers. comm./photos, N. Saunders pers. comm./photos). They are now being seen near Calgary, Alta., in summer. ***Saskatchewan***: "Harlan's" birds are photo-documented in boreal forest in this province (much farther southeast than the subspecies' core range) in what appears to be very atypical but undoubtedly historical habitat for at least light morphs on n. Great Plains of s. regions of the province (W. S. Clark pers. comm./photos; Sullivan and Liguori 2010; N. Saunders pers. comm./photos). (They have probably always been in this region but likely [and easily] mistaken for "Krider's.") ***North Dakota***: A light morph nested with a heavily marked "Eastern" bird in Turtle Mts. (Sullivan and Liguori 2010), which is the most southerly nesting record of this subspecies and 1st for U.S. These s. region light morphs may nest with other light morph "Harlan's" but often mate with "Eastern," "Western," and even "Krider's" counterparts. *Winter*.—Main historic "Harlan's" winter quarters are in large cyan block of s.-cen. U.S. on range map. A large number winter in scattered, isolated locations west of Rocky Mts. throughout Great Basin, and lesser numbers along Pacific coast lowlands. There are several records of adults in Baja California Sur, Mexico; and 1 for late Feb. in se. Alaska (S. Heinl, A. Piston pers. comm.).**"Fuertes."** *Year-round*.—This subspecies is common in all of its range. Separation from "Eastern" is difficult, because both subspecies can have minimal ventral markings or lack ventral markings, but Fuertes has longer wings that usually reach tail tip when perched. (Scapular color of adults is gray or tawny, whereas in "Eastern" adults, it is mainly white.) Fuertes intergrades broadly with "Western" along borders of its N.Mex. and Ariz. range, and most birds from se. Ariz. appear to be intergrades. As noted in "Eastern" status, n. border is uncertain because so many "Eastern" birds of s. Great Plains are lightly marked. Range map follows Johnsgard (1990), which seems accurate in terms of current range knowledge and DNA studies on "Eastern" in n. Tex. by Hull et al. (2010). **"Alaskan."** Data for plumages and illustrations are based on examples from isolated Graham Island of Haida Gwaii (Queen Charlotte Islands), B.C. (AMNH and RBCM; unpubl. images by M. Hearne and P. Hamel). This isolated population possess consistent plumage traits of this subspecies. Separating this subspecies from some vartiations of "Western" can be difficult in the field. *Summer*.—"Alaskan" birds are sparsely distributed and uncommon in near-coastal mainland areas west of Coast Mts. of B.C. and se. Alaska (S. Heinl, A. Piston pers. comm.), where they undoubtedly mix in some locations with "Western." The most unique-to-subspecies examples inhabit larger islands of this region, especially Haida Gwaii, B.C. Some adults of se. Alaska, including outermost islands, such as Baranof (C. Susie, unpubl. photos), appear to align more with "Western" subspecies and are perhaps intergrades with "Alaskan." Some individuals from Haines, Alaska, have pale supercilium patches and white ventral areas and do not appear like this subspecies but likely have "Harlan's" influence ("Harlan's" birds have nested just north of Haines [C. Susie, unpubl. photos; G. Vanvliet, unpubl. photos]); other birds are identical to classic light morph "Western" (W. S. Clark, unpubl. photos; C. Susie, unpubl. photos; S. Heinl, unpubl. photos). Birds from Juneau, Alaska, mainly appear as good examples of this subspecies (M. Schwan, unpubl. photo of juvenile; S. Heinl, unpubl. photos of adults). As noted previously, in "Western" subspecies section, birds on Vancouver Island, B.C., are very much unlike those of Haida Gwaii, B.C., or even coastal Alaska and are considered by the author to be "Western" subspecies. Birds from nearby San Juan Islands, Wash., align with typical lighter morphs of "Western" (pers. obs.). *Winter*.—"Alaskan" birds are rare in winter on coastal mainland (Ketchikan, Alaska); most vacate breeding areas and move south an undetermined distance (S. Heinl, A. Piston pers. comm.). Some island birds irregularly winter on larger islands, as Baranof Island, Alaska (noted in some Sitka Christmas bird counts). Haida Gwaii, B.C., population is sedentary. **NESTING: "Eastern."** Begins Jan.–May, depending on elevation and latitude. Even in n. states, nesting activities may begin in Jan.–Feb., but in Apr.–early May in n. Canada. Pairs remain together year-round in southerly and cen. latitudes and even in some areas in northerly latitudes. Large stick nest is built by both sexes and placed high in live or dead tree that has easy access and is situated with expansive view of surrounding area; typically isolated, tall dominant deciduous or coniferous tree, or similar size and type of tree on edge of woodlot or woodland. Nests may be placed on tall utility poles or on buildings in urban areas. Nests are placed in main fork or on inner or mid-part of sturdy horizontal branches. Nests may be reused for many years and can become quite large. Brood size is 2 or 3, rarely up to 5. Both sexes incubate. Nesting is completed Apr.–Sep. **"Krider's."** Season begins Mar.–Apr., possibly Feb. for s. pairs. Nest and egg data as in "Eastern," but nest located in mainly isolated live or dead deciduous tree or within shelterbelt grove of deciduous trees. **"Western."** Highly variable nesting-cycle timing depends on elevation and latitude inhabited. Southern low-elevation pairs may start in Dec.; high-elevation and/or northerly pairs may begin in Apr. All other data are similar to "Eastern," including nest site locations, except clutch size is smaller with high-elevation and/or northerly latitude birds. Birds of far s. areas may nest on tall cacti. **"Harlan's."** Nest cycle begins in early Apr.–early May in Y.T. and Alaska, but probably slightly earlier in more southerly locations. Nests are placed in upper part of White Spruce or Balsam Poplar; also on low cliffs (especially along Porcupine River in ne. Alaska; B. Ritchie pers. comm.). Brood size is 2 or 3. Cycle completed in Aug.–Sep. **"Fuertes."** Nesting may begin in Dec. Nests placed in tall live-oaks or tall mesquite; may use utility poles and especially Saguaro cactus (*Carnegiea gigantea*) in Ariz. **"Alaskan."** No data available. Nest sites are probably in tall live or dead Sitka Spruce or Western Hemlock, the region's dominant trees. **MOVEMENTS: "Eastern."** Nesting s. adults and many mid-latitude adults are resident. Northerly latitude breeding adults from n. U.S. and most of w. Canada are short-distance or moderately long-distance migrants (very few winter records in s. Canada [s. Alta.]). Canadian banding data show population nesting east of Rocky and Mackenzie Mts., moving mainly in sharp southeasterly direction from breeding grounds to winter grounds, staying east of Rocky Mts. Far lesser numbers move slightly southeast but still east of Rocky Mts. A very small percentage of Canadian birds move due south through e. Great Basin. Younger adults and juveniles engage in the most movement, with juveniles moving farthest. *Spring.*—"Eastern" adults move before returning juveniles. Adult movement is noted on Great Plains in Feb.–Mar. and peaks in mid–late Mar. Along w. Great Lakes, adults peak in early–mid-Apr. Young adults and returning juveniles move Apr.–Jun., peaking along Great Lakes in mid–late Apr., stragglers later. *Late-summer dispersal.*—"Eastern" juveniles and some nonbreeding adults engage in northward dispersal late Jul.–early Sep., prior to heading south. A juvenile from Okla. went north to s. Man. by early fall; a juvenile banded in n.-cen. N.Y. went northwest—also to Man. by early fall. *Fall.*—"Eastern" juveniles move 1st, the earliest starting in mid–late Aug., especially in n. Canada. Very large numbers of juveniles are noted in late Sep. on n. Great Plains. Along w. Great Lakes and on cen. Great Plains, juveniles peak in mid–late Oct. Adults start later, peak in late Oct., but continue into Nov. in small numbers. Preble (1908) noted several (undetermined age, subspecies) hunting "Varying" (Snowshoe) Hare on Sep. 30, 1903, at mouth of Great Bear River, near Mackenzie River, N.W.T. (near Tulita). **"Krider's."** *Spring.*—Data for all ages same as "Eastern. Fall.—Data for all ages same as "Eastern," but even breeding adults leave nesting grounds. **"Western."** Lower-elevation and southerly latitude birds are resident. Young birds disperse out of natal territory. High-elevation and northerly latitude breeders and offspring migrate short to moderate distances. Main population nests in Rocky Mts. and westward and remains in Intermountain West in winter. Based on Canadian banding data, birds from B.C. move due south from breeding/natal areas. Among the small contingent breeding east of Rocky Mts. (and likely Mackenzie Mts.), some head due south, but most, like Canadian "Eastern" population, head sharply southeast over Great Plains—or even farther east of the plains. *Spring.*—"Western" breeding adults move 1st, beginning in Feb. and peak in mid–late Mar. In w. Wash., adults begin in early–mid-Mar. and peak in early Apr. Young adults and juveniles move later, beginning in Mar. and peaking in mid–late Apr. *Summer dispersal.*—Following "Western" data are gathered from banding and telemetry data from s. Calif. sedentary breeding adult population in Bloom et al. (2015): Most southerly and mid-latitude juveniles and many nonbreeding adults (until they obtain nest mate and territory) engage in northward summer dispersal (similar to such movement documented for "Eastern" subspecies). Recently fledged birds initiate northward movement early May–early Aug. Movement northward may take only 3 days or up to 2 months, and birds arrive at maximum distance location by early Jul.–mid-Sep. Some immediately return to natal/summer location, others linger in north, arriving back at origin locales by early Sep.–early Nov. Subsequent summer movement begins earlier than previous year (if bird was a juvenile), as early as Apr. From s. Calif., movement is mainly northerly, but a few went eastward and very few southward. Most went north into n. Calif., but 9% radiated north, northeast, and east to cen. and e. Oreg., s. Idaho, sw. Mont. (1 crossed Continental Divide to Bozeman), Utah, and all parts of Ariz. Maximum dispersal distance was 904 mi. (1456km) from natal location. The interesting fact is that all that survived (in telemetry sample) returned to point of origin! This also illustrates that s. Calif. birds can show up anywhere west of Rocky Mts. during summer and fall. *Fall.*—"Western" juveniles and even adults begin moving out of high-montane elevations and down to valleys, with small percentage moving eastward onto e. foothills of Rocky Mts. and adjacent plains in mid–late Aug. Canadian birds that breed east of Rocky and Mackenzie Mts. mirror passage dates and pathways for all ages of "Eastern." U.S. intermountain movement is early. Juveniles begin moving before mid-Aug., move steadily throughout Sep., peak in 3rd week of Sep. Adult peak is in early Oct., with good numbers and possible 2nd peak in mid-Oct. (Much southward movement encompasses previously noted summer dispersal.) **"Harlan's."** This subspecies is a long-distance migrant. Majority follow historic passage to and from Y.T. and Alaska, likely through the narrow corridor above Rocky Mts. in B.C. and below Mackenzie Mts. in s. Y.T. and s. N.W.T.—and stay east of Rocky Mts. Far fewer numbers may use a narrow corridor between n. Mackenzie Mts. and s. Richardson Mts. in and out of n. Y.T. and n. Alaska (Porcupine River and Eagle Flats regions) and use Mackenzie River valley of N.W.T. A juvenile banded in w.-cen. Y.T. (Dunn et al. 2009) migrated along "normal" route via passage over and east of Rocky Mts. to s. Great Plains; 1 juvenile from s.-cen. Y.T. went due south through intermountain corridor to s. Ariz. Most of population that winters west of Rocky Mts. of U.S., including along West Coast, would seem to follow this pathway east of formidable Wrangell Mts. of Alaska and Coast Mts. of B.C. and into s. B.C. and spread throughout w. U.S. Island-hopping migrants of this subspecies have been noted on se. Alaska islands (C. Susie pers. comm.: adult at Kupreanof Island in fall). *Spring.*—"Harlan's" adults begin leaving winter grounds in late Feb., and most leave in mid–late Mar. Late adults may still be on winter quarters in Colo. in mid-Apr. (pers. obs.). Adults 1st arrive in Alaska in early Apr. and peak in mid-Apr. to sometimes 3rd week of Apr., at a rate of over 250 birds a day (Gunsight Mt. hawk watch, mile marker 118, Glenn Hwy.). Juveniles, as with other subspecies, start later and leave mainly in Apr., their arrival in Alaska spread out from late Apr. through May and into early Jun. *Fall.*—"Harlan's" juveniles migrate 1st, leaving core breeding grounds in mid–late Aug. Adults may begin moving in Alaska in late Aug.–early Sep., and most leave breeding grounds by late Sep. Juveniles, especially, are noted in fair numbers in N.Dak. by late Sep. and peak at w. Great Lakes hawk watches in mid–late Oct. Early season young adults arrive on winter grounds in se. Colo. by late Sep., and older adults by early Oct., but bulk arrive in late Oct. through mid–late Nov., with movement into Dec. **"Fuertes."** Sedentary. As with southerly latitude sedentary "Western" noted previously, younger birds may engage in northward summer dispersal. **"Alaskan."** *Spring.*—Late Mar.–early May; possible peak early–mid-Apr. (S. Heinl, A. Piston pers. comm.) *Fall.*—2 possible migrant juveniles seen near Sitka, Alaska, in late Aug.–early Sep. (C. Susie pers. comm.). Movement spans mid-Aug.–Nov. at Ketchikan, Alaska, with several seen in mid-Sep. (S. Heinl, A. Piston pers. comm.). **COMPARISON:** ***"Eastern" Red-tailed Hawk juvenile.*** **Red-shouldered Hawk (juvenile).**—Medium-brown eye, *vs.* tan, pale gray, or yellow eye. Yellow cere, vs. pale greenish cere but sometimes yellow cere (use caution). In flight, crescent-shaped panel/window on base of primaries, *vs.* pale, rectangular tan or white panel/window. Breast and belly are streaked, vs. unstreaked breast (only intermediate morph "Western" Red-tail has streaked breast but also has dark bellyband). **Broad-winged Hawk (juvenile)**.—Bright yellow cere, *vs.* pale greenish (but sometimes yellow) cere. Eye darker medium brown, *vs.* tan, pale gray, or pale yellow eye. When perched, wingtips are near tail tip, *vs.* wingtips distinctly shorter. Breast may be unmarked and belly marked, similar to Red-tail (use caution). In flight view, patagial area of wing's leading edge is unmarked, *vs.* dark patagial mark. **Swainson's Hawk (juvenile).**—Cere usually yellow but sometimes greenish, *vs.* greenish and sometimes yellow (use caution). Medium-brown eye, *vs.* tan, pale gray, or pale yellow eye. Dark patch on sides of neck, *vs.* streaked sides of neck. Wingtips nearly equal tail tip when perched, *vs.* wingtips much shorter. In dorsal view, secondaries are dark, *vs.* brown with dark barring. In ventral view, flight feathers are medium gray, *vs.* white. Uniformly marked underwing coverts, *vs.* dark patagial mark on wing's leading edge. ***"Krider's" juvenile.*** **Swainson's Hawk (pale juvenile, 1-year-old)**.—Share pale or white heads. Swainson's always has thin, dark line behind eye, but this marking is found on a few "Krider's." Dark patch adorns sides of neck, *vs.* streaking on sides of neck. When perched, wingtips equal tail tip, *vs.* wingtips much shorter than tail tip. Dorsal surface of secondaries is dark, *vs.* brown/tan with thin dark barring. Distinct dark patch on sides of breast, *vs.* sparse streaking. Gray flight feathers on underwing, *vs.* white flight feathers. ***"Harlan's" juvenile light morph.*** **Swainson's Hawk (pale juvenile, 1-year-old)**.—Similar to "Krider's" data, but, in addition, Harlan's tail often has wider dark bands or lack of bands and often is mottled. ***"Harlan's" and "Western" darker morphs.*** **Rough-legged Hawk (all ages).**—Forehead is white and throat is dark, *vs.* dark forehead and sometimes white throat ("Harlan's") or dark forehead and dark throat ("Western"). Legs are fully feathered, *vs.* bare lower legs. Tail patterns of females/juveniles can be identical to adult Harlan's. **Zone-tailed Hawk (juvenile).**—Barred pattern adorns base of outer primaries, *vs.* unmarked basal area. Tail patterns are similar. Dihedral wing position when soaring, with rocking motion, *vs.* wings held flat and stable flight. **Zone-tailed Hawk (adult).**—Underside of flight feathers is gray, *vs.* whitish. Tail has 1 or more broad, white inner bands, *vs.* lack of wide white bands.

Figure 1. Weld Co., Colo. (Jun.)
Nest site of "Eastern" in rural, semi-open to open grassland.

Figure 2. Weld Co., Colo. (Nov.)
Winter habitat for "Eastern," "Harlan's," and some "Western."

Figure 3. Jackson Co., Colo. (May)
Nesting habitat (with nest) for "Western " in montane valley at over 8,100' (2500m).

Figure 4. Codington Co., S. Dak. (Jul.)
Natural or disturbed pothole prairie is "Krider's" breeding habitat. Nests are placed in an isolated tree, in a small group of trees, or in a shelterbelt grove. Photo by Brian Sullivan

Figure 5. Matanuska River valley (Chugach Mts.), Alaska (Apr.)
"Harlan's" migration corridor near Glennallen.

Figure 6. Pima Co., Ariz. (Apr.)
Year-round habitat in Saguaro cacti for "Fuertes."

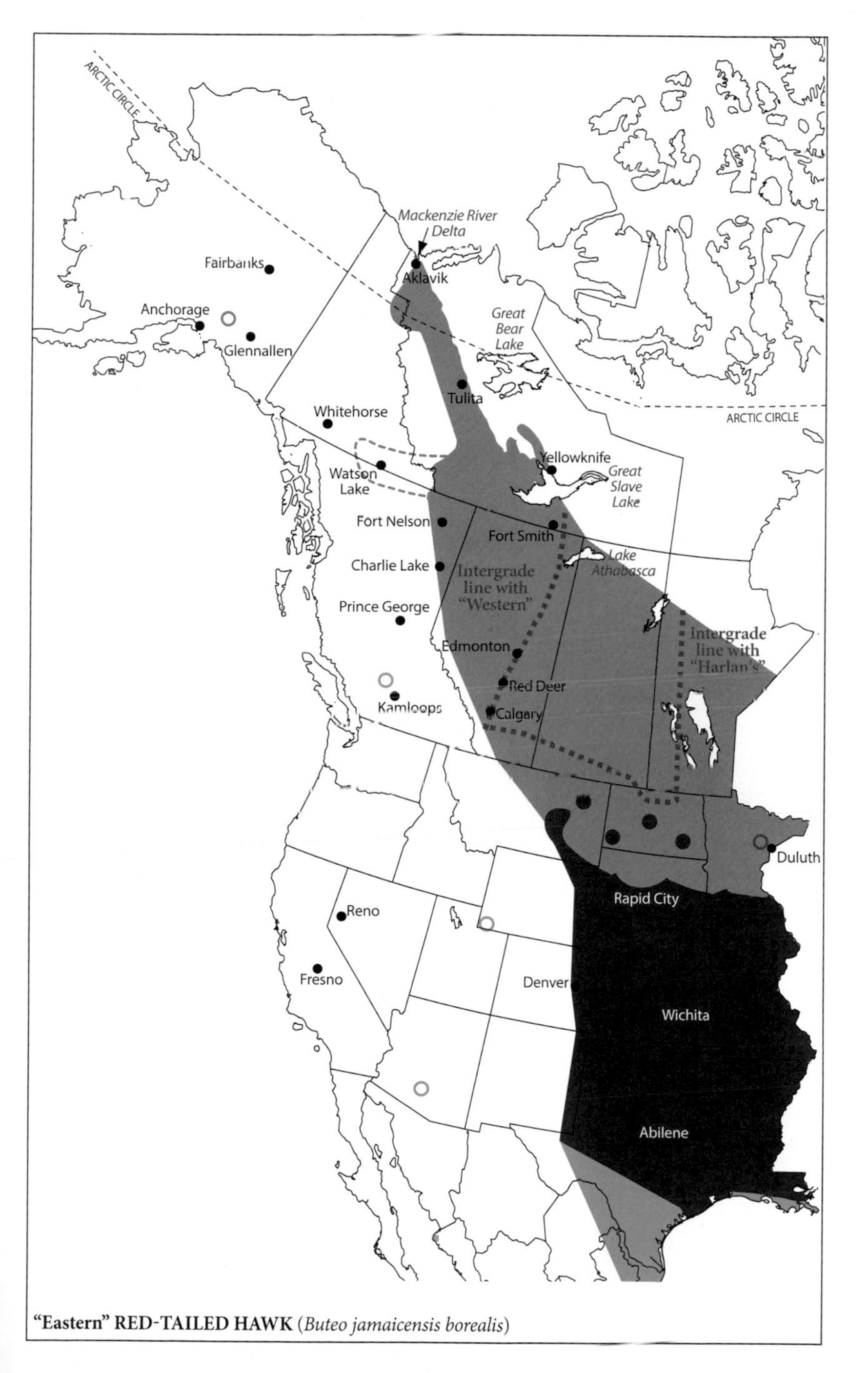

"Eastern" **RED-TAILED HAWK** (*Buteo jamaicensis borealis*)

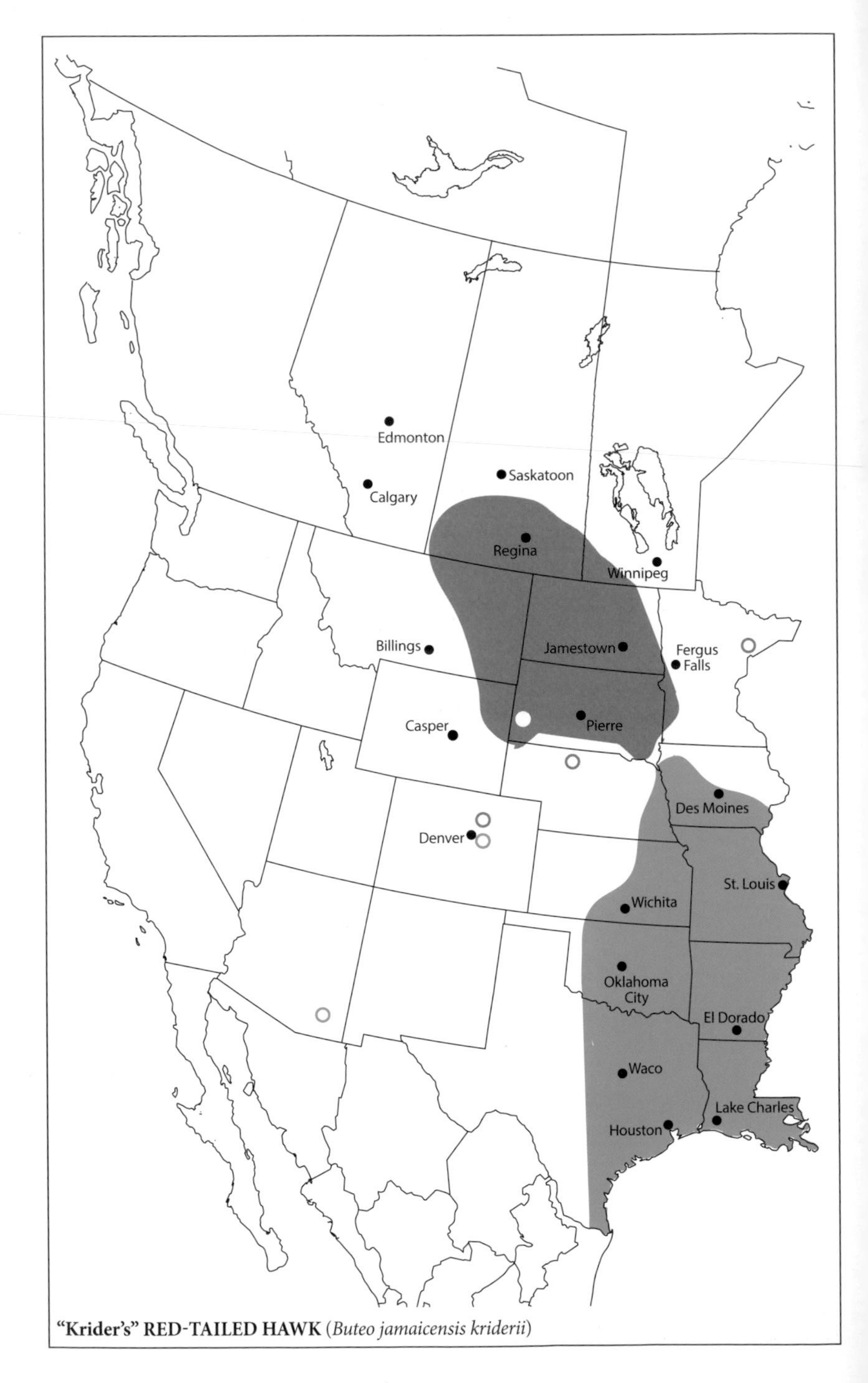

"Krider's" RED-TAILED HAWK (*Buteo jamaicensis kriderii*)

"Western" RED-TAILED HAWK (*Buteo jamaicensis calurus*)

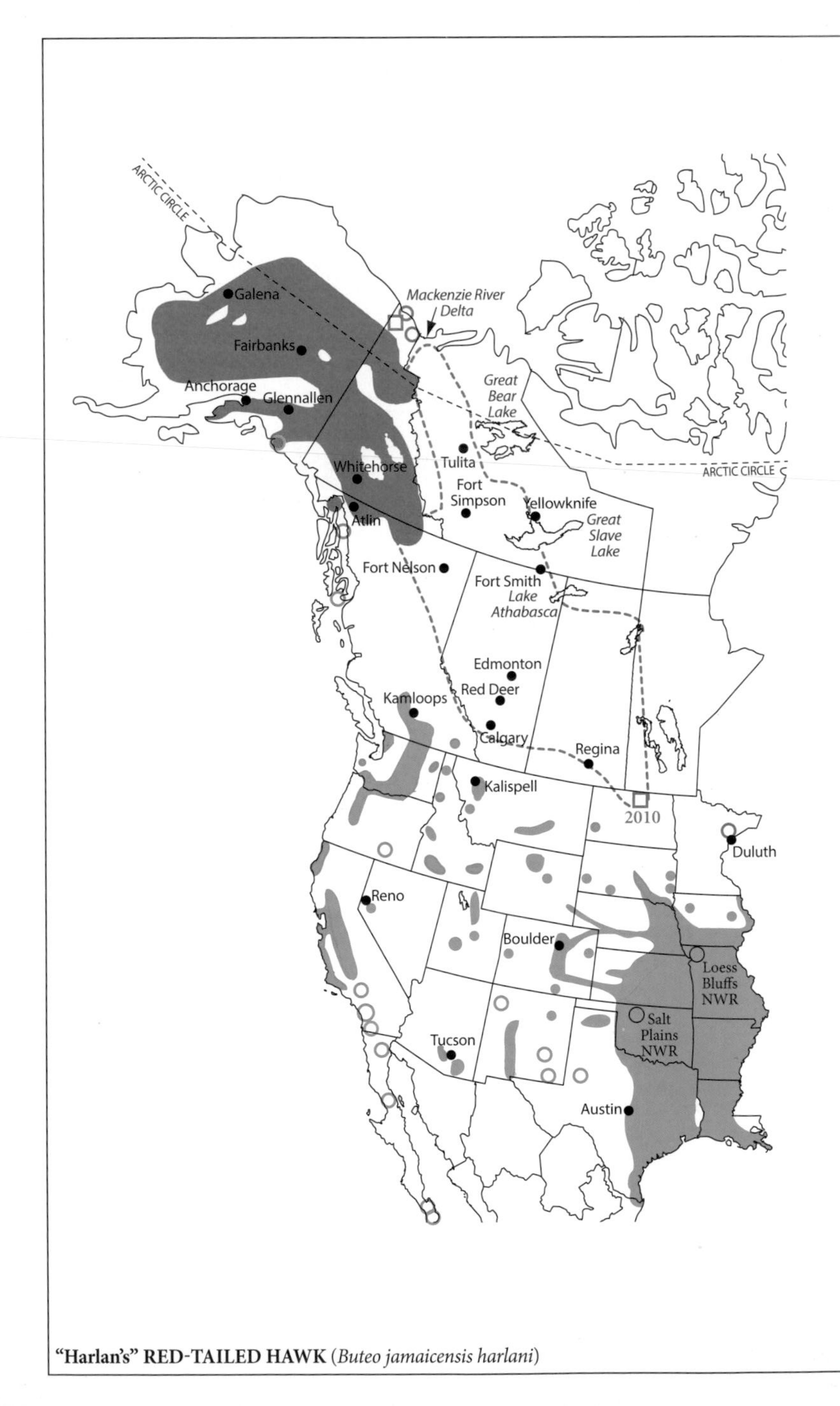

"Harlan's" RED-TAILED HAWK (*Buteo jamaicensis harlani*)

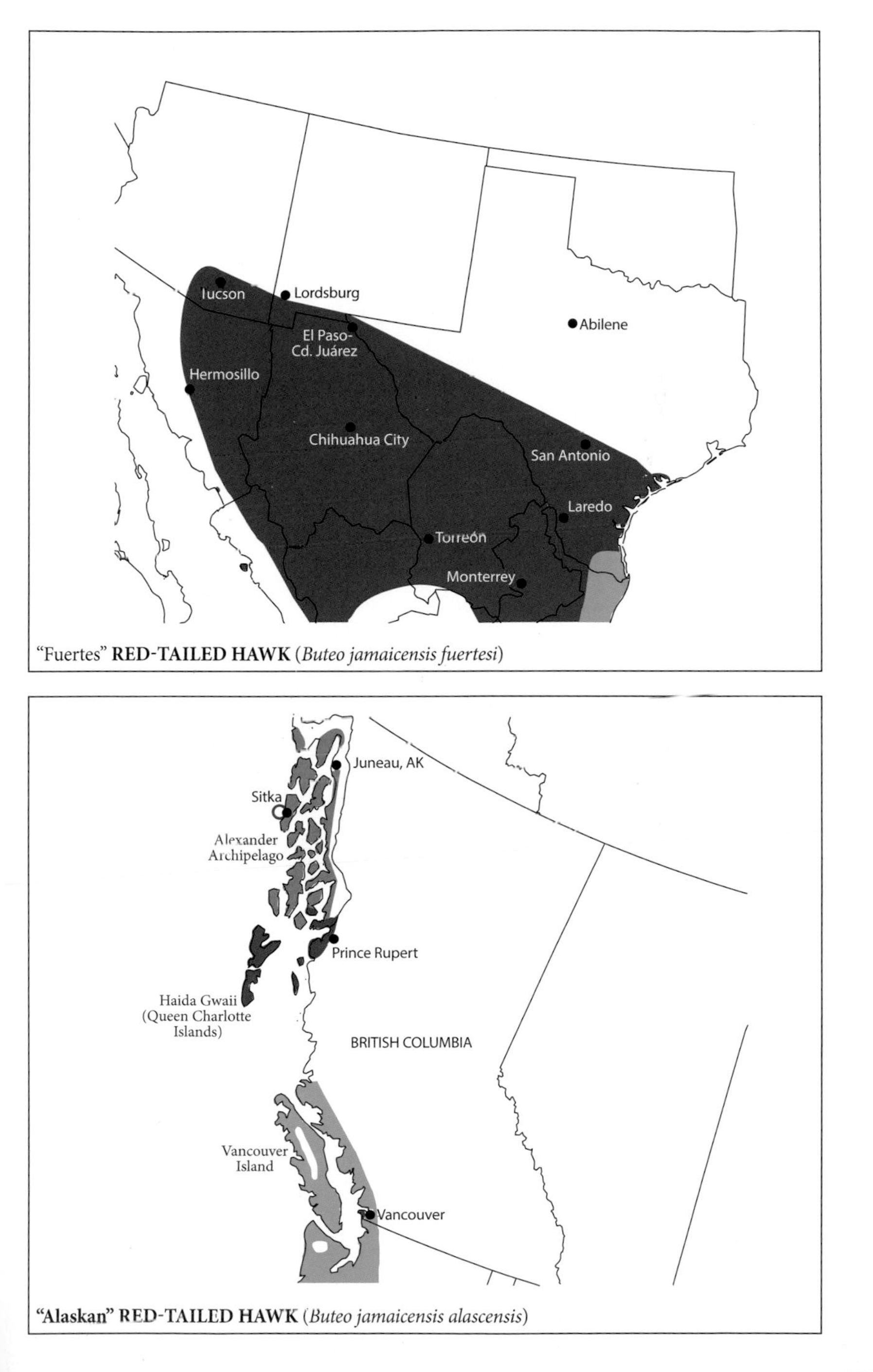

"Fuertes" **RED-TAILED HAWK** (*Buteo jamaicensis fuertesi*)

"Alaskan" RED-TAILED HAWK (*Buteo jamaicensis alascensis*)

Plumage Variation in "Eastern" Red-tailed Hawk

The "Eastern" subspecies of Red-tailed Hawk exhibits considerably more plumage variation than previously described or depicted in any publication. Some of these variations, such as barred legs and banded tails, which are more commonly found on the population in w. Canada—but also regularly seen on adults in the entire breeding range—are aligned more with "Western" plumage traits, thus the confusion regarding demarcation of "Eastern" and "Western" subspecies. Johnsgard (1990), Palmer (1988), Preston and Beane (1993), and Wheeler (2003b) give a basic synopsis of plumages of "Eastern" adults but do not address all plumage aspects found in this subspecies. Adults adorned with all-dark throats and heads, heavily marked breasts, blotchy or heavy belly and flank markings (bellybands), all-barred thigh feathers, heavily marked underwing coverts and/or barred axillaries, rufous uppertail coverts, and banded tails have been documented during breeding season in low to high frequency in the West and in low to moderate frequency in the e. U.S. and ne. Canada.

Plumage Documentation

The following plumage assessments are based on unpublished images and museum examples as well as published images that show a sampling of generic and "atypical" plumage variations to validate the assertion that this subspecies inhabits the w. Canadian region. Images showing plumage variations are cited in-text, and Internet URLs are provided in the Bibliography for verification. *Note:* The Internet is a constantly evolving medium, and some URLs will not remain intact over a period of time. An attempt was been made to utilize websites that will likely remain active for many years. For example, archive portions of the New York City (N.Y.C.) websites for East comparisons will retain images for a very long period. **PUBLISHED IMAGES (W. CANADA): "Pembina Valley."**—*Karl Bardon* (Bardon 2013). Sharp, in-flight images of 53 northbound migrants (2 in collection are probable non-"Eastern") photographed at the Pembina Valley hawk watch near Windygates, Man., in Apr. 2013; images provide, per Bardon, "good exposure of a random selection of birds." (Windygates is about 125 mi./200km south of the boreal forest; passing migrants head north or northwest into n. Man. and points farther north and west—as far as Alaska for "Harlan's.") **"Yukon."**—*Jukka Jantunen* (Jantunen 2014). Sharp, in-flight images of 3 heavily marked "Eastern"-like adults (1 adult, two 1-year-olds) from extreme s. Y.T. in summer (2 had less-than-typical markings on underside of flight feathers for "Eastern" and were likely intergrades). Also included are a close image of a perched, moderately marked "Eastern" juvenile and several superb images of rufous and dark "Western" and "Harlan's" from the same area. **"Yellowknife."**—*Reid Hildebrandt* (in Liguori and Sullivan 2014, Fig. *3*). A soaring adult photographed near Yellowknife, N.W.T., in Aug. **UNPUBLISHED IMAGES (W. CANADA): "Western Canada."**—Images and museum examples (ventral and dorsal photos) of 111 adult "Eastern" birds (80% of total of 139 randomly selected Red-tailed Hawk types) from *breeding grounds* in n. Alta. and ne. B.C. east of Rocky Mts. and in s. N.W.T. south of Great Slave Lake and east of Mackenzie Mts. This set of images/specimens includes all the following. **"Alberta Female."**—Heavily marked female museum example (dorsal/ventral photos; CMNH #101899) from Jun. from Peace River delta on w. side of Lake Athabasca in ne. Alta. (extreme n. edge of Red-tailed Hawk range in boreal forest). This bird was also noted as a proposed *abieticola* subspecies in Todd (1950); see "Analysis of the Proposed 'Northern' Subspecies of Red-Tailed Hawk (*B. j. abieticola*)." This bird is fairly similar to "Québec Female" (see "Unpublished Images [E. Canada]," below.) *W. S. Clark*. Field photos of 77 adults from

Jun.–Aug. expeditions in 2009, 2012, and 2013 to Alta. and B.C.; plus data from N.W.T. **"In-hand."**—Of Clark's 77 birds, 16 were captured for banding (14 Alta., 2 B.C.); a 17th, not included, was a light morph "Harlan's" × "Eastern" intergrade, road-killed sw. of Edmonton. **"Northwest Territories."**—*Gary Vizniowski*. Field images of 20 adults (and 1 close "Eastern"-like juvenile) from N.W.T. photographed from late Apr. to early Sep., over a period of a few years, the last 3 birds during day trips late Apr. and mid-May 2016 (along with 2 darker adult "Harlan's"). **"Nw. Alberta."**—G. Vizniowski also photographed 2 adult "Eastern" (and 1 adult rufous/intermediate morph "Western" and 1 undetermined subspecies), not included in the "Western Canada" 139 total, on Jun. 30, 2015, north of Manning, Alta., along a 19-mi. (30km) portion of Hwy. 35 in the extreme nw. part of the province. *Addendum:* G. Vizniowski supplied images of 3 more adults taken on Apr. 27, 2017, along Hwy. 5 near Ft. Smith, N.W.T., the only Red-tailed Hawks seen that day: All were quite "Eastern." (These are not included in any plumage assessment in the following text; they verify additional individuals of this subspecies.) In 2 of the 3 birds, the head showed well: 1 had a white throat with streaking, the other a dark patch at mid-throat; both had a pale supercilium. All 3 had white, unmarked breasts and moderate belly marks at best, with diamond-shaped marks on the 2 that showed the belly well. 1 had fully barred thighs; the others lacked visible thigh markings. All had minimal amount of underwing markings, including lack of distinct axillary barring. 2 of the 3 showed the tail area: Both had all-white uppertail coverts (1 had classic unmarked "Eastern," the other had bars), and 1 had multiple tail bands, while the other had faint bands only on basal region. **"Alaska Hwy."**—This author saw 7 "Eastern" adults along Alaska Hwy. in w. Alta. and ne. B.C. east of Rocky Mts. in late Jun. 1997; additionally, 1 adult dark morph "Western" near Red Deer, Alta., another adult dark morph in ne. B.C., and 1 rufous/intermediate morph in Alta. *Addendum*: A small sample of these birds are depicted on p. 20 and p. 191 **PUBLISHED IMAGES (EAST): "Ontario."**—*Raymond Barlow* (Barlow 2015). Sample of 29 adults of all plumage types (and 31 juveniles) photographed in winter near Hamilton, Ont. Based on Canadian banding data (Dunn et al. 2009), this is a wintering area for some of the breeding population originating from Que., Lab., and Maritimes. **"Connecticut."**—Jim Zipp (Zipp 2015). Adult and juvenile migrants photographed over a period of several years during Oct. near New Haven, Conn. This is a migrant and winter region for birds originating from e. Canada breeding locations. **UNPUBLISHED IMAGES (E. CANADA): "Québec Female."**—A heavily marked female museum specimen (dorsal/ventral photos; CMNH #57348) from May 1917 near Clarke City, Que. (n. side of Gulf of St. Lawrence) that was selected by W. E. C. Todd as the type specimen of the proposed *B. j. abieticola* subspecies in "Northern Race of Red-tailed Hawk" (Todd 1950). *Addendum*: A breeding adult from Happy Valley-Goose Bay, Nfld-Lab. is depicted in *Birds of Prey of the East*, p. 159.

Atypical Markings in "Eastern" Red-tailed Hawk

This section presents data on various plumage variations for each anatomical region, beginning on the head and working to the tail. Western-area birds are described 1st, then a comparison is made to adults of e. U.S. and e. Canada. **HEAD MARKINGS:** *Paler heads and throats:* Most "Eastern" birds have moderate brown or pale brown head; pale, often white supercilium; and white throat or white throat with black streaking, often with black collar at base of throat. Most adults with white throat also have white forehead. Such white-throated types are common in West to Rocky Mts., including all of Canada east of Rocky and Mackenzie Mts. Of the 111 "Western Canada" adults, a total of 81% had white throats; 88% of the 51 "Pembina Valley" birds had mainly white throats. Of the 2 definitive "Eastern" hawks in the "Nw. Alberta" sample, 1 bird was a classic white-throated type. Even the "Yukon" birds, which are quite heavily marked on their underparts, have all-white throats. A lightly marked adult male perched near his heavily marked mate northwest of Calgary near Cochrane, Alta., in May is white-throated (and very light-bellied) "Eastern" (Kube 2012). An in-flight white-throated adult from late Jul. near Calgary is lightly marked and identical to a large percentage of adults of e. U.S. (Sim 2011). A perched "Eastern" more indicative of this w. area—with barred legs and fully banded tail—is a white-throated type (with white forehead), photographed in Jul., also near Cochrane, Alta. (Lichter 2010). The white-throated bird with full tail bands depicted in the "Yellowknife" example from N.W.T. is identical to white-throated soaring adults seen north of Edmonton near Redwater, Alta., in Jun. (Edwards 2012); at Edmonton, Alta., in Jul. (Bauschardt (2012); and on prairies of se. Sask. in late summer (Herriot 2010). *Note:* The percentage of white-throated birds in w. Canada is probably lower than in the East but still is high; south of Canada the percentage likely equals the greater number in the East. White-throated and typical-plumaged birds are found in large numbers from Lab. (Mactavish pers. comm./photos) and southward. *Dark heads and throats:* Adults with dark heads and mostly dark or all-dark throats are also found throughout this subspecies' entire breeding range but in a much smaller percentage than pale-throated birds. Dark-headed birds typically have dark foreheads (as on "Western" subspecies). Many nw. Canadian birds with mostly dark heads and dark throats also have white supercilium patch above the eye; white supercilium seen on some in the East, but this mark is more typical of paler-headed birds. Dark-headed and dark-throated types can have lightly to heavily marked bellybands on their bellies and flanks, and darker head markings occur on either sex. A dark-headed but very lightly marked male was part of the "In-hand" group from near Edmonton, Alta. A dark-headed, moderately marked nesting female was photographed on Sturgeon River northwest of Edmonton, Alta. (Conlin 2015), and is similar to a large number of birds in "Western Canada" group. A dark-headed, heavily marked nesting female was photographed in flight just south of Edmonton near Millet, Alta., in Jun. (Eriksson 2011; pers. comm.). The heavily marked mate of the Cochrane, Alta., male is also all-dark-headed (Kube 2012). A heavily marked adult from Apr. near Edmonton, Alta, had an all-dark head (Borlé 2016d). "Alberta Female" is all-dark-headed (and darker-throated than "Québec Female"). Only 3 of 23 (13%) of the "Northwest Territories" and 9% of the "Pembina Valley" adults had dark heads and throats. Dark-throated types are least common on Great Plains and far outnumbered by white-throated types (pers. obs.). **Eastern U.S. and Eastern Canada:** Dark-headed types are relatively common in all e. provinces and states and have been noted on breeding/resident birds (of either sex) in the breeding season from Que. ("Québec Female") south to especially N.Y.C. and N.J. (Cosby 2014a), and farther south to S.C. (Miller 2012), and west to s. Ohio (Dice 2012). Only 1 of the "Ontario" winter group had a dark head and throat. The following are a few N.Y.C. birds with dark heads and throats: Bobby and Rosie, 2012 pair of Washington Sq. Pk. (Washington Sq. Pk. Blog 2012); Lima of Central Park (Karim 2011, Oct. 21, Dec. 27, Dec. 29); Norman of n. Manhattan (Schmunk 2014a); Charlotte of Central Pk., 2005–8 (Karim 2005); and a 2015–16 breeding resident of Washington Sq. Pk. (Yolton 2016). A Bergen Co., N.J., female from May (AMNH #8259) is identical, with dark throat and pale bellyband, to the "In hand" Edmonton, Alta., male noted above. **Summary:** White- or streaked-throated types are much more common than dark-throated types throughout the subspecies' range. Dark-throated birds, however, are found everywhere, and dark throats adorn both sexes, but possibly females more commonly. Dark throats may be more common in w. Canada than in the East. All-white-throated and especially dark-throated variants noted above from w. Canada can be matched exactly with plumage markings of breeding individuals of e. U.S. and e. Canada. **COLORED/MARKED BREAST:** Underparts of "Eastern" Red-tailed Hawks are typically uniformly white or pale tawny with a rufous patch on sides of neck/breast. A fairly high percentage of adults have colored or marked breasts. If colored, the rich tawny color of autumn's newly molted plumage often fades paler—to white—by late fall or winter. Color and/or markings may adorn breasts of white- or dark-throated types as well as lightly or heavily marked birds. *Unmarked, colored breasts:* Yellowish/tawny breasts that are otherwise unmarked or very lightly streaked are uncommon in the West, even in Canada. This feature was noted on a few birds of the "Western Canada" sample but not seen on any of the "Pembina Valley" birds (possibly because of fading). *Marked breasts:* A fair percentage of breeding birds from this subspecies' entire North American breeding range have tawny or white breasts with brown or rufous streaks or blobs on part of the breast (sides) or on all of the breast (heaviest on sides). Those with tawny color exhibit this in late summer and fall, but it fades thereafter, including any rufous markings, which bleach and become paler and narrower in width. In "Western Canada" sample, 12% of the birds sported breast markings; "Pembina Valley" sample had 9% with such markings. Of the 16 "In-hand" examples, 5 (31%) had distinct breast markings. 2 of "Alaska Hwy." sample had heavy breast streaking. 2 of the 3 "Northwest Territories" adults from 2016 had lightly streaked breasts. Neither the 2 "Nw. Alberta" birds nor "Alberta Female" had marked breasts. A fall migrant from cen. Alta. had a dark-streaked pattern on its yellowish-tinged breast, a white throat, and moderate belly markings (Lynch 2016). **Eastern U.S. and Eastern Canada:** The "Ontario" wintering group showed breast streaking on 14% of the 29 adults. Breast streaking is a fairly common trait on breeding birds in e. U.S. as well. It is notable on several N.Y.C. past and present residents: Bobby and especially Rosie, the 2012 pair; very distinct on Charlotte of Central Pk. (rich tawny in early Sep. [Karim 2005], faded whitish by mid-Nov. [Yolton 2008b]); and on white-breasted George of n. Manhattan, who has white throat (Schmunk 2010); also on many other residents. It is seen also on a Conn. resident pair, both of which have streaking on sides of their breasts (Hand 2003a); on a partially white-throated male of a 2014 pair in Atlantic Co., N.J., whose breast was rich tawny in Nov. but faded yellowish by Jun. (Cosby 2014b, 2015a, respectively); on Scarlette, resident of James City Co., Va., which also had a white throat (she was an educational-use bird with injuries sustained when her nest tree was felled when she was a nestling in 1989; she died in 2011 [Deal 2011; D. Lepkowski pers. comm.]); and on dark-throated residents in Brunswick Co., N.C. (Ennis 2010). It is a common feature on otherwise typical "Eastern" birds among the "Connecticut" migrants and on many fall migrants of unknown breeding origin seen at Hawk Mt. Sanctuary near Kempton, Pa. (pers. obs.). **Summary:** Birds with marked breasts seem to represent a fair percentage of adults throughout this subspecies' range in U.S. and Canada, though fewer such birds occur on Great Plains. This trait is found on both sexes and on any plumage type. **BELLY AND FLANK MARKINGS (BELLYBAND):** Most w. region adults of the "Eastern" subspecies, based on "Western Canada" and "Pembina Valley" data, have lightly or moderately marked mid-bellies and flanks, with diamond-shaped, streaked, or small to medium-size bloblike markings; flanks are often thinly or broadly barred but can lack barring and have small or large streaks or blobs. *Lightly marked type:* Light markings are fairly common in most areas, especially on males. Palest birds lack ventral markings, particularly on the belly, and can be as pale as "Fuertes" subspecies individuals. In "Pembina Valley" group, 14% had lightly marked bellybands, as did at least 2 birds from boreal forest region, 1 in n. Alta. and 1 in N.W.T., from "Western Canada" sample. Light-bellied male of pair at Cochrane, Alta., photographed by Kube (2012), noted above, is a pale type with minimal belly markings. In-flight adult from Calgary, also noted above, photographed in late Jul. by Sim (2011), has classic lightly marked pattern. Lightly marked types are more common than other types in Great Plains region. *Moderately marked type:*

This is the most common type in w. Canada. Among "Pembina Valley" adults, 51% had moderately marked bellybands. In "Western Canada" sample, 85% had lightly to mainly moderately marked bellybands. A moderately marked bellyband is noted on 1 of the 2 "Nw. Alberta" birds as well. An adult from s.-cen. Sask. in May exhibits a moderate bellyband (Saunders 2009).The Sturgeon River female near St. Albert north of Edmonton, Alta., noted above, is on the heavier side of the moderate bellyband category (Conlin 2015), as is the adult seen north of Edmonton at Redwater, Alta., in late Jun. (Edwards 2012). *Blotchy/bloblike bellyband pattern:* Birds with blotchy or blob-type bellybands seem to be fairly common in "Western Canada" and "Pembina Valley" samples, but not as common as those with smaller belly/flank markings. The "Yellowknife" bird has blotches, as does the Edmonton bird of Bauschardt (2012). *Very dense blotchy/streaked bellyband pattern:* Birds with this heavy bellyband pattern can be white- or streak-throated with moderate-colored heads, or dark-throated with dark heads. Some have breast markings. Where sex is known, heavy bellybands appear to be more common on females than on males (4 of 5 museum examples from boreal forest of this type examined by Todd [1950] were females). Heavy bellybands also appear to be more common in Canada than in U.S.; however, they are still uncommon, even in w. Canadian range. In "Western Canada" breeding season sample they occurred in 15% of the 111 adults, including "Alberta Female," in 2 of the 3 "Northwest Territories" adults of 2016, and in 2 of the 16 (12%) "In-hand" birds. This feature was prevalent, though still a minority type, in "Pembina Valley" sample, showing on 18 of the 51 (35%) birds. Even a female among a mainly paler "Eastern" and "Krider's" population from n. N.Dak. in Jul. sported a quite heavy bellyband (in Liguori and Sullivan 2010, Fig. *9d*), indicating such markings can show up in any region. Male of nesting pair from Pambrun, Sask., in arid sw. part of the province—also a "paler" Red-tail region—had a fairly heavy bellyband (Wainman 2011b). Female of Cochrane, Alta., pair (Kube 2012) and the other "Nw. Alberta" "Eastern" bird had heavy bellybands. In "Yukon" sample, all 3 adults had rather heavy bellybands. *Solid bellyband pattern:* A very small percentage of w. Canadian birds have solid black bellybands (1 of 51 [2%] in "Pembina Valley" and 5 of 111 [5%] in "Western Canada"). 1 of the 3 in "Yukon" sample had a nearly solid bellyband. None of the N.W.T. adults had solid bellybands. A bird with a solid bellyband was noted in "Alaska Hwy." group. It is likely these birds are intergrades with, especially, "Western" subspecies, in which solid bellybands occur on some light and especially on rufous/intermediate morph birds. (A "Western Canada" bird with a solid bellyband, that appeared to be "Eastern" but had long wings that reached its tail tip when perched—a trait of "Western" subspecies—was categorized as an intergrade.) **Eastern U.S. and Eastern Canada:** Lightly and moderately marked bellyband types prevail in the East as well. Some adults, especially males, lack belly markings. Female Red-tailed Hawks, as with most raptors, are on average, more heavily marked than males. *Blotchy/bloblike bellyband pattern:* Charlotte (Karim 2005); the male of a pair in J. Hood Wright Pk., N.Y.C. (Schmunk 2014b); and especially Dora of N.Y.C. (Yolton 2014) all had rather large, round, black blotches on their bellies and flanks. A 2011 Tompkins Sq. Pk., N.Y.C., resident (Karim 2011, Dec. 21) had such a heavy blob pattern that it nearly falls into the heavily marked bellyband category. 1 of the Atlantic Co., N.J., resident birds had a distinctly large-blobbed bellyband pattern (Cosby 2015b). *Very dense blotchy/streaked bellyband pattern:* U.S. breeders can also have rather heavy black bellybands and be *identically* marked to birds from w. or e. Canada. Heavy bellyband samples from N.Y.C. include Lima (Karim 2011, Oct. 21, Dec. 27, Dec. 29), George (Schmunk 2010), and a 2015–16 breeder in Washington Sq. Pk. (Yolton 2015, 2016)—all of which are identical to 2 of the 3 "Northwest Territories" birds of 2016. Other e. examples are "Québec Female"; a nesting female in May near Ottawa, Ont. (Schneider 2008a); and Big Red, resident female at Cornell University, Ithaca, N.Y. (Cornell Lab 2012). A 1-year-old in Pike Co., Pa.—based on molt sequence on inner primaries and mid-tail was in late spring–early summer stage (probable Jun.; image posted in Sep.)—had a heavy black belly (D[ennis] S. 2014). A nesting probable female from Pike Co., Pa., in early June had large dark marks on her flanks and likely a heavier bellyband (Pope 2012). Both sexes of a resident pair in 2003 of Sherwood Island St. Pk., Fairfield Co., Conn., had heavy bellybands; the female had a large black-blobbed pattern (Hand 2003a, 2003b). Of the above-listed birds, "Québec Female," Lima, and the 2015–16 Washington Sq. Pk. resident of N.Y.C. had dark throats. *Note:* No adult from the East has been recorded with solid belly and flanks (bellyband), as seen on a very few w. Canadian birds, although the Pike Co., Pa., 1-year-old (D[ennis] S. 2014) is nearly solid on its mid-belly but not on its flanks. "Ontario" group had 24% with dark bellybands, but none solid black. **Summary:** Birds with heavy bellybands seem to be more common in northerly latitudes but are still *far outnumbered* by more lightly marked types, especially moderately marked birds. *Note:* Pairs of mixed plumage types are common, and the combination of extreme opposites, of heavily marked female and lightly marked male, was exhibited in the West in the Cochrane, Alta., pair (Kube 2012) and in the East with the virtually identically marked Lima–Pale Male pair (Karim 2011, Oct. 21, Dec. 29). Pairs of similarly marked birds are common, too, whether moderately or heavily marked, as in a few pairs in "Western Canada" collection photographed by W. S. Clark. Similarly marked birds of a pair were noted in the East as well, with heavy bellybanded Sherwood Island St. Pk. pair of Fairfield Co., Conn., being a prime example (Hand 2003a). **THIGH (LEG FEATHER) MARKINGS:** Thigh feathers are typically unmarked white or pale tawny in "Eastern" birds. Dickerman and Parkes (1987) remarked that thighs can be "weakly barred with cinnamon." Preston and Beane (1993) noted that they can be a deeper color and that they may contrast with the rest of the underparts. Distinct thigh barring was previously associated with lighter morphs of "Western" subspecies or intergrades with them, which can also have solid dark tawny or rufous thighs. The typically unmarked white or pale tawny thigh feathers occur on "Eastern" birds in much of the subspecies' breeding range, including many birds from n. Canada of any plumage type: the moderately marked adult from Jun. at Redwater, north of Edmonton (Edwards 2012), and the heavily marked Apr. adult from Edmonton (Borlé 2016d) both lacked thigh markings. However, on n. Great Plains of U.S. and s. Canada and in boreal forest of Canada, partial or full thigh barring is a common trait. The barred pattern can be faintly or strongly marked brown or rufous and *identical* to that on any light morph "Western." Of the 51 birds in "Pembina Valley" sample, 18 (35%) clearly showed partial or full leg barring. "Western Canada" field-distance images were not assessed, because of angle of view and distance; however, 1 of the 2016 "Northwest Territories" adults had partial barring and the other 2 had fully rufous-barred thighs. Of the 16 "In-hand" birds, 2 (12%) had partial barring on upper thighs and 8 (50%) had fully barred thighs. The "Yukon" sample birds all had brown-barred thighs. Thigh barring adorned the fairly heavily marked female of the Pambrun, Sask., pair (Wainman 2011a). A bird from Bragg Creek, south of Calgary, from May had partial barring on the upper thighs (Martin 2012); the fence-post-perched single Cochrane, Alta., bird in Jul. shows lightly barred thighs (Lichter 2010). "Alberta Female" had brown-marked, lightly barred thighs. Thigh barring is common on breeders along e. foothills of Colo.; some of these birds are undoubtedly intergrades with "Western" (pers. obs.). **Eastern U.S. and Eastern Canada:** Thigh barring is uncommon in the East but seems to be much more common than previously thought. Lima of N.Y.C. had faint brown bars on some thigh feathers (Karim 2011, Nov. 12, Dec. 27). Pale Male (Karim 2011, Dec. 20) and 2015–16 Washington Sq. Pk. resident (Yolton 2016) of N.Y.C. had partial rufous barring, as did a 2016 resident of Atlantic City, N.J. (Cosby 2016). Thigh feathers were fully and distinctly barred with rufous on the following N.Y.C. residents: Pale Male Jr. (Karim 2006); Christo (Goggin 2015), which is almost identical to May 2016 "Northwest Territories" bird; and female from J. Hood Wright Pk. pair (Schmunk 2014d). This feature is also seen well on the 2014–15 resident male of Atlantic Co., N.J., on which thighs are totally rufous-barred (Cosby 2015a). Rufous-barred thighs on otherwise classic-plumaged adults are seen at Oakland Co., Mich., from May (Podrasky 2011) and on a nesting female from Toronto, Ont. (Brokelman 2010). Fully barred thighs are also seen on birds with more heavily marked plumage, such as marked breasts and heavier bellybands: George of N.Y.C. (Schmunk 2014e); Scarlette of Va. (Deal 2011); and a resident of Brunswick Co., N.C. (Ennis 2010). Barred thighs are seen regularly on moderately and heavily marked fall migrant adults of unknown breeding origin at Hawk Mt. Sanctuary, Pa. (pers. obs.), and on several in "Connecticut" sample. In "Ontario" sample, 24% had partial or full thigh barring; the percentage is probably much higher in Canada than in a comparable sample of e. U.S. breeders. "Québec Female" had faint brown bars on the very upper thighs. **Summary:** Thigh barring is common on population of n. Great Plains of U.S. and s. Canada and northward throughout boreal forest. It is uncommon to very uncommon elsewhere in U.S. This plumage feature adorns any plumage type, though found on lightly marked birds only in w. Canada. **MARKED UNDERWING/AXILLARY BARRING:** Majority of birds of "Eastern" subspecies have few if any markings on the underwing coverts, though a moderate percentage of moderately and heavily marked birds have some markings. In w. Canada, a small percentage of lightly marked types and a fairly high percentage of moderately and heavily marked types exhibit a fair amount of dark markings on the underwing coverts. Those with marked underwings are likely to also have partially or distinctly barred axillaries. Barring on axillaries, which is brown, rufous-brown, or black, was thought to occur mainly on Canadian birds—and lighter morphs of "Western" ("Western" can have identical barring but it can also be richly rufous, which has not been seen on "Eastern"). It is an especially common feature in "Western Canada" sample, occurring on a large number of birds (no percentage available). Among "Pembina Valley" migrants, all photographed in flight, 34 birds (66%) showed this trait. In "In-hand" collection barred axillaries showed on 8 of 14 (57%) of the Alta. birds and on the 2 B.C. birds. In the small "Yukon" group, all had some markings, but none had heavily marked underwings; axillary barring was thicker than average and flight-feather barring on 2 of the 3 was quite thick and may show influence of, especially, "Harlan's." The fairly heavily marked Pambrun female had barred axillaries and heavily marked "Western"-like underwing coverts, as did her dark-bellied mate (Wainman 2011a, 2011b). A heavily marked female in Turtle Mts. of N.Dak. (mated to a light morph "Harlan's") also had barred axillaries (in Sullivan and Liguori 2010, Fig. *6*). In w. Canada, even paler plumage types show barred axillaries as a fairly common trait. *Note:* A very *few* Canadian birds (e. and w. regions) have more pronounced underwing markings than the most heavily marked of U.S. breeders. Of the 51 "Pembina Valley" birds, only 3 (6%) had heavy underwing markings (and strongly barred axillaries), and these birds may perhaps have mixed genetics. These markings were also noted on a few of the in-flight birds of the "Western Canada" sample. **Eastern U.S. and Eastern Canada:** Moderately and heavily marked birds in the East also can have marked underwings. A lesser percentage have barring on their axillaries, which is an uncommon to very uncommon trait in this region. Barred axillaries adorn George of N.Y.C. (Yolton 2008a), show faintly on Lima (Karim 2011, Oct. 8), and appear on J. Hood Wright Pk. female (Schmunk 2014c; difficult to see, but 1 axillary protrudes under her right wing). This feature was also seen on a Brunswick Co., N.C., resident, which had black-barred axillaries that nearly rival those of any heavily marked Canadian bird (Ennis 2010); it also adorns a DeKalb Co., Ga., resident in Jul., which had distinct rufous-brown-barred axillaries (Polucci 2012). **Summary:** More heavily marked underwings and especially barred axillaries are quite common on Canadian birds, especially from w. Canada. These 2 traits occur on a smaller percentage of breeding adults from areas south of Canada and in the East. Very heavily marked underwings appear to be present in Canadian stock but are found in a very low percentage of birds. **RUFOUS UPPERTAIL COVERTS:** This marking is an uncommon trait for this subspecies, but one that seems to show up frequently throughout its entire breeding range, including in the East. The vast majority of "Eastern" birds—in all of this subspecies' range—have all-white coverts or white coverts with dark crossbar(s) on each feather. In those that have rufous coverts, each feather always has a white distal tip or a black crossbar behind the white tip (feather is *never* solid rufous, as on many "Western"). There are no uppertail covert data for "Western Canada" or "Pembina Valley" birds, since photo angles often did not show this part of the plumage. 2 of the 2016 "Northwest Territories" adults had rufous-type coverts. Of the 16 "In-hand" samples, 3 (19%) exhibited this trait. The 1 "Yukon" that showed the dorsal side of the tail had white coverts with black barring, as did 1 of the 2016 "Northwest Territories" adults. *Note*: Depicted on image on p. 20. **Eastern U.S. and Eastern Canada:** This trait was found on a few birds in the East: Ottawa, Ont., nesting female in May, photographed in flight carrying a vole (Schneider 2008b); Big Red of Cornell University, in N.Y., shown with her wings drooped while on her nest (Erickson 2012); Scarlette of Va., perched, in rear angle view (Wildlife Center of Va. 2011); and the bird of DeKalb Co., Ga., photographed flying away with a squirrel (Polucci 2012). Only 1 "Ontario" bird showed this trait, but uppertail coverts were obscured in most photos and may be more prevalent. **Summary:** Rufous uppertail coverts seem to be an uncommon trait on moderately and heavily marked "Eastern" birds from all regions. **TAIL MARKINGS:** Most "Eastern" adults have uniformly rufous tails on the dorsal side (pale pink or gray on ventral side, unless lit by translucent light, when shows rufous). A variable-width single subterminal band is most common, but this band can be absent. The majority of the birds of n. Great Plains and w. Canada, including boreal forest, have partially or fully banded tails, which is possibly due to a long period of genetic exchange with "Western" subspecies in this region. Whether banding pattern is partial or full, it is identical to the thinly banded pattern on "Western" birds of any color morph (darker morphs can have [much] thicker bands). *Partial banding pattern:* Palmer (1988) and Preston and Beane (1993) noted that partial banding occurred in the East, but no author made note of its high occurrence in the West. This pattern is most prevalent on outer tail feathers, especially on inner web of outer feathers (Palmer 1988). It also regularly adorns the basal region of some or many feathers, especially the central feathers. Some feathers may be nearly fully banded, while others lack banding. Of the 51 "Pembina Valley" migrants, 39 (76%) had some type of tail banding, whether partial or full bands. Of the 16 birds of the "In-hand" sample, 13 (81%) had partial or full bands (some had faint markings that would be difficult to see at field distances). A large percentage of "Western Canada" birds had partial banding and a fair percentage had full banding on their tails, which may be a partial influence by close proximity and/or intermixing with "Western" subspecies. The "Yellowknife" bird showed rather a typical moderately marked "Eastern" body with no thigh barring but had a fully banded tail. It was identical to Edmonton, Alta., Jul. adult (Bauschardt 2012); Redwater, Alta., Jun. adult (Edwards 2012); and se. Sask. Aug. adult (Herriot 2010). The fence-post-perched bird from Cochrane, Alta., had a fully banded tail that rivals any such pattern on a "Western" bird (it was otherwise very "Eastern"). All 3 "Yukon" adults and 2 of the 3 "Northwest Territories" adults from 2016 had fully banded tails (the other was unbanded). A very lightly marked "Eastern" from n. N.Dak. in Jul.—in "Krider's" range—had a fully banded tail (in Liguori and Sullivan 2010; Figs. *5a*, *5b*). A heavily marked type female from Turtle Mts., N.Dak., had a fully banded tail (in Sullivan and Liguori 2010; Fig. *6*). Birds with banded tails are common breeders on n. Great Plains and along e. edge of Rocky Mts. in Colo., though in this area some are undoubtedly intergrades with "Western" (pers. obs.). "Alberta Female" had single-subterminal-band pattern, as did very heavily marked Apr. adult from Edmonton (Borlé 2016d). **Eastern U.S. and Eastern Canada:** Partial (but not full) tail banding adorns a fair number of "Eastern" adults, which is noted by Palmer (1988) and Preston and Beane (1993). Only a moderate percentage of U.S. breeders exhibit multiple tail bands; however, the percentage is probably much larger than previously thought. *Partial banding pattern:* Pale Male of N.Y.C. has partial or full bands on many tail feathers (Michaels 2015a). Pale Male Jr., also of N.Y.C., had full bands on many of his tail feathers (Karim 2006). The resident adults from Brunswick Co., N.C. (Ennis 2010), and DeKalb Co., Ga. (Polucci 2012), had partially banded tails. *Full banding pattern:* A 1-year-old in early May on Long Island, N.Y., had a fully banded tail (Michaels 2015b); this bird was nearly identical to Edmonton, Alta., adult photographed by Bauschardt (2012) and "Yellowknife" bird. Scarlette of Va. (Wildlife Center of Va. 2011) and the female of the 2015 resident pair in Atlantic Co., N.J., in Jun. (Cosby 2015a), had nearly fully banded tails. Full tail banding was relatively common on otherwise typical paler adults among the fall migrants in the "Connecticut" group, as well as birds at other migration locales in the East (pers. obs.). **Summary:** Based on "Western Canada" data, especially on "In-hand" images of W. S. Clark (and pers. comm.), full or partial tail banding is a common trait in this w. part of "Eastern" subspecies' range—as it is on n. Great Plains—with a large number having at least partial banding. It may not be any more common in e. Canada than in e. U.S., showing on only 2 of 29 birds (7%) in "Ontario" sample (though heavy bellybands and barred thighs were more common in this small sample than in e. U.S. birds). **PLUMAGE VARIATION IN "EASTERN" JUVENILES:** Birds from all regions are quite similar, occurring in lighter and darker variations, as in adults; lighter types are prevalent on Great Plains. The throats of all juveniles of this subspecies are white, but some have thin, dark streaking and/or a dark collar (*none* have all-dark throats as on many "Western" juveniles). The "Yukon" and "Northwest Territories" juveniles are quite typical of the "Eastern" subspecies. A few "Eastern" juveniles have a dark crossbar on each undertail covert, a trait previously thought to be found only on some "Western" juveniles. ("Eastern" adults *never* show crossbars on undertail coverts.) These barred undertail coverts were 1st noted on an "Eastern" fledgling from Kettering in Montgomery Co., Ohio (Lincoln Pk. News Blogspot 2012); they were also seen on a few fall migrants in New Haven Co., Conn. (Zipp 2015), and a museum example from the East (AMNH).

Regional Status of Red-tailed Hawk Plumages

This section gives an overview of origins of plumage variations encountered. See also "Status" in the Red-tailed Hawk natural history text. **"Eastern" in Western Canada.** *Summer.*—"Eastern" is the dominant subspecies in w. Canada; based on "Western Canada" sampling it comprised about 80% of the Red-tailed Hawk population in n. Alta., ne. B.C., and N.W.T. The *revised* Canadian range of this subspecies is now shown to extend west and north to e. border of Rocky Mts. in boreal forest of n. Alta. and ne. B.C., then north into N.W.T., east of Mackenzie Mts. along narrow, fingerlike extension of band of larger-size White Spruce and Balsam Poplar habitat, west and north of Great Bear Lake along Mackenzie River to its s. delta. (This species was not seen by Edward Preble in 1903 [Preble 1908] north of Fabre Lake, which is halfway between Great Slave Lake and Great Bear Lake, until he got to mouth of Great Bear River near Mackenzie River.) Mackenzie River delta is the farthest north this species now regularly(?) nests in N. America. A pair (of unknown subspecies/color morph) seen by A. C. Twomey in spring of 1942 near Aklavik,

N.W.T., in s. Mackenzie River delta, was the 1st documentation of Red-tailed Hawk breeding at this latitude (in Todd 1950). A. E. Porsild (1943) did not see this species in the delta during expeditions in the 1930s. No photographs or museum examples exist of "Eastern" subspecies north of Yellowknife, N.W.T. On his 1903 expedition on Athabasca and Slave Rivers, in ne. Alta. and se. N.W.T., respectively, Preble (1908) had seen 7 light "normal" probable "Eastern" birds (and 2 rufous/intermediate morph "Western"). Along Athabasca River, E. Preble obtained a classic-looking "Eastern" juvenile in late Aug. 1904 (dorsal/ventral photos, USNM #195662). However, Preble later "saw the species ('melanistic' and 'light phases') nearly every day" farther north along Mackenzie River in N.W.T., from Ft. Simpson to Tulita (formerly Ft. Norman); the most northerly bird (light type) he saw was at Ft. Good Hope. On his return trip, Preble noted fewer birds, seeing only 5 from Ft. Simpson, N.W.T., south to Ft. McMurray, Alta. On numerous Peregrine Falcon expeditions for the Canadian government on the Mackenzie River, K. Hodson saw only "a handful of buteos [Red-tailed Hawks]" on all trips combined, which illustrates their low density (pers. comm.). A recently fledged juvenile in early Aug. from Saline River, a very small river near mouth of Great Bear River near junction of Mackenzie River, may have some "Eastern" in it, as indicated by dorsal markings, but also has a light morph "Harlan's" tail pattern (dorsal/ ventral photos, CMN #18549). *Yukon:* There is limited and probably irregular presence of this subspecies, or most likely intergrades with this subspecies, in s. Y.T. ("Yukon" sample, Jantunen 2014). In this region it extends westward in the lowland boreal forest corridor between n. Rocky Mts. and s. Mackenzie Mts. Of the 3 "Yukon" adult birds, 2 were 1-year-olds and possible nonbreeders, and 2 had markings on their ventral flight feathers that were thicker than normal for this subspecies and likely exhibited mixed genetics (especially "Harlan's"). Overall, the "Yukon" adults were identifiable as heavily marked "Eastern" subspecies and not "Western" or "Harlan's." *Note:* The juvenile from the "Yukon" sample was identical to any moderately marked bird of this subspecies found in e. U.S. *Additional data:* Breeding density declines with latitude, especially in the less-diverse habitat of the boreal forest. "Eastern" density is high in s. and cen. regions of the Canadian provinces but very low in northerly latitudes of the forest, especially in N.W.T. (W. S. Clark pers. comm.; G. Vizniowski pers. comm.). The "Western Canada" data are unique because of the substantial number of adults photographed during the breeding season. Based on this data, heavily marked birds appear to be more common in boreal forest than farther south, or in e. U.S. However, paler types, especially moderately marked birds, most with typical "Eastern" markings, far outnumbered more heavily marked types. There was a complete mix of adult types, from those with lightly marked (or unmarked) bellies and underwings to those with heavily marked bellies and underwings. Different types could occur in the same area—and even mate with one another. This mix was also seen in "Alaska Hwy." data, with extreme plumage variances only miles apart. Even among "Pembina Valley" migrants, which are undoubtedly heading near or into boreal forest, more heavily marked plumaged types are very much the minority; except for showing a high percentage of tail banding and thigh barring, most birds looked like any breeding in e. U.S., where such markings, though uncommon, are still widespread. **"Eastern" Extralimital.** *Summer.*—Birds of the "Eastern" subspecies are infrequently seen/ photographed in summer near Rocky Mt. Trench in se. B.C. (D. Leighton pers. comm./ unpubl. photos). A few are seen near Glennallen, Alaska (mile marker 118), at the Gunsight Mt. hawk watch during Apr. migration (B. Sullivan pers. comm.; J. Liguori pers. comm.; pers. obs.). A probable "Harlan's" × "Eastern" intergrade with mainly "Eastern" body plumage was seen at Delta Junction, Alaska, in late Jun. 1997 (pers. obs.). A moderately marked "Eastern" photographed near Palmer, Alaska, in mid-May 2012, had whitish throat and pale underwing but lacked thigh or axillary barring or tail banding, which are common traits on many w. Canada adults of "Eastern" subspecies (Iliff 2012). The northernmost breeding record was 3 nestlings (of unknown subspecies) banded in early Jul. 1973 in British Mts. on North Slope of Y.T. (P. Sinclair pers. comm.). **"Eastern" in Eastern Canada.** *Summer.*—There is minimal plumage data on Red-tailed Hawks from e. Canadian boreal forest during the breeding season, though there has been extensive and comprehensive coverage for Ont. (Atlas of the Breeding Birds of Ontario 2006) and Que. (Québec Breeding Bird Atlas 2010–14) in breeding bird atlases. As noted for w. Canada, both provinces have a high density of nesting Red-tails in their s. regions, but density decreases into boreal forest. "Québec Female," of the boreal forest, which was the type specimen used by Todd (1950), is a heavily marked bird that is virtually identical to the female of the pair from Cochrane, Alta. (Kube 2012), which is over 100 mi. (160km) south of the boreal forest, and to Lima of N.Y.C., 600 mi. (965km) south of the forest (the latter 2 were mated to rather typical pale-plumaged "Eastern" males). Images of 3 adults photographed over a period of 2 decades—1 from the 1990s (in flight) and 2 from Jun. 2007 (1 in close flight)—in Goose Bay, Lab., which is at the extreme ne. part of Red-tailed Hawk range in N. America (and boreal forest), are the very classic, moderately marked type of "Eastern" subspecies, with absolutely no "atypical" markings ((B. Mactavish; see Fig. *1*, p. 159 in *Birds of Prey of the East*). Though only 3 birds, this set matches data from the paler-plumaged breeding-season adults documented in "Western Canada" data set. *Winter.*—The 2 most northerly inland winter records of this species in the East, a juvenile and adult photographed near Québec City, Que., in Feb. and Dec., respectively, which is at the northernmost interior edge of the winter range and near the s. edge of the boreal forest zone, were both classic paler "Eastern" birds (Maire 2013a, 2013b). Other "typical" paler juveniles from near Québec City from Oct.–Feb. are also depicted by Maire (2016). *Note:* Of the "Ontario" winter sample of 31 juveniles, none of the juveniles had any different plumage features than found on resident birds in e. U.S. **"Western."** See "Status" in the natural history text for information on this subspecies' range. Status of plumage types is covered herein. *Light morph (adults):* Only 6 birds were possible light morph "Western" and/or intergrades in n. Alta. in "Western Canada" survey sample. A light morph (barred belly, dark head, all-barred legs, all-banded tail) was photographed in Jul. near St. Albert, Alta. (Borlé 2016e). 2 (1 with all-rufous uppertail coverts) were photographed in s. N.W.T. (G. Vizniowski, unpubl. photos). A classic bar-bellied adult was photographed at Atlin, B.C., at extreme n. part of range—and in historic s. "Harlan's" range (W. S. Clark, unpubl. photo). An adult was photographed near Calgary, Alta., in May (Martin 2012). *Rufous/intermediate morph (adults):* "Western" intermediate morphs are uncommon and sparsely distributed west and north of Edmonton, Alta., and very uncommon in N.W.T. south of Great Slave Lake, and are far outnumbered in all areas by "Eastern." In "Western Canada" sample, 7 were from Alta. (W. S. Clark = 4; Preble 1908 = 1 along Athabasca River in May 1903; pers. obs. = 1; G. Vizniowski = 1); not in this sample is an early Sep. adult from nw. Alta. (G. Vizniowski pers. comm.). The s. N.W.T. produced 3 in "Western Canada" group (G. Vizniowski = 2; Preble 1908 = 1, a nesting "melanistic" male mated to a "normally colored" [probable "Eastern"] female on the Slave River near Ft. Smith in 1903 [dorsal/ventral photos; USNM #193557]; based on this museum example, his "melanistic" is a rufous/intermediate morph). Also in N.W.T., 1 adult in early Sep. was seen south of Enterprise, just north of Alta. border (G. Vizniowski, unpubl. photo). Northernmost example of this subspecies is an adult male from Mackenzie River delta, N.W.T., on Aug. 5, 1955 (which has probable "Harlan's" traits of a few small white speckles on rufous breast; head is all-dark ["Harlan's" intermediate morph-type plumage has white throat] and tail all-rufous, as is typical of this plumage [dorsal/ventral photos; CMN #136425]). Preble (1908) noted, "The melanistic and normal phases seemed to be about equally represented" along Mackenzie

River from Ft. Simpson to Tulita, N.W.T.; on his return trip he saw "a melanistic below Tulita Jul. 25, 1903." 2 adults were in same area as "Eastern" in the "Yukon" sample. A perched adult was seen well at Palmer, Alaska, in mid-Apr. 2009 (pers. obs.). *Dark morph (adults):* A dark intermediate morph adult (all-dark except slight rufous cast on breast) was photographed at Fox Creek, Alta., (northwest of Edmonton) in early May 2010 (Russell 2010). 1 was seen near Red Deer, Alta., (soaring with a paler type "Eastern" adult) and 1 near Ft. Nelson, B.C., in late Jun. 1997, both east of Rocky Mts. (pers. obs.). Adult male nested with "Eastern" intergrade-type-plumaged female near Ft. St. John, B.C., in 2013 (W. S. Clark pers. comm./unpubl. photos). Easternmost dark morph (or dark morph "Harlan's) in Alta. was a juvenile near Peace River (W. S. Clark pers. comm.). No all-dark examples of this color morph have been seen or photographed in N.W.T. (W. S. Clark pers. comm.; G. Vizniowski pers. comm.). A dark morph was sighted near Beaufort Sea in n. Y.T. in early May 1974 (Salter et al. 1980). 3 adults from same area as the "Yukon" (light morph group) are classic types of this color morph (and in known "Harlan's" breeding range). A dark morph nested with an intermediate morph "Harlan's" near Klukwan, Alaska, in 2013 (C. Susie pers. comm./unpubl. photos; W. S. Clark, unpubl. photos). A few all-dark Red-tailed Hawk types with all-red tails were seen in n. interior Y.T. during numerous Gyrfalcon and Peregrine Falcon aerial surveys (D. Mossop pers. comm.). *Note:* Since "Harlan's" birds sometimes have quite rufous tails among their shared *jamaicensis* traits—and interbreed with other Red-tailed Hawk subspecies—it is difficult by any standard, even in the hand, to separate some darker Red-tailed Hawks to exact subspecies. **Mixing of Red-tailed Hawk Subspecies (Western Canada):** The balance of the 139 adults from the "Western Canada" selection north of cen. Alta. and west of Sask. comprised 28 (20%) "Western," "Harlan's" (mainly light morph), and probable intergrades: 15 (11%) were "Western," with 10 rufous/intermediate morph, 3 dark morph, and 2 light morph; 5 (4%) were "Harlan's" light morphs; 8 (5%) were light and/or dark intergrades of undetermined-subspecies. In their DNA research, Hull et al. (2010) note that the "Eastern" subspecies "has historic and contemporary contact" with "Harlan's" and "Western." "Eastern" interbreeds with the smaller percentage of "Western" and "Harlan's" that inhabit this region, especially from Calgary, Alta., and northward, east of Rocky and Mackenzie Mts. *Subspecies mixing in breeding season near Edmonton, Alta.*—The following birds were observed in the same area, presenting a high potential for interbreeding: dark-headed, heavily marked "Eastern" female from Millet in Jun. (Eriksson 2011); dark-headed, moderately marked "Eastern" from Sturgeon River near St. Albert (Conlin 2015); light morph "Western" (all-barred belly) from St. Albert in Jul. (Borlé 2016e); white-throated, moderately marked "Eastern" at Edmonton in Jul. (Bauschardt 2012); white-throated, moderately marked "Eastern" nesting bird in Jun. from Redwater (Edwards 2012); light morph "Harlan's" from Morinville in Jun. (Borlé 2016b); and an intermediate morph "Harlan's" nesting near Edmonton in Jun. (Borlé 2016a). Farther north, in boreal forest of nw. Alta., just south of N.W.T., G. Vizniowski, in "Nw. Alberta" group, saw/photographed 4 adults along 19-mi. (30km) stretch of Hwy. 35 in this order: white-throated, moderately marked "Eastern" with single-banded tail and no thigh barring or markings on underwing; classic rufous/intermediate morph "Western"; heavily marked "Eastern" with whitish throat (similar to Sherwood Island St. Pk., Conn., female; Hand 2003a); and heavily marked light morph bird of undetermined lineage (underwing had "Harlan's"-like barring). *Documented interbreeding of subspecies.*—M. Borlé (2016c) shows an image of a light morph "Eastern"-like juvenile in early Sep. near Edmonton, Alta., from a brood of 3 from a dark "Harlan's" male and an "Eastern" female; its 2 siblings were dark morphs. A dark morph male "Western" was mated to a mixed-lineage "Eastern"-like female at Charlie Lake (near Ft. St. John), B.C. (east of Rocky Mts.), in 2013, producing 2 nestlings similar to their unmarked-white-breasted, heavily marked mother

(W. S. Clark pers. comm./Fig. 2, p. 191). The N.W.T. rufous morph that E. Preble collected on the Slave River near Ft. Smith, N.W.T. (dorsal/ventral photos, USNM #193557 [cataloged as "Harlan's" but clearly a rufous "Western"]), was mated to a "normally colored" (likely "Eastern") bird. A partial albino "Eastern" was mated to a light morph "Harlan's" near Peace River, Alta. (W. S. Clark pers. comm./unpubl. photos). A heavily marked "Eastern" female was mated to a light morph "Harlan's" male in Turtle Mts., N.Dak. (Sullivan and Liguori 2010). *Note:* DNA studies published by Hull et al. (2010) suggest that "on average over the generations, gene flow has been greater from 'Eastern' [and 'Harlan's'] to 'Western' than from 'Western' to the other 2 subspecies"; and that there is "significantly greater gene flow from 'Eastern' to 'Harlan's' than vice versa." **Summary:** Based on the above data, the n. Great Plains and especially the boreal forest—particularly from cen. Alta. and northward—can be seen as a virtual melting pot of Red-tailed Hawk genetics. The 3 subspecies in this region have been interbreeding for eons, and it is remarkable that a "pure" example of any subspecies is encountered from cen. Alta. and northward. Adhering a subspecies label to many of these Red-tails is *impossible*, especially to offspring of darker morphs of "Harlan's" or "Western" × "Eastern" parents, which may look like either of their parents, but, interestingly, may not show a combination of both parents' traits. The previously noted "Eastern"-like juvenile of the "Harlan's" × "Eastern" pair may *look* "Eastern" but in reality is not a "pure" type. *Note:* Todd (1950) was perhaps the 1st author to note that w. Canada (mainly Alta.) possesses a conglomerate of subspecies with the distinct probability of interbreeding. *Addendum*: Sample of w. Canada adults depicted on p. 191.

Analysis of the Proposed "Northern" Subspecies of Red-tailed Hawk (*B. j. abieticola*)

This proposed subspecies has been advocated, as a split from the "Eastern" subspecies, by Todd (1950), Dickerman and Parkes (1987), and Liguori and Sullivan (2014). A "Northern" subspecies was not accepted by the American Ornithologists' Union in its last subspecies assessment in 1957. The initial criteria for this subspecies were that it is much more heavily marked and more richly colored on its underparts than typical "Eastern," and that it was thought to breed only in the White Spruce–Balsam Fir boreal forest of Canada. **Todd (1950).**—In "A Northern Race of Red-tailed Hawk," Todd based his findings on 5 breeding-season adult museum examples from the boreal forest biome: 1 from Que., noted in the preceding text as "Québec Female" and the type specimen for his publication, and 4 from n. Alta.—the previously described "Alberta Female" plus 2 females and 1 male from Lac La Nonne, a short distance northwest of Edmonton, Alta. All 5 have dark heads, mostly dark throats, and moderate to heavy bellyband markings. *Comparisons:* "Alberta Female" is from n. edge of boreal forest and is nearly identical to Lima, 5th mate (2011) of Pale Male of Central Pk., N.Y.C. (Karim 2011, Nov. 12, Dec. 27), found 2,400 mi. (3860km) from the Alta. bird. The similar-looking female from Cochrane, Alta., south of boreal forest, was, like Lima, mated to a very pale "Eastern" male (Kube 2012). The stately, very dark-headed and streak-breasted Charlotte, a 2005–8 resident of Central Pk., N.Y.C. (Yolton 2008b), was more heavily marked on the breast than "Alberta Female." The birds of 2003 resident pair from Sherwood Island St. Pk. in s. Conn. both had even heavier bellybands than "Alberta Female" but had whitish or white throats (Hand 2003a, 2003b). Images by W.

S. Clark and G. Vizniowski in "Western Canada" group clearly show that lightly marked birds greatly outnumbered heavily marked birds in n. Alta. and s. N.W.T. Published images of moderately marked nesting adults from around Edmonton, Alta., show paler birds intermixing with darker birds (Bauschardt 2012; Edwards 2012). Todd did note that Alta. may contain a confounding mixture of subspecies. He is very correct on this point. **Dickerman and Parkes (1987).**—In "Subspecies of Red-tailed Hawk in the Northeast," the authors expounded on Todd's proposed subspecies based on heavily marked features; however, none of their museum examples were from the boreal forest. The authors noted 25 birds (18 adults, 7 juveniles) from fall–spring collected in N.Y., N.J., Pa., Md., and Iowa; 7 other birds were from ne. Canada, but only 1 of 2 adults in the sample was from the breeding season, a female from P.E.I., which was evidently heavily marked and "inseparable from western *calurus*." This study *speculated* that more heavily marked birds were from n. regions but provided no proof, since only 1 bird of the sample was from breeding season and from a northerly latitude (but not from boreal forest, the focus of Todd's speculation). *Comparisons:* Adults are discussed above. The authors noted that if thigh markings were present, they were "sepia" (dark brown) on *abieticola* types rather than reddish brown, as on *calurus* ("Western") birds. "Alberta Female" did have dark brown partial barring on her thighs. However, this author has found reddish-brown- (rufous-) barred thighs, as well as dark-brown-barred thighs, on many birds of the "Western Canada" sample. The single perched adult from Cochrane, Alta., had rufous-barred thighs (Lichter 2010). The heavily marked female from Millet, Alta., had only faint brown upper-thigh barring (Eriksson 2011); the very heavily marked Edmonton bird from Apr. lacked thigh barring (Borlé 2016d). In the East, Christo of Tompkins Sq. Pk., N.Y.C. (Goggin 2015), shows fully rufous-barred thighs, as do the resident male of Atlantic Co., N.J. (Cosby 2015a), and many others noted in the prior essay. *Note:* In a personal communication with this author, Robert Dickerman reversed his opinion on the proposed "Northern" subspecies several years after the publication of Dickerman and Parkes (1987): "Birds that fit this description [*abieticola*] occur infrequently and do not support a recognized population" (initially printed in Wheeler 2003a). This author agrees fully with Mr. Dickerman's statement. **Liguori and Sullivan (2014).**—In "Northern Red-tailed Hawk (*Buteo jamaicensis abieticola*) Revisited," the authors tout the previous 2 publications on dark-headed and heavily marked individuals. However, they also present plumage characteristics of more lightly marked types that they believed came from more northerly breeding populations, but these characteristics—streaked breasts, partial tail banding, blob-type belly markings—have now been found to be relatively common in all of the "Eastern" subspecies' range. They also noted that *abieticola* birds that did not have dark throats had a "dark-bordered throat"—a dark mark next to the paler throat. Only 2 of their examples were from the breeding season: the "Yellowknife" bird from N.W.T. and a blob-patterned, moderately marked bird from Ottawa, Ont., from Aug. (Liguori and Sullivan 2014, Fig. *3* and Fig. *6*, respectively). Other examples exhibited mixed-genetic plumage features, especially such "Western" traits as rufous-barred underwing coverts and axillaries. *Comparisons: Head markings.*—The data Liguori and Sullivan published regarding a "dark-bordered throat" are misleading: All Red-tailed Hawks except all-dark-headed and all-white-headed ("Krider's"/light morph "Harlan's") birds have a black malar mark on each side of the paler/white throat, and this has no association with *abieticola* subspecies. *Breast markings.*—This author has found breast markings to be the most common of the atypical plumage markings adorning "Eastern" birds across all regions of the breeding range. They adorn both sexes. They show well on the cen. Alta. migrant (Lynch 2016) but on few other published images from Alta. This pattern, however, is well documented on numerous individuals from the East: Bobbie and Rose, 2012 residents of N.Y.C., both sported such markings (Washington Sq. Pk. Blog 2012), as did Charlotte (Karim 2005) and numerous others from N.Y.C., as well as adults from other areas farther south, including Va. resident Scarlette (Deal 2011) and a N.C. resident (Ennis 2010). *Tail banding.*—Partial to full tail banding is the most common atypical trait in w. Canada, where it is found in all regions and latitudes. The "Yellowknife" image presents a classic example. This trait is also found on the nesting bird of Redwater, Alta., near Edmonton (Edwards 2012), and on the late-summer adult from near Regina, Sask. (Herriot 2010). This pattern is also found in the East, though much less commonly; Scarlette of Va. (Wildlife Center of Va. 2011) and a 1-year-old from Long Island, N.Y., in May (Michaels 2015a) are 2 examples. *Blob-type belly/ flank markings.*—These markings are just variations from the more normal square-edged or diamond-shaped markings. They are found on many Canadian birds and are commonly seen on populations in the East. Above-noted Bobby and Rosie show good examples of this pattern, as do the male from J. Hood Wright Pk., N.Y.C. (Schmunk 2014c, 2014d), and the resident male of Atlantic Co., N.J. (Cosby 2015b). The pair from Sherwood Island St. Pk., Conn., presents a prime example of the heavy bellyband pattern: The male has more square-edged black markings, and the female has a very blobbed pattern (Hand 2003a); the female is identical to the bird with a heavy bellyband in Fig. *7* in Liguori and Sullivan (2014). **Red-tailed Hawk Subspecies of Western Canada:** As noted in "Mixing of Red-tailed Hawk Subspecies" in the previous section, the region west of Man. is a veritable melting pot of intermixing and interbreeding of subspecies. Sask. has "Krider's," "Eastern," and light morph "Harlan's" on Great Plains and the latter 2 subspecies in boreal forest. Alta. does not have "Krider's" (all pale birds thought to be "Krider's" have been light morph "Harlan's"), but north of Calgary and especially Red Deer and northward throughout N.W.T., there is a total mixing of dominant "Eastern," and "Harlan's" and "Western." "Pure" examples of subspecies surprisingly do occur, but there are a good number of unidentifiable birds. Data from "Western Canada" sample and "Pembina Valley" published images and other places (see "Plumage Variation") show that the array of plumages from n. Canada encompasses all types, from lightly marked to heavily marked. Heavily marked birds are a small minority. In this region of the "Eastern" subspecies' range, whether the plains or boreal forest, 2 plumage traits are exhibited in a high percentage of adults: banded tails and barred thighs. If one were inclined to create a subspecies, these w. Canadian birds, though very "Eastern" in plumage, regularly have banded tails, barred thighs, and barred axillaries, more often than in any other region of the subspecies' range. However, these traits are also seen throughout the East range in identical patterns, although in much smaller percentages. *Note:* Western Canada shows 1 trait that is found only in that region, and it is very likely due to "Western" genetics: solid black bellyband. **Summary:** (1) "Eastern" birds, as with light morph "Harlan's" and "Western," show a realm of lightly marked to heavily marked types with a variety of marking patterns. In "Harlan's" and "Western," however, the variations go farther, into polymorphism; but even in darker morphs of these subspecies there is much individual variation. (2) The boreal forest is not a habitat barrier for "Eastern," nor is it one for "Harlan's" or "Western" in w. Canada. Across this remote region, the forest is about 450 mi. (725km) wide in a north–south band, but Red-tailed density is low to very low in n. portions, especially in N.W.T. (W. S. Clark pers. comm.; G. Vizniowski pers. comm.). (3) Heavily marked types of "Eastern" are more common north of Great Plains, but in all areas, such birds with extra melanin in their feathers appear to always be a minority type. There will always be the odd bird with excessive markings, but these may well be individuals of mixed genetics and certainly do not comprise a subspecies. Virtually all plumage types from w. Canada—except intergrade variants that are found only in w. Canada in the breeding season—can also be found breeding in e. U.S. and e. Canada. (4) The intergrading of 3 subspecies in w. Canada, as noted in "Mixing of Red-tailed Hawk Subspecies," is no doubt tremendous. What may appear as a "pure" subspecies example may not be so at all. Based on the above data and the opinion of Robert Dickerman, this author concludes that the proposed *abieticola* simply represents a more heavily marked and more saturated variant of the "Eastern" subspecies, which as with other subspecies, possesses considerable individual variation. In w. Canada, all types intermix and interbreed—and have done so for many hundreds of years. Creating a subspecies in this region, in particular, would seem illogical. So-called "Northern" is not a subspecies, just a "*type*" (i.e., heavily marked, dark-headed, streak-breasted, dark-bellied, bar-legged, band-tailed, intergrade, etc.).

Figure 1. Bruce, Alta. (Jul. 2013)
A moderately marked dark-collared adult "Eastern." Majority of w. Canada adults have tail banding. Thighs are unmarked, but often barred in this region. Photo by W. S. Clark

Figure 2. Charlie Lake, B.C. (Jul. 2013)
A heavily marked dark-throated adult "Eastern"-like (mated to dark morph "Western"; see pp. 187, 188). She likely has "Western" lineage (very large patagial marks); lightly spotted thighs is an "Eastern" trait. Photo by W. S. Clark

Figure 3. Great Slave Lake, N.W.T. (May 2016)
A moderately marked dark- collared adult "Eastern." Its thighs are barred (hidden); tail is not banded. Photo by Gary Vizniowski

Figure 4. Great Slave Lake, N.W.T. (Apr. 2016)
A heavily marked darkish-throated adult "Eastern"-like. Its axillaries and thighs are barred; tail not banded. Photo by Gary Vizniowski

Plate 42. ROUGH-LEGGED HAWK (*Buteo lagopus*) Juvenile, Perching

Ages: Juvenile plumage is retained for 1st year. *Bill is thin and small, with small bluish basal area on both mandibles.* Eyes are pale brown or, less commonly, pale gray. *White outer lores and forehead form mask, which is more distinct on darker birds. Long, thin, yellow gape extends under eye.* Legs are fully feathered to toes. *Feet are very small. When perched, wingtips equal tail tip. 3 main undertail variations are shared by all variants of all color morphs. Light/intermediate morphs.*—Dorsal tail surface has variably sized white base; white area is always small in intermediate morph but can be small even on typical light morph. Undertail coverts are mainly pale. *Dark morph.*—Dorsal side of tail is dark. Tail can have pale bands. Tail can be partially or fully banded on either sex. Undertail coverts are mainly dark. Dark morph plumages are divided into 2 main types: *Brown type* has all-brown head and body or has paler head and often paler breast (pale-headed type). This plumage variation adorns either sex. *Male type* is a darker blackish-brown plumage that adorns only some males. **Subspecies:** *B. l. sanctijohannis* in N. America; 3 others in Eurasia. **Color morphs:** Cline from light to dark in juveniles of *both* sexes. **Size:** L: 18–23" (46–58cm). Females average larger. **Habits:** Tame to wary in West. Perches on exposed, often high objects and on ground. **Food:** Aerial and perch hunter of small rodents. **Voice:** Typically silent on winter grounds. If agitated, emits plaintive *keeaah*.

42a. Light morph: Pale head with thin, dark eye line. Inner lores are often whitish and blend with white outer lores and forehead. *Note:* Head of intermediate morph is similar.

42b. Dark morph, pale-headed brown type: Head is moderately dark, with thin, dark eye line. Inner lores are dark; *white on outer lores and forehead.*

42c. Dark morph, brown type: Dark inner lores; *distinct white outer lores and forehead.* Thin, dark eye line. Supercilium is paler than rest of brown head.

42d. Dark morph, male type: *Very distinct white outer lores and forehead.* Head is uniformly dark. Eye is pale gray. *Note:* This eye color is mainly on black-type male.

42e. Light morph [left] leg: Leg feathers are unmarked or have light streaking on paler birds. Legs are fully feathered.

42f. Light morph [left] leg: Leg feathers are moderately streaked on heavily marked birds. Legs are fully feathered.

42g. Dark morph [left] leg: Dark brown, fully feathered legs.

42h. Unbanded tail (ventral): Inner ½ has white basal band; distal ½ has gray band.

42i. Single-banded tail (ventral): As on *42h*, except has ill-defined, darker subterminal band.

42j. Banded tail (ventral): Similar to *42i*, except has partial narrow inner tail bands. *Note:* Uncommon for multiple bands to show on ventral side unless seen in translucent light. All banded pattern can adorn either sex.

42k. Light morph, lightly marked type: Dark streaking on side of neck; clear breast. *Flanks and belly are dark brown and form a solid bellyband. Primaries have white patch on their base.* Basal part of tail is white. Legs are fully feathered.

42l. Light morph, moderately/heavily marked type: Dark streaking on side of neck; streaked breast. *Dark brown flanks and belly form dark band.* Streaked or spotted leg feathers. *Small white patch on base of primaries.* Base of tail has small patch of white. *Note*: It is common for tails to have rufous wash on dorsal surface, as on this figure.

42m. Intermediate morph: Heavily streaked breast. Dark brown flanks and belly form dark band. Legs are heavily streaked. Small white area on base of primaries. Small white area adorns base of tail. *Note:* Can appear like a paler dark morph, except having white on base of tail.

42n. Light morph flank feather: Solid brown. *Note:* Compare to adult female (*44k, right*).

42o. Dark morph, pale-headed brown type: *Head as on 42b.* Rufous-streaked breast. Mostly solid brown flanks, belly, and legs. Legs are fully feathered. Dark tail is unmarked or banded.

42p. Dark morph, brown type: *Head as on 42c.* Breast has minimal amount of pale tawny or rufous streaking. Legs are fully feathered. Dark tail is unmarked or banded.

42q. Dark morph, male type: *Head as on 42d.* Plumage is very dark brown. Legs are fully feathered. *Note:* Male-only plumage. Tail is typically banded.

a
b
c
d
e
f
g
h
i
j
k
l
m
n
o
p
q

Plate 43. ROUGH-LEGGED HAWK (*Buteo lagopus*) Juvenile, Flying

Ages: Juvenile plumage is retained for 1st year. *White outer lores and forehead form mask, which is more distinct on darker birds.* Legs are fully feathered to toes. *Feet are small. On dorsal surface, primaries have pale, rectangular panel that contrasts with dark primary coverts.* Pale primary panel creates rectangular window on underwing when seen in translucent light. *On ventral surface, outer primaries have large, black tip. 3 main undertail variations are shared by all variants of all color morphs. Light/intermediate morphs.—Large, square, black carpal patch. Outer primaries have small white area on base of dorsal side. Broad, dark brown band adorns belly and flanks. Uppertail coverts are white or tawny.* Axillaries are pale and streaked on light morph but dark on intermediate morph. On dorsal side, tail has variably sized white base; white area is always small in intermediate morph but can be small even on typical light morphs. Tail can be partially or fully banded on either sex. Undertail coverts are mainly pale. *Dark morph.*—Dorsal surface of tail is dark. Tail can have pale bands. Undertail coverts are mainly dark. Axillaries are dark brown. Dark morphs are divided into 2 main types: *Brown type* has all-brown head and body or has paler head and often paler breast (pale-headed type). Dark carpal patch may not show on all-brown birds. This plumage variation adorns either sex. *Male type* is a darker blackish-brown plumage that adorns only some males; dark carpal patch is not visible. **Subspecies:** *B. l. sanctijohannis* in N. America; 3 others in Eurasia. **Color morphs:** Cline from light to dark in juveniles of *both* sexes. **Size:** L: 18–23" (46–58cm), W: 48–56" (122–142cm). Females average larger. **Habits:** Tame to wary in the West. **Food:** Aerial and perch hunter of small rodents. **Flight:** Slow wingbeats with irregular glide sequences. Glides in modified dihedral and soars in low dihedral. Regularly hovers and kites. **Voice:** Typically silent on winter grounds. If agitated, emits plaintive *keeaah.*

43a. Light morph, lightly marked type: *Dark carpal patch. Broad, dark band on belly and flanks.* Primaries show translucent window. Tail is unbanded.

43b. Light morph: *Pale, rectangular panel on primaries; white base on outer few primaries. Primary coverts are dark.* Base of tail is white. Shows partial banded pattern of *43l.*

43c. Light morph, lightly marked, split-bellied type: Very pale birds have incomplete dark band across belly. *Black carpal patch is often incomplete.* Tail has 1 dark band. *Note:* Very uncommon.

43d. Light morph, heavily marked type: Streaked breast. *Carpal patch is black. Bellyband is solid brown.* Tail is banded. *Note:* Uncommon for inner bands of tail to show unless in translucent light.

43e. Intermediate morph: *Black carpal patch and tawny-rufous underwing coverts.* Breast is heavily streaked, and body is dark brown. Tail is not banded. *Note:* Head is as pale as on many light morphs.

43f. Dark morph, pale-headed brown type: *Black carpal patch and rufous underwing coverts.* Breast is streaked with rufous, and body is dark brown. Tail has darker subterminal band.

43g. Dark morph, brown type, 1st prebasic molt (May): *Black carpal patch subtly darker than rest of coverts; slight rufous on some coverts.* Body is dark brown. Tail is banded. *Note:* Innermost primary is dropped.

43h. Outer primary feather: All of tip (finger) is black. *Note:* Compare to Ferruginous Hawk (*48b*).

43i. Dark morph, male type: Blackish-brown body and underwing coverts. Tail is banded. *Note:* Male-only plumage.

43j. Dark morph, brown type: Pale rectangular primary panel contrasts with solid dark coverts. Tail lacks bands.

43k. Light morph, unbanded tail (dorsal): Basal ½ is white, distal ½ is dark. White uppertail coverts.

43l. Light morph, partial-banded tail (dorsal): Variable portion of basal area is white; some inner banding.

43m. Light morph, partial-banded tail (dorsal): Small amount of basal area is white; very partial banding.

43n. Light/intermediate morph, unbanded tail (dorsal): Minimum amount of basal white for light morph. Uppertail coverts of light morph are tawny on such dark-tailed types; coverts are dark on intermediate morph.

43o. Light/intermediate morph, banded tail (dorsal): Small basal white area; dark uppertail coverts.

43p. Dark morph, partial-banded tail (dorsal): 2–4 thin gray bands on center of tail. Uppertail coverts are dark.

43q. Dark morph, banded tail (dorsal): 3–4 narrow, complete, gray or whitish bands.

a
b
c
d
e
f
g
h
i
j
k
l
m
n
o
p
q

Plate 44. ROUGH-LEGGED HAWK (*Buteo lagopus*) 1-Year-Old and Adult, Perching

Ages: 1-year-old and adult (includes 2-year-old). *Bill is thin and small, with small bluish basal area. Long, thin, yellow gape extends under eye. White outer lores and forehead form mask.* Lower breast may have pale "necklace" band above bellyband. Both ages have wide, black subterminal tail band. Legs are fully feathered to toes. *Feet are small. When perched, wingtips extend past tail tip. 1-year-old.*—Eye is pale brown. A few juvenile feathers are retained on wings and body. Males exhibit adult-female-like plumages (Clark and Bloom 2005). Ventral can also be adult male-like (pers. obs.). *2-year-old.*—Adult-plumaged; not separable when perching. *Adult.*—Eye is dark brown. Plumages are sexually dimorphic: Males are grayish-mottled on upperparts; females are brown. *Light/intermediate morphs.*—Females have dark flanks, and males have spotted/barred flanks. Thin dark eye line. Leg feathers are spotted or barred. Dorsal side of tail has variably sized white base, smallest on intermediate morph and some light morphs. *Dark morph.*—Banded tail patterns are shared by both sexes; unbanded pattern (not shown) only on some females. 1-year-olds and adult females are *brown*; adult males are *black*, sometimes brownish on head and breast. **Subspecies:** *B. l. sanctijohannis* in N. America; 3 others in Eurasia. **Color morphs:** Light and dark; cline from light to dark *only* in females in adults (see Flying [dorsal view], Plate 46). **Size:** L: 18–23" (46–58cm). Females average larger. **Habits:** Tame to wary in West. Perches on exposed objects and on ground. **Food:** Aerial and perch hunter of small rodents. **Voice:** Silent on winter grounds; emits plaintive *keeaah* if agitated.

44a. Light morph 1-year-old: Eye is pale brown. *White mask. Lores are gray. Tawny head has thin, dark eye line. Note:* Adult female (*44c*) is identical but has dark brown eye.

44b. Light morph male: Dark brown eye. *White mask. Most light morph males have black inner lores that connect to black ring under eye and to black malar.*

44c. Light morph female: Eye is dark brown. Head is pale, with dark malar. Throat is pale. *White mask.* Inner lores are grayish. *Note:* Lightly marked males can be identical (as on *44g*).

44d. Dark morph 1-year-old, pale-headed type: Pale brown eye. *White mask is distinct.* Thin, dark eye line. Supercilium and auriculars are paler tawny. *Note:* Pale-headed types have tawny/rufous breast.

44e. Dark morph female/some males: *White mask with dark inner lores. Note:* Body is dark brown.

44f. Dark morph male: *Jet-black head with distinct white mask. Note:* Body is jet black.

44g. Light morph male, lightly marked type: Tawny ventral plumage. Side of breast is lightly streaked; lightly barred flanks and leg feathers. Belly may be lightly spotted. Gray-mottled dorsal plumage. *Note:* Palest type of male.

44h. Light morph male, lightly marked type: *Dark bib. White ventral plumage is lightly barred on flanks and leg feathers; unmarked or lightly spotted on belly.* Gray-mottled dorsal plumage. 2 thin, dark tail bands.

44i. Light morph male, moderately marked type: *Similar to 44h but has moderately barred belly and flanks.*

44j. Light morph male, heavily marked type: *Darker than 44i. Flanks and belly are heavily barred.*

44k. Light morph flank feathers: *Male (left):* barred. *Female (right):* solid black, often light-tipped.

44l. Light morph female, lightly marked type: Lightly streaked sides of breast. *Dark flanks with light-tipped feathers.* Belly is unmarked or lightly speckled. Leg feathers are spotted. Tawny/rufous-edged dorsal feathers.

44m. Light morph female, split-bellied type: *Black bellyband has narrow, tawny split down center.* Tawny necklace adorns breast. Leg feathers are barred, typical of adult. Tawny/rufous-edged upperparts.

44n. Light morph female, heavily marked type: *Solid black bellyband; tawny necklace.* Leg feathers are barred.

44o. Intermediate morph female: Pale head. Belly and flanks are solid brown. *Thin necklace separates breast and belly.* Leg feathers are mostly dark. *Small amount of white adorns base of tail.*

44p. Dark morph 1-year-old/female, pale-headed type: Moderately pale head (as on *44d*). Breast is tawny/rufous and contrasts with dark brown belly, flanks, and legs. *Note:* Paler retained juvenile feathers on wing.

44q. Dark morph 1-year-old/female: Head and body are uniformly dark brown. *Note:* 3-banded type tail.

44r. Dark morph male, speckled type: Black body, but brownish on head and breast. Has sparse, white speckling on body, especially on necklace. *Note:* Possible intermediate morph, but *no* cline from light to dark in males.

44s. Dark morph male: Jet-black head and body. Tail is 3-banded type. *Note:* Adult-male-only plumage.

a
b
c
d
e
f
g
h
i
j
k
l
m
n
o
p
q
r
s

Plate 45. ROUGH-LEGGED HAWK (*Buteo lagopus*) 1-Year-Old and Adult, Flying (ventral view)

Ages: 1-year-old and adult (includes 2-year-old). *White outer lores and forehead form mask.* Both ages have wide, black subterminal tail band. *Feet are small.* Wings are long, with 5 fingers; broader than on juveniles. *1-year-old.*—A few juvenile feathers are retained on flight feathers (retains up to p7–10, s3, s4, s8, s9) and sometimes on tail. 1-year-old males exhibit adult-female-like plumage (Clark and Bloom 2005). Can also be adult male-like on ventral areas (pers. obs.; Wheeler 2003b, Fig *482*). *2-year-old.*—Separable from adult only in dorsal view in flight (see *46c, d, i*). *Adult.*—Plumages are sexually dimorphic: Males are grayish-mottled on dorsal plumage; females are brown. *Light/intermediate morphs.*—Females have dark flanks, and males have spotted/barred flanks. *Underwing has black carpal patch at wrist.* Leg feathers are spotted or barred. Dorsal side of tail has variably sized, often small, white basal area. *Dark morph.*—Banded tail patterns are shared by both sexes; unbanded pattern only on some females. Dorsal side of tail is *always* either all-dark, or black with white/gray banding. 1-year-olds and adult females are *brown*; adult males are *black*, but sometimes brownish on head and breast. **Subspecies:** *B. l. sanctijohannis* in N. America; 3 others in Eurasia. **Color morphs:** Cline from light to dark exists *only* in adult females and juveniles. There is *no* cline between light and dark males in tail pattern or markings on underwing coverts or undertail coverts. Some males, however, do appear to have somewhat intermediate plumage markings, but without clinal blend seen in females (and juveniles). **Size:** L: 18–23" (46–58cm), W: 48–56" (122–142cm). Females average larger. **Habits:** Tame to wary in West. **Food:** Aerial and perch hunter of small rodents. **Flight:** Slow wingbeats with irregular glide sequences. Glides in modified dihedral and soars in low dihedral. Regularly hovers and kites. **Voice:** Silent on winter grounds; plaintive *keeaah* if agitated.

45a. Light morph [left] leg feathers: *Spotted type (left):* Adorns some 1-year-olds and adult females. *Barred type (right):* Most common type on all adult ages. Legs are fully feathered to toes.

45b. Tail (ventral): All color morphs and both sexes have fairly wide to wide black terminal band; may have multiple thin, dark inner bands.

45c. Light morph male, lightly marked type: Unmarked or lightly speckled white/tawny underparts. *Black carpal patch is mottled and confined mainly to primary coverts.* Tail has 1 or 2 bands. *Note:* Palest type of male.

45d. Light morph male, lightly marked type: *Dark bib. White belly; flanks are darker and barred. Black carpal patch often mottled but still distinct. Note:* Paler types typically have fewest tail bands.

45e. Light morph male, moderately marked type: Dark bib. *White necklace separates breast and belly. Belly is spotted or barred; flanks are darker than belly and always barred.* Tail has up to 5 bands.

45f. Light morph male, heavily marked type: *Dark bib. Necklace is pale and separates bib from darker, barred flanks and belly. Carpal patch is black.* Marked undertail coverts. Tail has up to 5 bands.

45g. Light morph female, lightly marked type: Tawny underparts are lightly streaked on breast. Belly is unmarked or lightly spotted. *Flanks are black, often spotted. Carpal patch is black.* Tail has 1 black band.

45h. Light morph female, split-bellied type: *Lightly or heavily marked breast has necklace demarcation. Flanks are solid black but can be spotted; belly has split down center. Carpal patch is black.* Tail has 1–3 bands.

45i. Light morph 1-year-old female/adult female, moderately/heavily marked type: Thin, pale necklace between streaked breast and black bellyband; some lack necklace. *Flanks and belly are solid black. Carpal patch is black.* Underwing coverts are heavily marked on more heavily marked birds. Retains faded, narrower juvenile flight feathers (p8–10; s3, s4, s8, s9).

45j. Dark morph, pale-headed type: Rufous-marked breast; dark brown body. *Rufous underwing coverts may be speckled with tawny or white; black axillaries and carpal patch.* Axillaries can be solid (shown) or barred.

45k. Dark morph female tail (ventral): 1 wide, black subterminal band. (Dorsal side is gray with black band.)

45l. Dark morph, dark-headed type: Uniformly dark brown body. Underwing coverts are faintly rufous, with black carpal patch. Axillaries can be solid or partially barred (shown). Tail is 3- to 4-banded type.

45m. Dark morph male, speckled type: Black body, but brownish on head and breast. Has sparse, white speckling on body, especially on necklace. *Note:* Possible intermediate morph, but *no* cline exists from light to dark in males.

45n. Dark morph male: Jet-black body and wing coverts. Axillaries can be partially barred with white.

a
b
c
d
e
f
g
h
i
j
k
l
m
n

Plate 46. ROUGH-LEGGED HAWK (*Buteo lagopus*) 1-Year-Old and Adult, Flying (dorsal view)

Ages: 1-year-old and adult (includes 2-year-old). *White outer lores and forehead form mask.* Both ages have wide, black subterminal tail band. *Feet are small.* Wings are long, with 5 fingers; broader than on juveniles. *1-year-old.*—A few juvenile feathers are retained on wings, especially visible on flight feathers (up to p7–10, s3, s4, s8, s9), and sometimes on tail. Males exhibit adult-female-like plumages (Clark and Bloom 2005). Can also be adult male-like on ventral areas (pers. obs.; Wheeler 2003b, Fig *482*). *2-year-old.*—Adult-plumaged; age is separable in dorsal flight view. *Adult.*—Plumages are sexually dimorphic: Males are grayish-mottled on dorsal plumage, females are brown. *Light/intermediate morphs.*—Females have dark flanks; males have spotted/barred flanks. Dorsal side of tail has variably sized, often small, white basal area. *Dark morph.*—Banded tail patterns shared by both sexes; unbanded pattern only on some females. Dorsal side of tail is *always* all-dark or black with white/gray banding. 1-year-olds and adult females are *brown*; adult males *black*; can be brownish on head, breast. **Subspecies:** *B. l. sanctijohannis* in N. America; 3 others in Eurasia. **Color morphs:** Cline from light to dark *only* on females (and juveniles). No cline exists in males between light and dark in tail pattern or markings on underwing coverts or undertail coverts. Some males appear to have somewhat intermediate plumage. **Size:** L: 18–23" (46–58cm), W: 48–56" (122–142cm). Females average larger. **Habits:** Tame to wary. **Food:** Aerial and perch hunter of small rodents. **Flight:** Slow wingbeats with irregular glide sequences. Glides in modified dihedral and soars in low dihedral. Regularly hovers and kites. **Voice:** Silent on winter grounds; emits plaintive *keeaah* when agitated.

46a. Light morph 1-year-old: Retains brown juvenile flight feathers (p8–10, s3, s4, s8, s9). *Tail has 1 black band; basal ½ is white.*

46b. Light morph 1-year-old: Retains brown juvenile flight feathers (p9, p10, s4, s9) and some wing coverts. *Tail is 5-banded type; basal ½ is white.*

46c. Light morph 2-year-old male, lightly/moderately marked type: Gray upperparts. Small white patch on outer primaries (p6–8), which are faded 1-year-old feathers; rest of wing is adult feathers. *Tail is 3-banded type.*

46d. Light morph 2-year-old female: Brown upperparts. Small white patch on outer primaries (p6–8), which are faded 1-year-old feathers. *Tail is 3-banded type* (typically greatest number of bands for light females).

46e. Light morph male, heavily marked type: Dorsal areas are darker gray than on *46c*. *Tail is 5-banded type.*

46f. Light morph female: Full adult plumage. Upperparts are brown. *Tail is 1-banded type with small amount of white on base* (as on *46n*). *Note*: Small amount of white on base of tail is common.

46g. Intermediate morph female: Head is pale. Upperparts are brown. Tail has very small amount of white on basal area, visible when fanned. *Tail has 1 band* (can have 3 dark bands).

46h. Dark morph 1-year-old, pale-headed type: Dark brown upperparts. Retains pale, faded brown juvenile flight feathers (p8–10, s3, s4, s8, s9). Unmarked, female-only tail pattern.

46i. Dark morph 2-year-old, dark-headed type: Dark brown head and upperparts. Small gray patch on outer primaries (p7, p8), which are 1-year-old feathers. Tail is partial-banded type (as on *46q*).

46j. Dark morph female, brown type: Dark brown head and upperparts. Tail is 4-banded type, with thin, white or gray bands. *Note:* Dark morph females commonly have up to 4 pale tail bands.

46k. Dark morph male, speckled type: Black; gray-mottled scapulars are similar to those of light morph males. *Tail is 4-banded type.*

46l. Dark morph male: Plumage is jet black but often mottled gray on scapulars. *Tail is 3-banded type.*

46m. Light morph 1-year-old tail (dorsal): Shows retained brown juvenile feathers (r2, r5 sets) among black-banded adult feathers.

46n. Light morph, minimal-white tail (dorsal): Smallest amount of white on tail base for light morph.

46o. Light morph, 3-banded tail (dorsal): Typically, maximum number of bands for light morph females; typical number for light morph males.

46p. Light morph, 5-banded tail (dorsal): Maximum number of bands for light morph males; rare on females.

46q. Dark morph, partial-banded tail (dorsal): 2 or 3 partial, pale bands; found mainly on 1-year-old and female dark morphs.

46r. Dark morph, 3-banded tail (dorsal): 3 thin, white/gray bands. This pattern is classic.

46s. Dark morph, 4-banded tail (dorsal): 4 thin, white/gray bands. Found on dark morphs of both sexes, more commonly females, which are larger and have longer tails.

a
b
c
d
e
f
g
h
i
l
j
k
m
n
o
p
q
r
s

ROUGH-LEGGED HAWK (*Buteo lagopus*)

Habitat: *Summer.*—Arctic tundra that is interspersed with cliffs, embankments, and rock outcrops. Southern areas may have sparse growths of White Spruce. *Winter/migration.*—Semi-open and open areas of airports, prairies, meadows, agricultural locales (especially grass hay, alfalfa), fresh- and saltwater marshes, and large swamps. **STATUS:** Fairly common. Sparsely distributed in breeding season, but can be locally common and gregarious in prey-rich areas during winter. Population fluctuates widely with summer cyclic abundance of especially Brown Lemming (*Lemmus trimucronatus*). When lemming population is high, the hawk's breeding success is high; when lemming population crashes, every 3–4 years, hawks either do not breed or have low nesting success, and smaller numbers of mainly adults are seen during winter. *Color morph status.*—Adheres to Gloger's rule: Arid climate in Arctic west of Hudson Bay produces mainly light morph, and only 5–7% of wintering birds on Great Plains are dark morph. Spring migrants heading to moist w. Alaska breeding grounds—especially Aleutian Islands—exhibit a high percentage of dark birds, which rival or surpass percentage in e. U.S. in winter (Apr. 2009 in s. Alaska: 146 of 348 [42%] were dark morph [pers. obs.]; 29% dark morph in Apr. 2002 at same location [P. Fritz, in Wheeler 2003b]). A fairly high percentage of dark morphs are seen west of Rocky Mts. and especially west of Cascade Mts. in winter. **NESTING:** Begins in May–early Jun. and ends as late as early Sep. Nests may be reused for many years and can become quite large. Male brings nest material, and female builds nest. Nests are built on cliffs, embankments, tops of elevated rocky outcrops or knolls, or on flat ground. Nests may be placed in tops of spruce trees in s. areas. Female lays 2–7 eggs; higher numbers when lemmings are abundant. **MOVEMENT:** Moderate-distance migrant. *Spring.*—Adults move 1st, in late Feb.–Mar., and peak on Great Plains early–mid-Mar. Bulk of female population migrates before males. Juveniles peak in early–mid-Apr., but movement continues into May–early Jun. *Fall.*—Birds leave breeding/natal grounds late Aug.–Sep.; juveniles and young adults, especially females, arrive on Canadian border areas by late Sep.–early Oct. Numbers peak in late Oct.–early Nov.; smaller numbers in Nov.–Dec.; late-season push on Great Plains in late Dec.–early Jan. Adult males move in latter part of season. **COMPARISON:** ***Rough-legged Hawk light morph.*** **Northern Harrier (all ages).**—Share similar white uppertail coverts. Flight mode, with wings held in dihedral and hovering, is similar. Harriers have all-white belly with rufous (adult male), streaked (adult female), or orange flank markings (juvenile), *vs.* white belly with black spots/bars on flanks or black/brown bellyband on females and juveniles. **Red-tailed Hawk (juvenile).**—Can have similar darker band on belly, but it is rarely solid. Ventral wing has dark patagial mark, *vs.* lack of such mark. When perched, wingtips are shorter than tail tip, *vs.* equal tail tip. Tail has many thin, dark bands, *vs.* possible few dark bands. White uppertail coverts can be similar. ***Rough-legged dark morph.*** **"Western" and "Harlan's" Red-tailed Hawks (juvenile darker morphs).**—Winter only. These juvenile Red-tails have multiple thin, dark tail bands, *vs.* maybe a few dark bands. Forehead is dark, *vs.* white. When perched, wingtips are distinctly shorter than tail tip, *vs.* wingtips equal to tail tip. In flight, wings held on flat plane, show pale dorsal panel extending to front edge of primary greater coverts, *vs.* wings in dihedral and pale rectangular panel only on primaries. **"Harlan's" Red-tailed Hawk (adult dark morph).**—All-black adult "Harlan's" is similar to juvenile and adult female Rough-legged, with tail dark and unmarked on dorsal side and whitish with darker terminal band on ventral side. Forehead is dark, *vs.* white forehead.

Figure 1. Weld Co., Colo. (Mar.)
Winter habitat of open areas with scattered trees for perching.

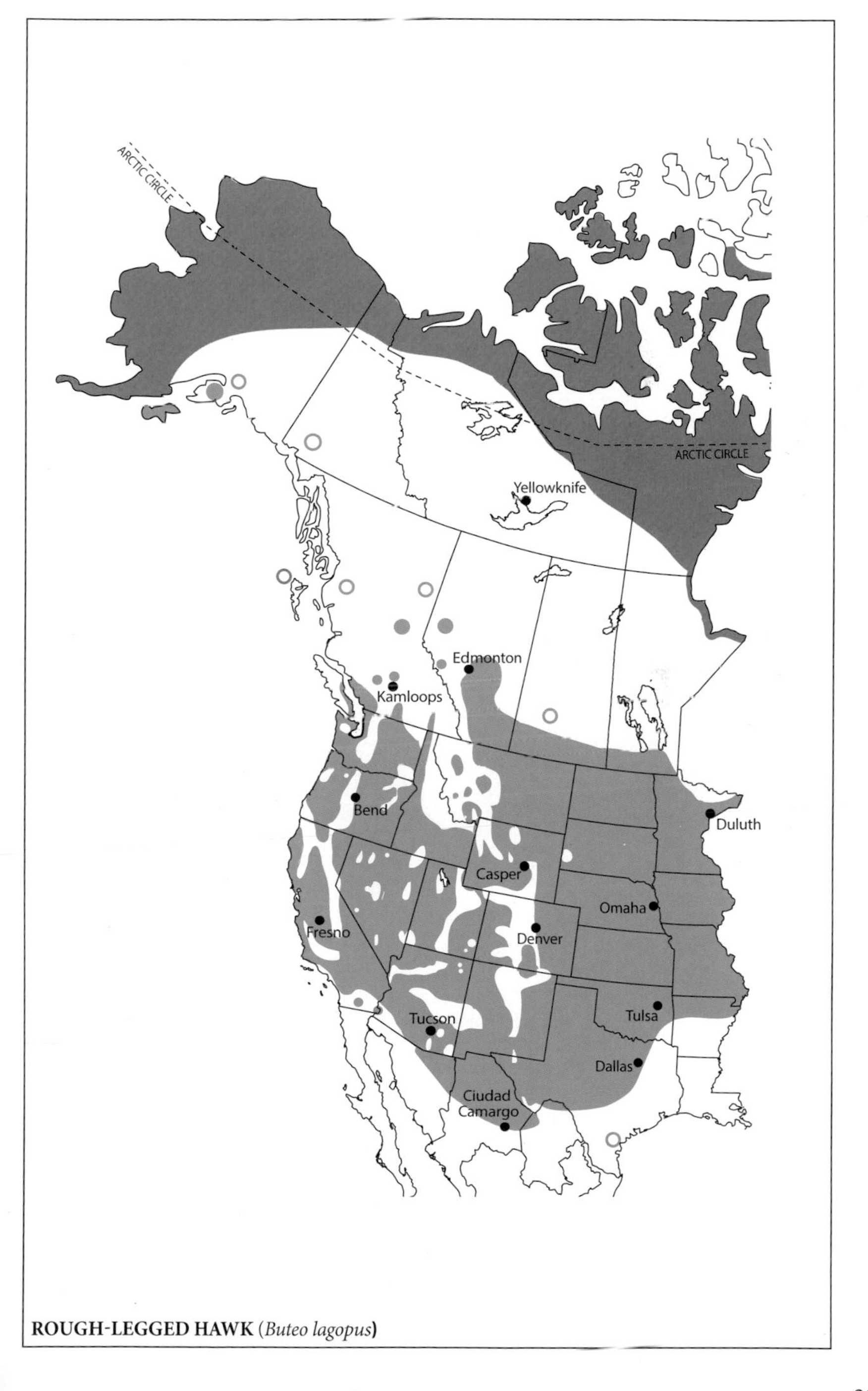

ROUGH-LEGGED HAWK (*Buteo lagopus*)

Plate 47. FERRUGINOUS HAWK (*Buteo regalis*) Juvenile, Perching

Ages: Juvenile plumage is held for 1st year. Plumage wears and fades considerably by spring. Eye color varies from pale gray when fledgling/recently fledged to pale yellow or tan in fall–winter and to tan-only by spring. *Black bill has small, pale blue spot on lower basal area of upper mandible. Yellow cere; long yellow gape extends to mid-eye. Legs are feathered to toes. Primaries show pale gray (silver) outer edge when folded (only raptor to have such marking).* When perched, wingtips are distinctly shorter than tail tip. Juveniles (more so than adults) use their tail as a brace against the ground to obtain leverage when feeding; this breaks off distal part of tail by late winter–spring (pers. obs.). *Light morph.*—2 variations of underparts on juveniles: lightly and moderately marked types. *Narrow, pale patch extends under eye. Thick, black eye line extends behind eye and under pale patch and touches gape. No dark malar mark. White uppertail coverts have large dark spot on each feather. Grayish (fresh) or brownish (worn) tail has 2–4 moderately wide partial bands, wider dark subterminal band, and variable amount of white on dorsal surface of base. Darker morphs.*—Uppertail coverts are dark. Tail is grayish or grayish brown and partially or fully banded, with moderately wide, often offset dark bands; wider, dark subterminal band. **Subspecies:** Monotypic. **Color morphs:** Light, intermediate, and dark; cline occurs only between intermediate and dark morphs. **Size:** L: 20–26" (51–66cm). Females average larger. **Habits:** Recently fledged birds are wary but become quite tame by fall and winter. Uses exposed perches. **Food:** Perch and aerial hunter of ground squirrels, prairie dogs, rabbits, hares, and some terrestrial birds; takes insects and reptiles in warm seasons. Feeds on carrion, especially in winter. **Voice:** Soft *kreeaah* is emitted if agitated.

47a. Light morph (recently fledged): Eye is pale gray. *Long gape is often orange. Dark eye line is thick and touches gape.* Head is tawny. Breast is tawny and forms bib but fades to white by late summer–fall.

47b. Light morph (fall–winter): *Pale head with thick, dark eye line.* Eye is pale yellow.

47c. Light morph (spring): Head bleaches to white. *Dark eye line is thick and touches gape.* Eye is tan.

47d. Light morph tail (summer–fall, dorsal): *Gray with white base; 3 or 4 dark bands. Coverts have dark spots.*

47e. Light morph tail (fall–winter, dorsal): Gray sheen wears off. Moderate-width, partial, offset dark bands on only middle feathers. *Note:* Variant with small amount of white on base of tail.

47f. Light morph tail (spring, dorsal): *White base; partial dark banding on middle feathers. Tips are broken.*

47g. Secondary flight feather: Gray with 3 broad, dark bars. *Note:* Compare to other large buteos.

47h. Light morph tail feather (summer–winter): *3 or 4 dark bands and wider subterminal band.*

47i. Light morph tail feather (spring): Similar to *47h*, but white tip is broken off.

47j. Light morph [left] leg: Feathering on lower leg extends to toes. On darker birds, feathering is dark brown.

47k. Light morph, lightly marked type (recently fledged): Tawny head and bib on breast, which fades to white by fall. White underparts have a few dark spots on flanks and part of leg feathering. Tawny-rufous edges on dorsal feathers.

47l. Light morph, moderately marked type (fall–winter): White underparts are moderately marked with dark spots on flanks, belly, and legs. Upperparts are edged with tawny. *Primaries are silver-edged.*

47m. Light morph, moderately marked type (spring): Bleached white head. Similar to *47l* on rest of body, but upperparts are faded to paler color. *Primaries are silver-edged.* Tail tip is broken off.

47n. Dark morph: *Long gape is distinct.* Eye is pale yellow. Head and body are all dark brown.

47o. Intermediate morph: *Long gape is distinct.* Eye is tan. Head and breast are tawny-rufous.

47p. Dark morph: Uniformly dark head and body, including feathered legs. *Note:* Very uncommon type. Primaries are silver-edged, as on *47q*.

47q. Intermediate morph: Tawny-rufous head and breast. Rest of body is dark brown, including feathered legs. *Silver-edged primaries are distinct.* Tail is gray and partially banded.

47r. Darker morph, partial-banded tail (summer–winter, dorsal): Faint banding; subterminal band.

47s. Darker morph, banded tail (summer–winter, dorsal): Irregularly banded; subterminal band widest.

a
b
c
d
e
f
g
h
i
k
l
m
j
r
n
q
s
p
o

Plate 48. FERRUGINOUS HAWK (*Buteo regalis*) Juvenile, Flying

Ages: Juvenile plumage is held for 1st year. Plumage wears and fades considerably by spring. *Legs are feathered to toes. Primaries have large, pale gray (silver), rectangular panel on dorsal surface that contrasts sharply with darker secondaries and wing coverts, including greater primary coverts (similar to juvenile Rough-legged Hawk;* Plate 43). *On underwing, small gray mark adorns tips of otherwise unmarked 5 outer primaries (fingers).* White rectangular panel also shows, as pale window, on underwing in translucent light (shared by juvenile Red-tailed and Rough-legged Hawks). *Long wings are pointed and show outward bulge between inner primaries and outer secondaries.* Juveniles (more so than adults) use their tail as a brace against the ground to obtain leverage when feeding; this breaks off distal part by late winter–spring (pers. obs.). *Light morph.*—2 variations of underparts on juveniles: lightly and moderately marked types. *Narrow, pale patch extends under eye. Thick, black eye line extends behind eye and under pale patch and touches gape. No dark malar mark. White uppertail coverts have large dark spot on each feather. Grayish (fresh) or brownish (worn) tail has 2–4 moderately wide partial bands, wider dark subterminal band, and variable amount of white on dorsal surface of base. Darker morphs.*—Uppertail coverts are dark. Tail is grayish or grayish brown and partially or fully banded, with moderately wide, often offset dark bands; wider dark subterminal band. **Subspecies:** Monotypic. **Color morphs:** Light, intermediate, and dark; cline exists only between intermediate and dark morphs. **Size:** L: 20–26" (51–66cm), W: 53–60" (135–152cm). Females average larger. **Habits:** Recently fledged birds are wary but become quite tame by fall and winter. **Food:** Perch and aerial hunter of especially ground squirrels and prairie dogs but also rabbits and hares, and some terrestrial birds; insects and reptiles in warmer seasons. Feeds on carrion, especially during winter. **Flight:** Soars in high dihedral and glides in modified dihedral. Moderately slow wingbeats. Hovers and kites. **Voice:** Mainly silent. Emits soft *kreeaah* if agitated.

48a. Light morph, lightly marked type (fall–winter): *Small dark gray tips on 5 outer primaries.* Dark spotting on flanks forms dark patch; unmarked belly. *Gray, partial-banded tail has white base.*

48b. Outer primary feather: *Small, dark gray tip on paler gray finger of otherwise white primary feather (5 outermost primaries).* (Compare to juvenile Rough-legged Hawk; *43h*).

48c. Light morph, moderately marked type (spring): Wingtips become very pointed when gliding; *small, gray wingtip marks.* Flanks and belly are heavily spotted; dark patch on flanks. Tail has broken tip.

48d. Light morph, lightly marked type (recently fledged): Tawny bib on breast, which fades to white by fall. White underparts have lightly spotted belly and flanks; dark patch on flanks. *Small gray wingtip marks. Tail has wide subterminal band, 2 indistinct inner bands, and small, white basal area.*

48e. Light morph (fall–winter): *Silver, rectangular primary panel contrasts with darker primary coverts.* Scapulars and back are brown. *White uppertail coverts have dark spots. Tail has 3 dark bands, white base.*

48f. Light morph (spring): *White head with thick dark eye line. Silver, rectangular primary panel contrasts with darker primary coverts. White uppertail coverts have dark spots. Tail has 3 partial dark bands, small area of white at base. Note:* Tip of tail is worn off, as on *47f.*

48g. Intermediate morph (fall–winter): Tawny-rufous head and breast contrast with dark brown underparts and underwing coverts. *Small, dark gray wingtip marks.* Tail is partially banded.

48h. Dark morph (fall–winter): Uniformly dark brown head, body, and underwing coverts. *Small, dark gray wingtip marks.* Tail is partially banded. *Note:* Very uncommon plumage.

48i. Intermediate morph (spring): *Silver, rectangular primary panel contrasts with rest of darker wing.* Faintly banded, grayish-brown tail has broken tip.

48j. Flight silhouettes: *Soaring (top):* high dihedral. *Gliding (bottom):* modified dihedral.

a
b
c
d
e
f
g
h
i
j

Plate 49. FERRUGINOUS HAWK (*Buteo regalis*) Adult, Perching

Ages: Adult plumage, including all flight feathers, is attained as 1-year-old in fall of 2nd year, which is a rapid molt for a large raptor. Eyes are medium brown. *Black bill has small, pale blue spot on lower basal area of upper mandible. Large yellow cere; long yellow gape extends to mid-eye.* When viewed at close range, males tend to have grayish heads and females brownish heads. *Legs are feathered to toes. Primaries show pale gray (silver) outer edge when folded, in all ages, both sexes.* When perched, wingtips are barely shorter than tail tip. Tail typically lacks dark subterminal band, but partial band may adorn any color morph. *Light morph.—Dark eye line is thick and touches gape.* 3 main variations on underpart markings: lightly, moderately, and heavily marked. Any variation can have tawny wash on breast, especially in fall. Males may have brownish head streaking and be similar to females. *Light morph lacks dark malar mark.* 3 main dorsal tail patterns with varying degrees of white, gray, and rufous; all have some rufous, have white on basal area, and often are mottled or speckled on base. Ventral tail surface is white but shows variable amount of rufous in translucent light. *Rufous upperparts fade by spring. Darker morphs.*—Lores are pale, and forehead is dark. Tail is mainly gray on dorsal surface and white on ventral surface. Dark morph lacks any hint of rufous in plumage. **Subspecies:** Monotypic. **Color morphs:** Light, intermediate (rufous), and dark morphs; cline exists from intermediate to dark, with dark intermediate (dark rufous) morph slotted between them. **Size:** L: 20–26" (51–66cm). Females average larger. **Habits:** Wary and aggressive when nesting but becomes tame in fall and winter. Uses exposed perches. **Food:** Perch and aerial hunter of ground squirrels, prairie dogs, rabbits, hares, and some terrestrial birds; large insects and reptiles in warmer seasons. Feeds on carrion, especially in winter. **Voice:** Emits soft *kreeaah* when agitated, mainly at nest sites.

49a. Light morph male: Whitish or grayish head with thick, dark eye line. Eye is medium brown. Gape extends to mid-eye. *Note:* Female of lightly marked type (*49h*) is more brownish but not always separable.

49b. Light morph female: Brownish head with thick, dark eye line. Eye is medium brown. Gape is long.

49c. Light morph, white-rufous tail (dorsal): White, with rufous feather edges.

49d. Light morph, gray-rufous tail (dorsal)**:** Gray, with rufous feather edges and white base. Coverts are white.

49e. Light morph, rufous tail (dorsal): Rufous, with variably sized white base. Coverts are rufous. *Note:* Rufous-gray tail on heavily marked types; may have irregular, black subterminal band.

49f. Light morph, gray-rufous tail (spring, dorsal): Faded gray, with pale rufous feather edges. Coverts are white.

49g. Light morph, lightly marked [left] leg: Main leg feathers (on tibia) are white and barred; rufous feathers extend to toes on lower leg (tarsus).

49h. Light morph male/female, lightly marked type: Pale, streaked head. White belly; light spotting on white flanks. Main leg feathers are white and fully barred.

49i. Light morph male/female, moderately marked type (spring): *Upperparts are faded rufous.* Underparts are moderately spotted and barred on flanks and belly. *Legs are rufous and barred.* Tail as in *49f.*

49j. Light morph female, heavily marked type: Head as on *49b. Tawny, streaked breast and rufous flanks and belly are heavily marked; white lower belly is barred. Rufous legs are barred.* Tail as in *49e.*

49k. Dark morph [left] leg: Brown leg feathers extend to toes. Intermediate morph has rufous feathers.

49l. Intermediate morph male: Gray head. *Base of bill has small blue spot; pale lores; long gape.*

49m. Intermediate morph female: Brown head. *Base of bill has small blue spot; pale lores; long gape.*

49n. Intermediate morph male, all-rufous type: Gray head. Dorsal feathers are rufous-edged. Rufous underparts have dark markings. *Note:* Ventral pattern also found on some females.

49o. Intermediate morph female, white-breasted type: Brown head. Breast is mottled with white. *Outer edge of primaries is silver.* Tail is all-gray type. *Note:* Ventral pattern also found on some males.

49p. Intermediate morph, gray-rufous tail (dorsal): Medium gray with rufous on feather edges. Coverts are rufous.

49q. Dark morph tail: Dark gray; may have dark, partial subterminal markings. Coverts are dark brown. *Note*: Dark morph lacks rufous tones.

49r. Dark intermediate morph male: Gray head. Rufous breast; body and wings are dark brown.

49s. Dark morph female: Uniformly dark brown. *Outer edge of primaries is silver.* Tail as on *49q.*

a
b
c
d
e
f
g
h
i
j
k
l
m
n
o
p
q
r
s

Plate 50. FERRUGINOUS HAWK (*Buteo regalis*) — Adult, Flying

Ages: Adult plumage, including all flight feathers, is attained as 1-year-old, in fall of 2nd year; this is a rapid molt for a large raptor. *Yellow cere; long yellow gape extends to mid-eye.* When viewed at close range, males tend to have grayish heads and females brownish heads. *Legs are feathered to toes. Primaries have large, pale gray (silver), rectangular panel on dorsal surface that contrasts sharply with darker secondaries and wing coverts. On underwing, small black marks adorn tips of outer 5 primaries (fingers).* Also on underwing, white window shows in translucent light. *Long wings are pointed and show outward bulge from inner primaries to outer secondaries. Gray band on trailing edge of wing; few if any markings on rest of flight feathers.* Tail typically lacks dark subterminal band, but partial band may adorn any color morph. *Light morph.*—3 main plumage variations on underpart markings: lightly, moderately, and heavily marked types. Any variation can have tawny wash on breast, especially in fresh fall plumage. Males may exhibit brownish head streaking and be similar to females. *Light morph lacks dark malar mark.* 3 main dorsal tail patterns with varying degrees of white, gray, and rufous; all exhibit some rufous, are white on basal area, and typically lack dark subterminal band. Ventral tail surface is white but will show variable amount of rufous in translucent light. *Rufous upperparts fade by spring. Darker morphs.*—Forehead is dark, and lores are whitish. Tail is mainly gray on dorsal surface and white on ventral surface. Dark morph lacks any hint of rufous on plumage. **Subspecies:** Monotypic. **Color morphs:** Light, intermediate (rufous), and dark morphs; cline exists from intermediate to dark, with dark intermediate (dark rufous) morph slotted between them. **Size:** L: 20–26" (51–66cm), W: 53–60" (135–152cm). Females average larger. **Habits:** Wary and aggressive when nesting, but becomes tame in fall and winter. **Food:** Perch and aerial hunter of ground squirrels, prairie dogs, rabbits, hares, and some terrestrial birds; large insects and reptiles in warmer seasons. Feeds on carrion, especially in winter. **Flight:** Soars in high dihedral and glides in modified dihedral. Moderately slow wingbeats. Hovers and kites. **Voice:** Emits soft *kreeaah* at nest sites if agitated.

50a. Light morph male, lightly marked type: Pale gray head. *Small black marks on wingtips.* Underwing coverts are lightly marked with rufous. Underparts are white and unmarked. *Moderately distinct V is formed by barred, white upper legs and rufous feathering to toes.*

50b. Outer primary feather, all color morphs: Black-tipped finger, with up to 1 or 2 thin, inner black bars.

50c. Light morph female, moderately marked type: Brown head. *Small black marks on wingtips.* Underwing coverts are moderately marked with rufous. Underparts are white and sparsely marked on flanks and belly. *Distinct V formed by rufous legs. Note:* This is a classic-plumaged bird with rufous leg feathers and paler underparts.

50d. Light morph male/female, heavily marked type: Dark head. *Small black marks on wingtips.* Underwing coverts are heavily marked with rufous. Tawny or white breast is streaked. Rufous flanks and belly are *heavily barred. V formed by rufous legs.* Wing window does not show in direct light.

50e. Light morph male (spring): Pale gray head. Back, scapulars, and upperwing coverts fade to pale rufous. *Primaries show as silver, rectangular panel. Tail is faded white-rufous type. Note:* Moderately and heavily marked types tend to have more rufous on tail.

50f. Intermediate morph male: Gray head. Back, scapulars, and upperwing coverts are edged in rufous. *Primaries show as silver panel.* Uppertail coverts are rufous. Tail is plain medium gray.

50g. Dark morph male: Grayish head. Upperparts, including uppertail coverts, are dark brown. Primaries show as silver panel. Tail is dark gray with faint darker subterminal band.

50h. Intermediate morph male, all-rufous type: Gray head. Underparts, including underwing coverts, are rufous. Breast is speckled with white. *Small black marks on wingtips. Tail is white and unmarked.*

50i. Dark intermediate morph male: Gray head. Rufous breast; rest of underparts, including underwing coverts, are dark brown. *Small black marks on wingtips. Tail is mainly white and unmarked.*

50j. Dark morph female: Uniformly dark brown underparts, including underwing coverts. *Small black marks on wingtips. Tail is white, unmarked. Note:* Only color morph with distinctly barred secondaries.

a
b
c
d
e
f
g
h
i
j

FERRUGINOUS HAWK (*Buteo regalis*)

HABITAT: *Summer.*—Open, arid, short-grass prairies and sagebrush steppe regions isolated from human disturbance. Terrain can be flat or have small or large hills, often with rocky ridges or with rock or badland-type soil formations. Single deciduous or coniferous trees or small groups of trees or large boulders may dot the open landscape. A few locales may be near parcels of agricultural lands, sometimes within sparsely settled ranches. Elevation varies from 7,000' (2100m) in Wyo. to 7,200' (2200m) in Mont. and up to 7,500' (2300m) in w.-cen. N.Mex. *Winter.*—Inhabits similar regions as during breeding season but at lower latitudes and at elevations up to 6,000' (1800m), but to 6,600' (2000m) in N.Mex. This species readily acclimates to human-occupied locales during this season and is regularly found in semi-open rural pastures, light agricultural areas, and even suburban areas that have ample prey base. *Migration.*—Migrants may temporarily occupy intermountain valleys of above-listed habitat to 8,100' (2470m); otherwise habitat is similar to that of winter. **STATUS:** Uncommon. Population is deemed overall stable but guarded. Listed as sensitive species by U.S. Bureau of Land Management, Species of Special Concern in Ariz. and Colo., and endangered in Oreg.; Canada lists it as vulnerable species. Massive habitat alteration in Great Plains and Great Basin regions since late 19th century has greatly reduced breeding habitat. Gas, oil, and mineral exploration and construction of wind farms in prime breeding habitat may have uncertain effects on breeding populations in these regions. Habitat alteration has also greatly reduced prey species, especially Black-tailed Prairie Dog (*Cynomys ludovicianus*) of Great Plains, which has undergone intense historic and current decimation. Breeding population has *not* increased in any region in their range. Some peripheral breeding areas have seen continual population declines in recent years due to habitat and prey loss. Between 2003 and 2010, the number of breeding pairs in Wash. decreased from 78 to 36 (Wash. Dept. of Fish and Wildlife, 2012). With expansive amount of breeding habitat and ample prey base, Wyo. has the largest breeding population, which is estimated to be over 1,100 pairs (Wyo. Game and Fish Dept., 2011). *Color morph status.*—Light morph makes up nearly 95% of population. Percentage of nesting darker birds is higher in Canadian populations (9% in Alta., 7% in Sask.). Rufous (intermediate) plumages account for 5% and pure dark morphs only 1% of wintering birds in e. Colo. and n. Tex. **NESTING:** Starts in late Feb.–late Mar. and ends mid-Jun.–late Aug. Timing is determined by elevation and latitude. Breeding does not occur until 3 years of age. Birds may build their own nest, build atop nest of Black-billed Magpie (*Pica hudsonia*), or refurbish nests of Swainson's or Red-tailed Hawks. New nests can be shallow but have wide diameter; nests are often reused for many years (and decades) and can become several feet deep, with a columnar shape, and attain the size and mass of an eagle nest. Nests are most commonly placed up to 55' (17m) high in a live or dead tree that is single, in a small cluster, or in a shelterbelt. Nests are also placed on top of large bushes (especially tall sagebrush), large rocks, rock pinnacles, low rock or soil escarpments, wooden utility poles, abandoned buildings, haystacks, and artificial platforms. Nests are also placed in lower portions of tall trees and metal power-transmission structures. Especially in N.Dak., nests may be placed on the ground. Dung from horses and cattle (formerly from American Bison, *Bos bison*) is usually added to nest, as is greenery. Clutch is 2–4 eggs, incubated by both sexes. Nearly fledged youngsters ("branchers") leave nest before attaining flight and often disperse on the ground. *Note:* This species does not tolerate human disturbance at nest sites during early part of nesting season, until eggs have hatched. **MOVEMENTS:** Population nesting east of Rocky Mts. stays east of the mountains; population originating west of Rocky Mts. (to e. Wash.) may engage in lateral easterly–westerly movements, crossing back and forth over Sierra Nevada and Great Basin in the same season—and even over Continental Divide and onto Great Plains (*see* Wheeler 2003b). *Spring.*—Adults move northward mid-Feb.–Mar. Returning juveniles may move in Mar., but most move in Apr.–May. Movement may be in diagonal northeasterly or northwesterly direction. *Summer dispersal.*—All ages of U.S. population may disperse in northeasterly direction onto n. Great Plains of U.S. and Canada after nesting activities are completed and/or fledging occurs; juveniles move in any direction. (Northern Great Plains has abundance of young ground squirrels, notably Richardson's Ground Squirrel [*Urocitellus richardsonii*]; adult ground squirrels enter estivation in Jul. and escape late mid–late summer predation.) *Fall.*—1st migrants are noted on cen. Great Plains in late Aug.–early Sep., building to large movement in Oct.; lesser movement continues into Dec. **COMPARISON:** ***Ferruginous juvenile light morph.*** **Swainson's Hawk (juvenile and 1-year-old light morph).**—Similar white head, but Swainson's has dark malar mark, *vs.* unmarked malar region. In flight, gray flight feathers on ventral surface and all-dark dorsal surface of wing, *vs.* all-white underwing and whitish rectangular panel on dorsal side of primaries. **"Krider's" Red-tailed Hawk (juvenile).**—Similar white head, especially in spring, but Krider's has either all-white head or thin dark eye line, *vs.* white head with thick dark eye line connecting to gape. Cere is usually green but sometimes yellow, *vs.* always yellow. Thin, dark distal tail bands, *vs.* a few thick, irregular dark bands; both share white basal tail area. In flight, has faint dark patagial mark on leading edge of underwing and large whitish square panel on outer half of upperwing wing, *vs.* unmarked underwing and rectangular whitish panel only on primaries. **"Eastern"/"Western" Red-tailed Hawk (juveniles).**—Lack dark eye line or have thin dark eye line, *vs.* thick eye line connecting to gape. Cere is typically greenish, sometimes yellow, *vs.* always yellow. Underwing has dark patagial mark on leading edge, and upperwing has large, square-shaped tan panel on primaries and their greater coverts, *vs.* unmarked white underwing and whitish rectangular panel only on dorsal side of primaries. Tail is all-brown/rufous with multiple, thin dark bands, *vs.* white base and a few thick, irregular dark bands. ***Ferruginous adult light morph.*** **"Krider's" Red-tailed Hawk (adult).**—Share similar whitish head, but "Krider's" often has dark malar mark and thin dark eye line, *vs.* lack of malar mark and thick dark eye line connecting to gape. Share similar rufous-edged feathers on upperwing coverts, but "Krider's" has large white patch on scapulars, *vs.* all-rufous scapulars. Lower legs are bare yellow, *vs.* rufous-feathered lower legs on paler birds—and all-rufous legs on most adults. White underwing has rufous or brown patagial mark from body to wrist, *vs.* uniformly lightly to heavily rufous-marked underwing. ***Ferruginous all-rufous intermediate and dark morphs (both ages).*** **Swainson's Hawk (adult rufous/intermediate and dark morphs).**—All-rufous type of intermediate morph and dark morph are similar to respective color morphs of Ferruginous Hawk, but Swainson's has white forehead and white/tawny undertail coverts, *vs.* dark forehead and dark or rufous undertail coverts. Lower legs are bare yellow skin, *vs.* feathered. In flight, underside of flight feathers is darker gray, *vs.* white. Tail is thinly banded, *vs.* sparse, wide bands (juveniles) to unbanded (adults). **"Western" and "Harlan's" Red-tailed Hawks (juvenile dark morphs).**—All-dark body, including undertail coverts, similar to that of very uncommon darker morphs of Ferruginous Hawk (both ages), but Red-tails usually have greenish cere, thin yellow gape, and distinctly banded tail, *vs.* yellow cere, large yellow gape, and irregular, offset tail bands. In flight, all fingers of outer primaries are black on underside, *vs.* small gray/black tips. Lower legs are bare yellow skin, *vs.* feathered. **"Western" Red-tailed Hawk (adult intermediate and dark morphs).**—Similar dark head and rufous breast (intermediate morph) or all-dark body (dark morph), but Red-tail has thin yellow gape, and tail is all-red, *vs.* thick yellow gape and gray/white tail. **"Harlan's" Red-tailed Hawk (adult darker morphs).**—Very similar if breast is rufous-tinged and has white markings like those that adorn many adult all-rufous Ferruginous Hawks. Tail can be identical to either age of Ferruginous Hawk: gray with dark subterminal smudge (juvenile) or all-whitish or all-grayish, sometimes with darker mottling (adult). On underwing, white flight feathers are strongly barred and have large, black tips, *vs.* white with little or no barring and small black tips. **Rough-legged Hawk (juvenile dark morph).**—Similar, with tawny-rufous breast or all-dark body, but has white forehead, *vs.* dark forehead. Gape is long and extends under eyes as on Ferruginous Hawk but is thin, *vs.* thick. Both species have all-feathered legs. When perched, Rough-legged's wingtips extend to or beyond tail tip, *vs.* shorter than tail tip. In flight, underside of wing shows square, black carpal patch, *vs.* uniformly dark coverts. Fingers of outer primaries are all-black on underside, *vs.* small gray or black tips. Tail pattern can be identical grayish, partially banded pattern.

Figure 1. Albany Co., Wyo. (Jul.)
Nesting habitat on short-grass prairie at 7,200' (2200m). This stick nest atop a boulder has been used intermittently for decades. (The Medicine Bow Mts. are on the horizon.)

Figure 2. Boulder Co., Colo. (Aug.)
Black-tailed Prairie Dog, a major prey species especially during winter on the Great Plains.

Figure 3. Albany Co., Wyo. (Jul.)
Richardson's Ground Squirrel, a major summer prey species in n. Great Basin and n. Great Plains.

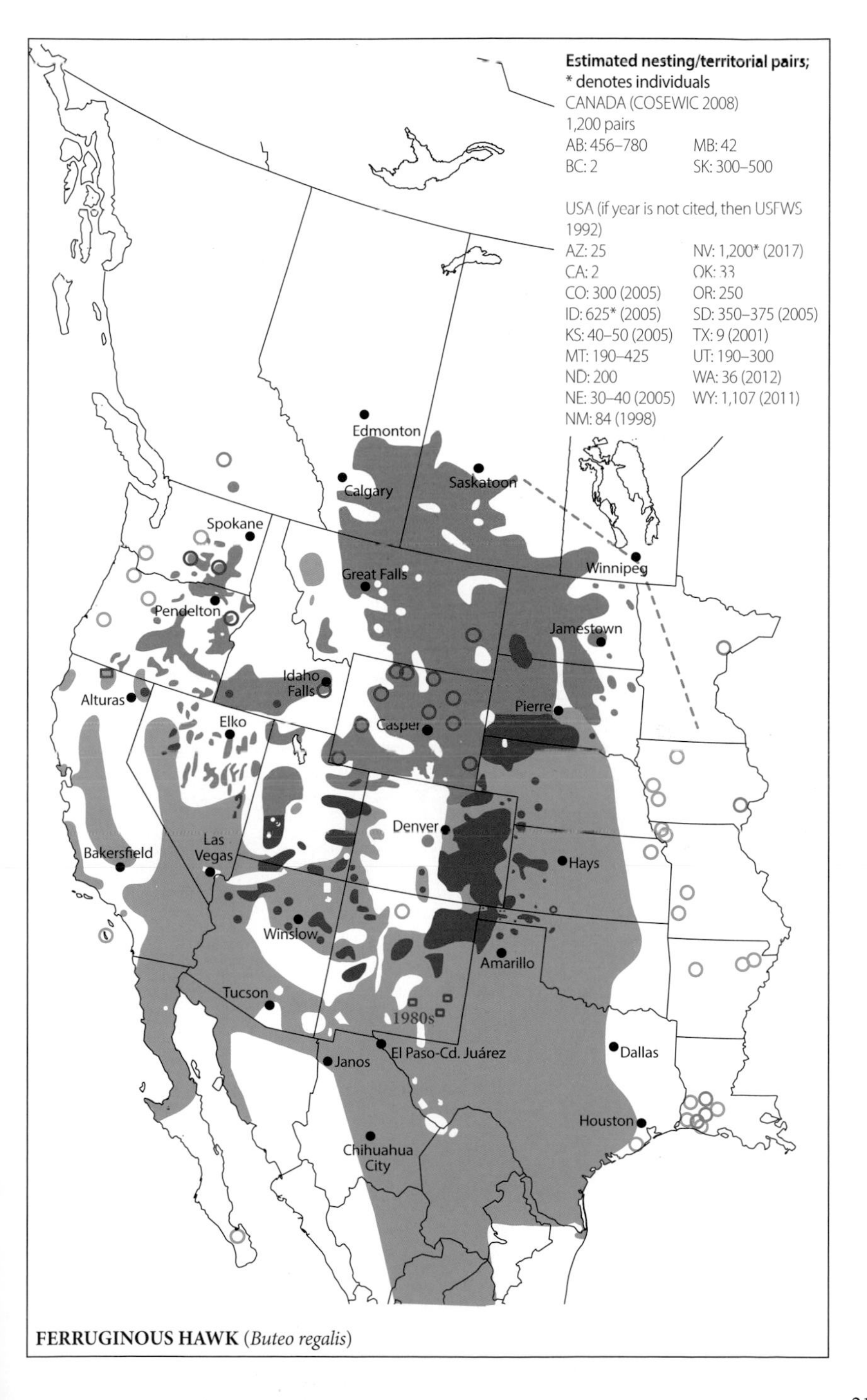

FERRUGINOUS HAWK (*Buteo regalis*)

Plate 51. BALD EAGLE (*Haliaeetus leucocephalus*) Head Portraits

Ages: Juvenile, 3 immature annual ages (banded, known age), and adult. Juvenile and 1-year-olds are separable from older ages by head pattern and bill color; older ages share traits and are difficult to age. Adult-like pure yellow bill and cere and all-white head can be attained by 3-year-old but are usually not attained until age 4 or older. Dark markings on bill and white head can be found even on old adults. Eye color gets paler with age but is also not reliable age feature. Bill is very large. Often-shaggy-looking head and neck feathers are long and pointed and are raised or lowered with temperature and mood. **Subspecies:** "Northern" (*H. l. alascanus*) and "Southern" (*H. l. leucocephalus*), with split at 40° N latitude. Subspecies are based on size, which is difficult to see in the field. **Color morphs:** None. **Size:** L: 27–37" (69–94cm). Females are larger. There is a clinal range in size, with largest farthest north and smallest farthest south. Based on respective sex, "Southern" birds can be 7" (18cm) smaller than "Northern" birds. "Southern" females average smaller than "Northern" males. **Habits:** Typically wary in nesting season but can be surprisingly tame in other seasons. Solitary in nesting season but often gregarious in nonbreeding season, especially "Northern" birds in prey-rich winter areas. Favored perches are large exposed branches, but birds often sit on other open objects and ground. They may sit for long periods. They sometimes perch on utility poles. "Southern" birds often seek shaded perches during heat of day. **Food:** Skilled aerial and perch hunter. Feeds mainly on live prey in nesting season but is opportunistic and often scavenges in other seasons. Fish are main prey in all seasons and regions, but also feeds on waterfowl and on jackrabbits and large rodents, particularly prairie dogs, in open areas away from water. Scavenges all sizes of animal life, especially ungulates; also feeds in garbage dumps. **Voice:** Highly vocal. Most calls are high-pitched, crisp, metallic notes. Staccato *whee-he-he-he* in decrescendo cadence is most common call. Others are series of *chirps* and *chitter* notes or *ca-ack, kah* notes.

51a. Juvenile (all year): *Uniformly black bill and cere; gape is pale yellow.* Eye is dark brown. *Throat is whitish.* Head and neck are dark brown. *Often shows white streaks on sides of neck.*

51b. 1-year-old, dark type: *Blackish bill and cere have pale yellowish patches on basal regions of bill, especially below nostril, on front edge of cere, and around nostril. Broad, dark band extends from lower jaw and rear of eye across side of head.* All other dark head feathers are speckled with pale tips. Eye varies from dark brown to pale brown. Throat is whitish.

51c. 1-year-old, pale type: *Blackish bill and cere have pale yellowish patches on basal regions of bill, especially below nostril, on front edge of cere, and around nostril. Broad, dark band extends from lower jaw and rear of eye across side of head. Rest of head and neck areas are whitish with thin, dark streaks.* Eye varies from dark brown to tan.

51d. 2- or 3-year-old, moderate/dark type: *Brownish-black bill has large, pale yellowish patches on lower base of upper mandible and portions of lower mandible. Cere is mainly yellowish but with dark area on ridge and below nostril. Moderate-width Osprey-like dark band on side of head extends from gape and rear of eye to nape.* Eye varies from tan to pale yellow. *Note:* Pattern on most 2-year-olds and some darker types of 3-year-olds.

51e. 2- or 3-year-old, pale type: *Orange-yellow bill and cere have dark smudges, especially on ridge and bill tip.* Moderate-width, Osprey-like dark patch extends from gape and rear of eye to nape; rest of head is dirty white, with forehead darker. *Note:* Pattern common on either age class.

51f. 3-year-old/older: *Orange-yellow bill often is covered with dark smudges, but cere is likely to be yellow. Mainly white head has thin, dark stripe on auriculars and down side of neck.*

51g. Adult (3-year-old/older): *Orange-yellow bill shows small amount of dark smudging next to cere and on tip; cere is yellow. Head is mainly white, with scattered dark flecks on crown, auriculars, and sides of neck. Eye is pale yellow.*

51h. Adult (3-year-old/older): *All-orange-yellow bill and cere. Head and neck are pure white. Eye is pale yellow.*

a
b
c
d
e
f
g
h

Plate 52. BALD EAGLE (*Haliaeetus leucocephalus*) **Perching**

Ages: Juvenile, 3 immature annual ages (banded, known age), and adult. Juvenile and 1-year-olds are separable from older ages by head pattern and bill color; older ages share traits and are difficult to age by these features. Adult-like pure yellow bill and cere and all-white head can be attained by 3-year-old but are usually not attained until age 4 or older. Dark markings on bill and white head can be found even on old adults. Eye color gets paler with age but is also not reliable age feature. Bill is very large. Often-shaggy-looking head and neck feathers are long and pointed and are raised or lowered with temperature and mood. **Subspecies:** "Northern" (*H. l. alascanus*) and "Southern" (*H. l. leucocephalus*), with split at 40°N latitude. Subspecies are based on size, which is difficult to see in the field. **Color morphs:** None. **Size:** L: 27–37" (69–94cm). Females are larger. There is a clinal range in size, with largest farthest north and smallest farthest south. Respective-sex "Southern" birds can be 7" (18cm) smaller than "Northern" birds. "Southern" females average smaller than "Northern" males. **Habits:** Typically wary in nesting season, but can be surprisingly tame in other seasons. Solitary in nesting season but often gregarious in nonbreeding season, especially "Northern" birds in prey-rich winter areas. Favored perches are large exposed branches, but birds often sit on other open objects and ground, and may sit for long periods. Unlike Golden Eagles, they do not regularly perch on top of utility poles. "Southern" birds often seek shaded perches during heat of day. **Food:** Skilled aerial and perch hunter. Feeds mainly on live prey in nesting season but is opportunistic and often scavenges in other seasons. Fish are main prey in all seasons and regions but also feeds on waterfowl and on jackrabbits and large rodents, particularly prairie dogs, in areas away from water. Scavenges all sizes of animal life, especially ungulates; also feeds in garbage dumps. **Voice:** Highly vocal. Most calls are high-pitched, crisp, metallic notes. Staccato *whee-he-he-he* in decrescendo cadence is most common call. Others are series of *chirps* and *chitter* notes or *ca-ack, kah* notes.

52a. Juvenile (faded plumage): *Black bill and cere. Head, neck, and breast are dark brown and form bib, which contrasts with tan belly.* May have white streaks on base of dark bib at junction of paler belly. Upperparts and median and lesser upperwing coverts are medium brown. Greater wing coverts and secondaries are dark brown. Leg feathers are dark brown and contrast with tan belly. Tail varies from black to whitish with black tip. *Note:* Plumage fades by winter and spring when near end of age class.

52b. Juvenile (recently fledged/fresh plumage): *Black bill and cere.* Head, neck, and rest of upperparts are dark brown and barely contrast with slightly paler *medium-brown belly.* Leg feathers (not shown) are dark brown.

52c. 1-year-old, pale type: *Broad, dark patch on side of whitish head. Neck, back, and belly are white with dark markings and blend with inverted white triangle on back; belly can be pure white or speckled. Breast has mottled dark patch that forms small, dark bib. Large white marks on wing coverts; small patch of tan, faded retained juvenile feathers on lesser coverts.*

52d. 1-year-old, moderately/heavily marked type: *Broad, dark patch on side of fairly dark head; breast is dark brown and forms bib, which contrasts with white, blotchy belly. White inverted triangle on back has large black markings.* Upperwing coverts often have remnant faded brown juvenile feathers.

52e. 2-year-old, moderate/pale type: *Moderate-width dark patch on side of whitish head; neck and inverted triangle on back are also whitish. Breast and leg feathers are dark brown. Belly is white with variably sized dark blotches.* Scapulars and upperwing are dark brown.

52f. 2- or 3-year-old, dark type: *Moderate-width dark patch on side of tawny head. All of body and upperwing are dark brown, with small amount of white speckling on back and belly. Upperwing coverts may have some white edging. Note:* Dark-type 2- and 3-year-olds are similar, but uppertail coverts and tail usually have more white in 3-year-old.

52g. 3-year-old/older: *Whitish head and neck have dark patch on forehead and thin, dark stripe on auriculars and side of neck.* Body and upperwing are dark brown; may have small white flecks on back and belly. Whitish tail has dark band on tip and often on outer edge.

52h. Adult (3-year-old/older): *Head, neck, tail, and upper- and undertail coverts are pure white. All of body and wings are dark brown.* Advanced 3-year-old is similar, but underwing coverts always have white blotches (not visible when perched). *Note:* Performing head-toss/sky-point behavior.

52i. Adult (3-year-old/older): White head and tail can have sparse dark markings; these can adorn even old birds.

a
b
d
c
f
e
i
h
g

Plate 53. BALD EAGLE (*Haliaeetus leucocephalus*) Flying (ventral view)

Ages: Juvenile, 3 immature annual ages (banded, known age), and adult. Progression of wing molt on flying birds, with retention of definitive sequence of juvenile feathers, is reliable ageing tool on 1- to 3-year-olds. 1 or more juvenile flight feathers can be held for 3 years (through 2-year-old class). Juveniles have longer, more pointed flight feathers, making their wings broader than older ages; retained feathers in immature ages appear spikelike. Juvenile tail feathers are longer than those of older birds. (*Note:* It is difficult or impossible to age birds beyond age 3.) Bill is large, and neck is long. *All have pale throat.* Outermost primaries form 6 fingers. **Subspecies:** "Northern" (*H. l. alascanus*) and "Southern" (*H. l. leucocephalus*), with split at 40°N latitude. Subspecies are based on size, which is difficult to see in the field. **Color morphs:** None. **Size:** L: 27–37" (69–94cm), W: 71–96" (180–244cm). Females are larger than males. There is a clinal range in size, with largest farthest north and smallest farthest south. Respective-sex "Southern" birds can be 7" (18cm) smaller than "Northern" birds. "Southern" females average smaller than "Northern" males. With longer flight and tail feathers, juveniles appear larger and longer than adults. **Habits:** Typically wary in nesting season but can be surprisingly tame in other seasons. Solitary in nesting season but often gregarious in nonbreeding season, especially "Northern" birds in prey-rich winter areas. **Food:** Skilled aerial and perch hunter. Feeds mainly on live prey in nesting season, but is opportunistic and often scavenges in other seasons. Fish are main prey in all seasons and regions, but also feeds on waterfowl, jackrabbits, and rodents, especially prairie dogs. Scavenges all sizes of animal life, especially ungulates; also feeds in garbage dumps. **Flight:** Can be quite aerial. Soars with wings held on flat plane. Powered flight has high upstrokes and shallow downstrokes. **Voice:** Highly vocal. Most calls are high-pitched, crisp, metallic notes. Staccato *whee-he-he-he* in decrescendo cadence; also, series of *chirps* and *chitter* notes or *ca-ack, kah* notes.

53a. Juvenile primary feather: Tip is pointed. Dark and neat in juvenile age; as retained outermost feathers of 1- and 2-year-olds become faded, worn, and frayed (as depicted).

53b. Immature/adult primary feather: Tip is round. Dark-colored when fresh; can become brownish when worn.

53c. Juvenile secondary feather: Tip is pointed. Pale markings often extend to tip.

53d. Immature secondary feather: Tip is rounded. Pale markings end before tip, creating broad, dark band.

53e. Juvenile, pale-type underwing (recently fledged/fresh plumage): *Pale throat. Head and breast are dark brown and contrast with medium-brown belly. Axillaries, underwing coverts, and all but outer 3 secondaries are white. Inner 2 or 3 primaries form white stripe.* White-type tail has thin, dark terminal band.

53f. Juvenile, dark-type underwing (worn plumage): *Pale throat. Head and breast are dark brown and contrast sharply with faded tan belly. Axillaries are mainly white; white bar on underwing coverts and small patch of white on innermost secondaries. Tail is black type; occurs on birds with lesser amounts of white on underwing.*

53g. 1-year-old, pale type: *Broad, dark patch on whitish head and neck. Breast has thin, dark necklace/bib. Belly is white with small dark spots. Axillaries and most underwing coverts are white.* Outer 5 primaries (p6–10) are frayed juvenile feathers. Secondaries have 2 groups (s2–4, s7–11) of retained juvenile feathers. *Note:* Average molt stage.

53h. 1-year-old, moderate/dark type: *Pale throat. Dark head and breast contrast with white-speckled belly. Axillaries are white; white bar on underwing coverts.* Outer 5 primaries (p6–10) are old juvenile feathers. Most secondaries (s2–13) are juvenile; outermost (s1) and innermost 1 or 2 (s14/s15) are new. *Note:* Slow molting stage.

53i. 2-year-old, pale/moderate type: *Pale head with thin, dark auricular patch. Dark breast forms bib. White belly is speckled. Axillaries are white.* Outer 2 primaries (p9, p10) are frayed juvenile feathers. Underwing is mainly dark, with 3 retained spikelike juvenile secondaries (s4, s9, s10, last feathers of 3 molt units). *Note:* Average molt stage.

53j. 2-year-old, dark type: Whitish head with thin, dark stripe. Body is adult-like. Dark underwing has some white speckling. Outer 2 primaries (p9, p10) are retained worn juvenile feathers; s4 is only retained juvenile secondary (same-age Golden Eagle almost never retains only s4). *Note:* Average molt stage. 3-year-old can be similar, with wings as on *53k*.

53k. 3-year-old/older: *White head and tail have dark marks.* Outer 1 or 2 primaries (p9, p10) are new (3-year-old only). *Note:* 3-year-old can have pure white head and tail. All 3-year-olds and some older birds have white blotches on wing coverts and axillaries. *Note:* Average molt stage.

53l. Adult (older than 3-year-old): White head and tail. Body and underwing are dark.

a
b
c
d
e
f
g
h
i
j
k
l

Plate 54. BALD EAGLE (*Haliaeetus leucocephalus*) — Flying (dorsal view)

Ages: Juvenile, 3 immature annual ages (banded, known age), and adult. The progression of wing molt on flying birds, with retention of definitive sequence of juvenile feathers, is reliable ageing tool on 1- to 3-year-olds. 1 or more juvenile flight feathers can be held for 3 years (through 2-year-old class). Juveniles have longer, more pointed flight feathers, making their wings broader than older ages; retained feathers in immature ages appear spikelike. Juvenile tail feathers are longer than those of older birds. (*Note:* It is difficult or impossible to age birds beyond age 3.) Bill is very large, and neck is long when in flight. Outermost primaries form 6 fingers. **Subspecies:** "Northern" (*H. l. alascanus*) and "Southern" (*H. l. leucocephalus*), with split at 40° N latitude. Subspecies are based on size, which is difficult to see in the field. **Color morphs:** None. **Size:** L: 27–37" (69–94cm), W: 71–96" (180–244cm). Females are larger than males. There is a clinal range in size, with largest farthest north and smallest farthest south. Respective-sex "Southern" birds can be 7" (18cm) smaller than "Northern" birds. "Southern" females average smaller than "Northern" males. With longer flight and tail feathers, juveniles appear larger and longer than adults. **Habits:** Typically wary in nesting season, but can be surprisingly tame in other seasons. Solitary in nesting season but often gregarious in nonbreeding season, especially "Northern" birds in prey-rich winter areas. **Food:** Skilled aerial and perch hunter. Feeds mainly on live prey in nesting season but is opportunistic and often scavenges in other seasons. Fish are main prey in all seasons and regions, but also feeds on waterfowl, jackrabbits, and larger rodents, particularly prairie dogs. Scavenges all sizes of animal life, especially ungulates; also feeds in garbage dumps. **Flight:** Can be quite aerial. Soars with wings held on flat plane. Powered flight has high upstrokes and shallow downstrokes. **Voice:** Highly vocal. Most calls are high-pitched, crisp, metallic notes. Staccato *whee-he-he-he* in decrescendo cadence is most common call. Others are series of *chirps* and *chitter* notes or *ca-ack, kah* notes.

54a. Juvenile (recently fledged/fresh plumage): *Blackish-brown head and neck (with pale throat) contrast with slightly paler dark brown upperparts.* Flight feathers are black. Tan blotch on inner greater coverts on upperwing. Tail is black type.

54b. Juvenile (faded): *Dark brown head and neck (with pale throat) and dark brown flight feathers contrast with paler medium-brown dorsal body and coverts.* Tan area on inner greater coverts on upperwing. Tail is white type.

54c. 1-year-old, pale type: *Broad, dark patch on side of whitish head. Inverted white triangle adorns back.* Inner 5 primaries (p1–5) are new, dark; outer 5 primaries are retained frayed juvenile feathers. Most secondaries are retained juvenile feathers; s1, s5, and s13–14/15 (mainly hidden) are new, dark immature feathers. Tail is white type. *Note:* A slow molt stage; shows 1st stage of molt at 3 molt centers on secondaries: s1, s5, and innermost (s14/15).

54d. 1-year-old, moderate/dark type: *Broad, dark patch on side of tawny head. Whitish inverted triangle on back.* Inner primaries (p1–5) are new, dark; outer 5 are retained faded juvenile feathers. 2 blocks of secondaries (s3, s4, s7–11) are retained juvenile feathers; s1, s2, s5, s6, s12–s14/15 are new. Tail is black type. *Note:* Average molt stage.

54e. 2-year-old, pale type: *Whitish head and neck with moderate-width dark patch on side of head. Whitish inverted triangle on back blends with pale head.* Most flight feathers are new and dark except outer primaries (p10 [shown] or p9 and p10). 2 small separate blocks of secondaries (s4 and s9, s10) are retained faded juvenile feathers. *Note:* Average molt stage. Can also retain just s4 and s9 or s4 and s10; also, only s4.

54f. 2-year-old, dark type: *Tawny head with moderate-width dark patch on side of head.* Upper body is mainly dark brown with sparse white spotting. Flight feathers retain faded juvenile feathers only on outer primaries (p10 [shown] or p9 and p10); secondaries are all molted to immature. Tail and uppertail coverts are black. *Note:* Advanced molt stage.

54g. 3-year-old: *Whitish head and neck with moderate-width dark patch.* Body is dark brown; may have white spotting on back. Lower back, uppertail coverts, and tail are white with dark marks. All flight feathers are shorter adult length. Outer 1 or 2 primaries (p10 shown) are new, darker. *Note:* Head as on *54e*, but tail is more whitish.

54h. 3-year-old/older: *White head and neck with dark stripe. Lower back, uppertail coverts, and tail are white; tail may have dark smudges on tip. Note:* Classic plumage for 3-year-old (3-year-old exhibits new, dark p10). Body molt more advanced than on *54g*.

54i. Adult (3-year-old/older): *Immaculate white head and neck, and lower back, uppertail coverts, and tail.*

54j. Flight silhouette: *High upstroke* (pale blue), *shallow downstroke* (dark blue) in powered flight.

a
b
c
d
e
f
g
h
i
j

Plate 55. BALD EAGLE (*Haliaeetus leucocephalus*) — Tails

Ages: Juvenile, 3 immature annual ages (banded, known age), and adult. This plate illustrates the immense variability of tail markings in all age classes, although tail patterns are not reliable age factors. It also shows similarities with immature ages of Golden Eagle (see Plate 59). Patterns are most visible when in flight, especially if tail is fanned when soaring or banking. Tail tip is wedge-shaped on juveniles, rounded on older ages. All juvenile feathers are replaced by shorter, round-tipped immature-age feathers in a quick molt in 2nd year as 1-year-old. **Subspecies:** "Northern" (*H. l. alascanus*) and "Southern" (*H. l. leucocephalus*), with split at 40° N latitude. Subspecies are based on size, which is difficult to see in the field. **Color morphs:** None. **Size:** L: 27–37" (69–94cm). Females are larger. There is a clinal range in size, with largest farthest north and smallest farthest south. Respective-sex "Southern" birds can be 7" (18cm) smaller than "Northern" birds. "Southern" females average smaller than "Northern" males. Juveniles have longer flight feathers and tails than older ages, which makes them longer and larger. **Habits:** Typically wary in nesting season but can be surprisingly tame in other seasons. Solitary in nesting season but often gregarious in nonbreeding season, especially "Northern" birds in prey-rich winter areas. **Food:** Skilled aerial and perch hunter. Feeds mainly on live prey in nesting season but is opportunistic and often scavenges in other seasons. Fish are main prey in all seasons and regions, but also feeds on waterfowl, jackrabbits, and larger rodents, especially prairie dogs, in arid areas away from water. Scavenges all sizes of animal life, especially ungulates; also feeds in garbage dumps. **Flight:** Can be quite aerial. Soars with wings held on flat plane. Powered flight has high upstrokes and shallow downstrokes. **Voice:** Highly vocal. Calls are high-pitched, crisp, metallic notes. Staccato *whee-he-he-he* in decrescendo cadence is most common call; also, *chirps* and *chitter* notes or *ca-ack, kah* notes.

55a. Juvenile, dark (ventral): *All of tail is black.*

55b. Juvenile, dark (dorsal): *Tail is black with small amount of white marbling on center feathers.*

55c. Juvenile, moderate (ventral): White marbling on basal region; outer edge and tip are black.

55d. Juvenile, moderate (dorsal): White marbling on center feathers of otherwise black tail.

55e. Juvenile, pale, closed (ventral): Thin, black outer edge and broad, black terminal band on otherwise white tail. *Note:* Very similar to tail of juvenile Golden Eagle (*59d, g*), but coverts are shorter.

55f. Juvenile, pale (ventral): Thin, black outer edge and irregular black band on tip of white tail.

55g. Juvenile, pale (dorsal): Mainly whitish tail with dark outer edges and thin, dark band on tip.

55h. 1- and 2-year-old, dark (ventral): Thin, black outer edges and irregular broad, black band on tip; inner portion is marbled whitish.

55i. 1- and 2-year-old, dark (dorsal): *Mainly black but may have some white marbling on very center.* Uppertail coverts are black. *Note: 3-year-old similar but usually has whitish uppertail coverts.*

55j. 1- and 2-year-old/dark 3-year-old (ventral): White, with thin, dark outer edges and thin, irregularly formed dark tip. Dark marks on undertail coverts. *Note:* 3-year-olds may have dark coverts.

55k. 1- and 2-year-old moderate/pale (dorsal): Dark uppertail coverts. Tail is mainly whitish on central part, with dark outer edges and dark band on tip.

55l. 2- and 3-year-old, dark (dorsal): *Whitish uppertail coverts. Tail is black with some white marbling on central feathers. Note:* Some advanced 2-years-olds may have white on uppertail coverts.

55m. 3-year-old/older (dorsal): *Dark marks on uppertail coverts. Tail has irregular, dark terminal band.*

55n. 3-year-old/older (ventral): *Dark marks on undertail coverts. Tail has irregular, dark terminal band.*

55o. 3-year-old/older (ventral): *White undertail coverts. White tail has irregular, dark terminal band.*

55p. 3-year-old/older (dorsal): *White uppertail coverts. White tail has irregular, dark marks on feather tips.*

55q. 3-year-old/older, closed (dorsal): *Uppertail coverts and tail are immaculate white.*

55r. 3-year-old/older (ventral): *Undertail coverts and tail are immaculate white.*

BALD EAGLE (*Haliaeetus leucocephalus*)

HABITAT: *Summer.*—Found in areas with large bodies of fresh or salt water, which may be lakes, marshes, rivers, reservoirs, swamps, or coastal waters, with adjacent stands of tall, live or dead trees. Found in arid areas far from water but with stands of tall trees (Plains Cottonwood) on parts of Great Plains. Most areas have little to moderate human disturbance. *Winter.*—Inhabits areas identical to summer—including locations at northerly latitudes—as long as there is open water or ample food supply. Large concentrations seasonally occur at salmon-spawning rivers and areas with large numbers of waterfowl. Lesser numbers inhabit areas far from water if there is adequate food supply (primarily large mammal carrion; also Black-tailed Prairie Dogs). **STATUS:** Overall uncommon but locally common in parts of some states and provinces, especially coastal Alaska and B.C. The species' population is dramatically increasing: U.S. estimate in 2009 was of 72,400 birds in the contiguous states and 70,500-plus in Alaska (unknown number in Canada); U.S. number is projected to surpass 200,000 birds, with most growth occurring in coterminous states, including those of Southwest, which had about 650 birds and should increase to 1,800 birds (USFWS 2016). Population has steadily grown since ban of DDT in 1972. The species was delisted from U.S. Threatened and Endangered Species List in contiguous 48 states in Jun. 2007. **NESTING:** Pairs remain together year-round as long as both partners survive. Ariz. birds begin nesting in late Dec. Pairs in mid-latitude states and coastal n. states and provinces begin in Jan.–Mar., interior Canadian birds in May. Nests are composed of large sticks and are typically placed beneath the canopy in tall, live tree, or sometimes above canopy of live tree or in dead tree. Cliff sites are also used, especially in Ariz. Nest site is usually close to water but can be far from water. Along rocky coastlines, nests may be placed on rock outcrops. Nests become very large with reuse. New nests are 1–3' (0.3–1m) deep and 3–6' (1–1.8m) in diameter; old nests can be 12' (3.5m) deep and 8' (2.4m) in diameter. Typical clutch is 1–3 eggs. Female does most of incubation. Fratricide occurs when food is scarce. **MOVEMENTS:** Mated pairs are resident in low-elevation s. and mid-latitude regions and in all Pacific coast areas, even in Alaska. Bald Eagles are short- to moderate-distance migrants from high-elevation and n. interior regions. *Spring.*—Adults leave winter grounds in Feb.–early Mar. and peak in n. U.S. and s. Canadian hawk watches in mid–late Mar.; movement of young birds continues through May. *Northward spring–summer dispersal.*—Nonbreeding adults and younger ages from southerly and middle latitudes disperse northward in spring and summer. From Ariz., Calif., and La., birds may go as far north as n. U.S. and s. Canada (B.C. to Man.); but 1 Calif. bird was tracked to N.W.T. Most dispersing birds reach northernmost dispersal areas by late summer–early fall and then return south in mid–late fall. Some Calif. birds remain in mild-climate n. Pacific regions for winter and then head south in late winter through spring (when most eagles are heading north to breeding grounds). Northward dispersal movements occur annually until breeding territory is established. *Fall.*—Southern- and mid-latitude-born spring- and summer-dispersing birds leave n. reaches of their dispersal movement and head back to their more southerly natal areas in early–mid-fall; northern-born populations leave n. interior regions later. At Hawk Ridge in Duluth, Minn., there is a peak of all ages in mid-Sep. of northward-dispersed birds returning south, and another peak in late Oct.–mid-Nov. of mainly northerly latitude population heading south. A similar passage is seen at Windy Point in sw. Alta., with early peak in late Sep.–early Oct. followed by later peak. **COMPARISON:** ***Bald Eagle juveniles and immatures (adults are distinctive).*** **Turkey Vulture (juvenile).**—Small gray head lacks feathers *vs.* large, fully feathered head. In flight, ventral surface of flight feathers is uniformly gray, *vs.* white patch on axillaries. **California Condor (immatures).**—Dark head has bare skin or is covered with gray downy feathers. In flight, shares white patch on axillaries, but white legs and toes extend nearly to tip of tail, *vs.* yellow feet extending far short of tail tip. **Osprey.**—Dark eye line, pale head, and pale yellow eye are similar to those of immature Bald Eagle. But Osprey has black bill and cere, with bluish fringe around nostrils, *vs.* solid blackish or yellowish bill and cere. **Golden Eagle (juvenile, immatures).**—Bill is black on outer half and bluish on inner half and cere is yellow, *vs.* all-black bill and cere or yellowish bill and cere. Throat is always dark, vs. throat is always white. Nape/hindneck is sharply defined pale golden yellow, *vs.* dark but often pale-tipped feathers in hindneck region. In flight, axillaries are dark, *vs.* white on immature Bald Eagles. Any white on flight feathers is on inner primaries or forms thin strip on base of secondaries, *vs.* 2 separate patches of white on inner primaries and many feathers of inner secondaries. Tail patterns can be identical. **Golden Eagle (adult).**—Bill, cere, and distinct pale nape as noted above. Flight feathers and axillaries are always dark. **Great Blue Heron (*Ardea herodias*), flight.**—Similar size; long legs, long bill, and tucked neck may not be readily visible at a distance or at certain angles. Wingbeats are slow, with even upstroke and downstroke, *vs.* similar slow beats but with distinctive high upstroke and shallow downstroke.

Figure 1. Johnstone Strait, B.C. (Aug.)
Year-round coastal saltwater habitat along Vancouver Island (Coast Mts. of mainland in distance).

Figure 2. Johnstone Strait, B.C. (Aug.)
An Orca (*Orcinus orca*) bull shares nearshore waters with eagles along Vancouver Island.

Figure 3. Loess Bluffs N.W.R., Holt Co., Mo. (Nov.)
Fall migration/winter area for eagles, which feed on the Snow Geese (*Chen caerulescens*) that seasonally visit the refuge by the hundreds of thousands.

Figure 4. Matanuska River and Chugach Mts., Palmer, Alaska (Apr.)
Open water of the river provides year-round foraging.

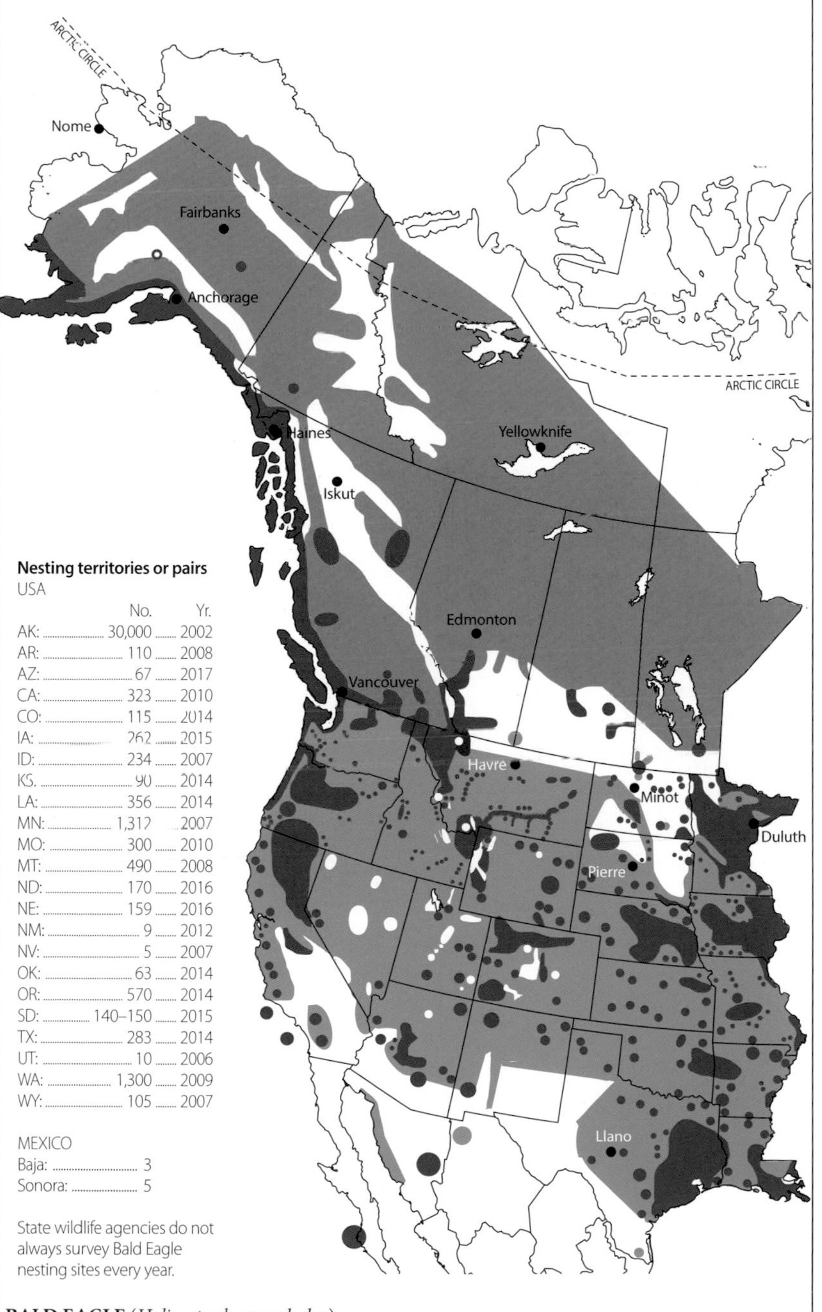

BALD EAGLE (*Haliaeetus leucocephalus*)

Plate 56. GOLDEN EAGLE (*Aquila chrysaetos*) Perching

Ages: Juvenile, 3 immature annual ages (based, in part, on Bloom and Clark 2001), and adult. Juveniles and 1-year-olds are easy to age when perched. Older immatures are difficult or impossible to age or separate from adults when perched. *Bill is moderate length and deep, distinctly bicolored, with black outer ½–¾ and pale blue or gray basal area; cere is always yellow. Large pale patch on nape and hindneck are golden yellow. Recently fledged birds are rusty brown on these areas until rapid fading occurs. Large feet, with legs feathered to toes; however, pale feathering is difficult to see in the field. Juvenile.*—Eye is medium or dark brown. Tail is longer than on older ages. Amount of white on base of tail is variable. *Immatures.*—Eye is medium brown or paler orange-brown, rarely yellow. Birds retain juvenile feathering on wings and tail, which includes some white in tail, until 3 or 4 years old. *Pale, broad bar shows on inner greater coverts and outer median coverts on mid-wing on birds older than 1 year.* Immature traits may extend past age 4. *Adult.*—Eye color is medium brown or paler, regularly orange-brown or yellow; typically paler on older birds. *Pale, broad bar shows on inner greater coverts and outer median coverts on mid-wing.* Pale gray barring/marbling adorns dorsal surface of flight feathers and inner greater coverts. *Tail:* Males lack pale bands/marbling or have 2 or 3 thin to moderate-width pale gray bands; females have moderate-width to wide pale banding/marbling. **Subspecies:** *A. c. canadensis* in N. America; 5 other subspecies in Palearctic of Eurasia. **Color morphs:** None. However, rare type called Barthelemyi variant has variably sized white patch on front of scapular tract at any age. **Size:** L: 27–33" (69–84cm). Females are larger. Juveniles are longer because of their longer tails. **Habits:** Varies from wary to very tame; shy at nest sites. Commonly perches on utility poles along roadsides in West, especially in fall and winter. Also perches on exposed ground, on rocky outcrops, or on larger exposed branches. Pairs remain together year-round; unmated birds are solitary. **Food:** Perch and aerial hunter. Mated pairs hunt cooperatively. Hunts mammals, from size of rabbits to small deer; also large waterfowl and upland game birds, and birds as large as Great Blue Heron (*Ardea herodias*). Primarily feeds on carrion in most areas fall–winter, especially on hunter-killed game that was not retrieved and on roadkill. **Voice:** Silent.

56a. Juvenile (recently fledged): *Bicolored bill and yellow cere. Nape and hindneck are medium rusty brown.* Eye is medium to dark brown. *Note:* Rusty-brown color quickly fades to golden yellow within a few weeks.

56b. Immature/adult: *Bicolored bill and yellow cere. Nape and hindneck are pale golden.* Birds older than juvenile may have medium to pale brown eye color. *Note:* Nape feathers are raised.

56c. Adult: *Bicolored bill and yellow cere. Auriculars, nape, and hindneck are pale golden.* Birds older than juvenile are often paler brown due to fading and molt. *Note:* Nape feathers are compressed.

56d. Juvenile: *Golden nape and hindneck.* Body is nearly uniformly dark brown, with same-aged feathering. Slightly paler bar often shows across middle of wing. Broad white band covers basal ¼–¾ of tail.

56e. Foot (right): *Pale feathering on lower leg (tarsus).* Pale feathers blend with yellow feet at field distances. Talons are very large, especially on innermost and rear toes (shown).

56f. 1-year-old: *Golden nape and hindneck.* Wing comprises mainly pale brown worn coverts and secondaries; secondaries lack a black band on trailing edge. Patches of white on the basal area of some coverts may show due to molting, missing feathers. Newly grown (darker) feathers adorn the tertials and respective greater coverts and scattered median and lesser coverts. *Note*: Late fall-spring.

56g. 2- and 3-year-old: *Golden nape and hindneck.* Body is mix of pale and dark brown feathers due to molt. *Pale band shows across tertials and mid-wing.* Black band on trailing edge of secondaries. Variable amount of white on outer base of tail. *Note:* Figure is indicative of a 3-year-old, with lack of faded retained juvenile secondaries, and primary tips showing new, darker feathers; may show faded coverts as on *56h*, *56i*. 2-year-olds have a few retained faded juvenile secondaries and 1 or 2 retained faded brown juvenile primaries on tip of wing.

56h. Adult: *Golden nape and hindneck.* Body is mix of pale and dark brown feathers due to molt. *There is a very distinct, pale mid-wing bar. Adults often have extensive gray barring and marbling on flight feathers and inner greater coverts.* Black band on trailing edge of secondaries. *Dark tail has pale gray, thin bands.*

56i. Barthelemyi variant, adult: *White blotch on front of scapulars.* Occurs on all ages; plumage is otherwise identical to respective age. Tail has wide, pale gray outer band, which is most typical of females.

a
b
c
d
e
f
g
h
i

Plate 57. GOLDEN EAGLE (*Aquila chrysaetos*) Flying (ventral view)

Ages: Juvenile, 3 immature annual ages (based in part, on Bloom and Clark 2001), and adult. *Golden feathers of hindneck and side of neck create pale yellow-orange patch on back of head, except when viewed in flight from directly underneath. Undertail coverts are long (longer than on Bald Eagles) and can conceal tail pattern when tail is closed. Large feet; legs feathered to toes, but feathering is impossible to see in flight. Broad primary tips show as 6 fingers.* Front edge of wing from body to wrist shows golden line merging into golden spot on wrist. *Short bill and neck make head look small. Juvenile.—Secondaries are longer, more pointed than on older ages, and wings appear broad.* On underside, flight feathers are plain gray, or have some barring on inner primaries and outer secondaries; variable amount of white on base of primaries and/or secondaries. Tail is longer, with more pointed feathers than on older ages. Amount of white on base of flight feathers and tail is highly variable. *Note*: Birds with much white on wing also have much white on tails. *Immatures.*—Retain juvenile feathers on wings and tail through 2-year-old age class: 1- to 3-year-olds can be aged in flight by wing molt sequence. Black bar adorns tips of older-aged flight feathers. All have at least some white on tail. *Adult.*—Black band adorns trailing edge of all flight feathers. *Tail:* Pale gray bands/marbling are somewhat sexually dimorphic but overlap: Males lack pale markings or have up to 2 or 3 thin or moderate-width pale gray bands; females have moderate to wide pale gray bands. Immature-looking traits may extend past 3-year-old class. **Subspecies:** *A. c. canadensis* in N. America; 5 other subspecies in Palearctic of Eurasia. **Color morphs:** None. However, rare type called Barthelemyi variant has variably sized white patch on front of scapular tract at any age (not shown here, but see *56i*, *58h*). **Size:** L: 27–33" (69–84cm), W: 72–87" (183–221cm). Females are larger. Juveniles are longer because of their longer tails. **Habits:** Varies from wary to very tame; shy at nest sites. Pairs remain together year-round; unmated birds are solitary. Commonly found along roadways in West. **Food:** Perch and aerial hunter. Mated pairs hunt cooperatively. Hunts mammals, from size of rabbits to small deer; also large waterfowl and upland game birds, and birds to size of Great Blue Heron. Primarily feeds on carrion in most areas fall–winter, especially on hunter-killed game that was not retrieved and on roadkill. **Flight:** *Soars with wings held in low or high dihedral. Wings are flapped with even up-and-down strokes from horizontal plane.* **Voice:** Silent.

57a. Juvenile, extensive white wing: *Large white patch on base of inner primaries, and thin white stripe on base of secondaries. Rest of flight feathers are gray, but distal part of inner primaries often has faint gray barring. Moderate-width black band on tail.*

57b. Juvenile, moderate white wing: *Small to medium-size white patch on base of inner primaries and sometimes outer secondaries; rest of flight feathers are mainly gray.*

57c. Juvenile, all-dark wing: Underwing is all-gray and lacks white patch; faint, pale gray barring may adorn distal part of inner primaries. White tail has wide black band; white may show only when tail is fanned.

57d. 1-year-old, extensive white wing: Much white on wing (as on *57a*). Inner 3 primaries (p1–3) are new, darker immature-age type. All visible secondaries are still juvenile (tertials would be new, but not visible from underside). Body shows mix of faded old and dark new feathers. Outer (r6) and middle (r1) tail feather sets are new, darker, and shorter. *Note:* Slow molt stage; can appear as a juvenile at a distance.

57e. 2-year-old, extensive white wing: Amount of white on wing is reduced and is irregularly formed due to molt. Outermost primary (p10) is worn and frayed juvenile feather. Wing is mainly immature feathers, with adult-like black trailing band. 2 inner secondaries (s9, s10) are classic retained juvenile feathers. Tail retains white base. *Note:* Average molt stage; may molt all juvenile secondaries. 3-year-old similar but has new, dark p9 and/or p10 (as on *57g*).

57f. Flight feathers: 1. *Juvenile outer primary:* pointed tip; faded and frayed on 1- and 2-year-olds. **2.** *Adult-type primary:* rounded tip. **3.** *Juvenile secondary:* pointed tip. **4.** *Adult-type secondary:* rounded, dark tip.

57g. 3-year-old, extensive/moderate white wing: Underwing is adult-like, with pale barring/mottling, but white spots may remain on base of some feathers. Outer primary (p10) is new, darker (can also include p9). Tail is mainly adult-like but has some white on base. *Note:* Average molt stage. 4-year-old can be similar but without distinction of new, darker outermost primaries (p9, p10). Inner 1 or more primaries molt annually.

57h. Adult: *Gray flight feathers have broad black band on trailing edge. Black tail has variable-width, pale gray barring or marbling. Tail can be all-black on some males* (not shown).

a
b
c
d
c
f
1
2
3
4
g
h

Plate 58. GOLDEN EAGLE (*Aquila chrysaetos*) Flying (dorsal view)

Ages: Juvenile, 3 immature annual ages (based, in part, on Bloom and Clark 2001), and adult. *Golden feathers of hindneck and side of neck create pale yellow-orange patch on back of head. Undertail coverts are long (longer than on Bald Eagles) and conceal tail pattern when tail is closed. Broad primary tips show as 6 fingers.* Front edge of wing from body to wrist shows golden line merging into golden spot on wrist. *Short bill and neck make head look small. Juvenile.—Secondaries are longer, more pointed than on older ages, and wings appear broad.* On underside, flight feathers are plain gray, or have some barring on inner primaries and outer secondaries; variable amount of white on base of primaries and/or secondaries. Tail is longer, with more pointed feathers than on older ages. Amount of white on base of flight feathers and tail is highly variable. *Note*: Birds with much white on wing also have much white on tails. *Immatures.*—Retain juvenile feathers on wings and tail through 2-year-old age class. 1- to 3-year-olds can be aged in flight by wing molt sequence. Black bar adorns tips of older-aged flight feathers. All have at least some white on tail. Immature traits may extend past 3-year-old class. *Adult.*—Black band adorns trailing edge of all flight feathers. *Tail:* Pale gray bands/marbling are somewhat sexually dimorphic but overlap: Males lack pale markings or have up to 2 or 3 thin to moderate-width pale gray bands; females have moderate to wide pale gray bands. **Subspecies:** *A. c. canadensis* in N. America; 5 other subspecies in Palearctic of Eurasia. **Color morphs:** None. Rare type called Barthelemyi variant has variably sized white patch on front of scapular tract; can be found on any age. **Size:** L: 27–33" (69–84cm), W: 72–87" (183–221cm). Females are larger. Juveniles are longer because of their longer tails. **Habits:** Varies from wary to tame; shy at nest sites. Often seen along highways feeding on carrion. Pairs remain together year-round; unmated birds are solitary. **Food:** Perch and aerial hunter. Mated pairs hunt cooperatively. Hunts mammals, from size of rabbits to small deer; also, large waterfowl and upland game birds, and birds to size of Great Blue Heron. Primarily feeds on carrion in most areas fall–winter, especially on hunter-killed game that was not retrieved and on roadkill. **Flight:** *Soars with wings held in low or high dihedral. Wings are flapped with even up-and-down strokes from horizontal plane.* In areas of strong updrafts, may kite with wings on flat plane. **Voice:** Silent.

58a. Juvenile, extensive white wing: *Large white patch on base of inner primaries.* Tail has broad white band on base and moderate-width black band on tip. *Note:* Juveniles regularly show pale barring on inner primaries.

58b. Juvenile, moderate white wing: Small to moderate-size white patch on base of innermost primaries. Tail is ½ white, ½ black. *Note:* Distal white area on tail is regularly speckled with black.

58c. Juvenile, all-dark wing: Upperside of flight feathers is solid dark but can be slightly paler on middle coverts. Tail is ½ white, ½ black.

58d. 1-year-old, moderate white wing: Large pale-brown patch on coverts of old, faded juvenile feathers. Some show white on base of innermost primaries. Outer 4 primaries (p7–10) and 2 blocks of secondaries (s3, s4, and s7–12) are retained faded juvenile feathers. New center (r1) and outer (r6) tail feather sets are more adult-like. *Note:* Average molt stage.

58e. 2-year-old: Outermost primary (p10) is very old, faded juvenile feather (can include p9). 2 inner secondaries (s9, s10) are long juvenile feathers; rest of secondaries are new adult-type feathers. Innermost (p1, p2) and middle (p6–9) primaries are molted for 2nd time (darker). Some white on tail base. *Note:* Average molt stage.

58f. 3-year-old: Adult-like tawny bar on middle coverts. All flight feathers are now adult-like, barred and mottled with pale gray. 3-year-olds have new, darker outermost 1 or 2 primaries (p10, as shown, or p9 and p10). Usually have small amount of white on base of tail. *Note:* Slow-molting birds can have wing as on *58e*, but with new, dark p10 or p9 and p10. *Note:* Average molt stage.

58g. Adult: Pale tawny bar on middle coverts. Trailing edge of wing has wide dark band with pale gray barring and mottling on inner part of flight feathers. Dark tail has 2 or 3 thin, pale gray bands, with distal pale band usually but not always wider. *Note:* Thin banding is mainly on males.

58h. Barthelemyi variant, adult: Plumage as on respective age, but with small white patch on front of scapular tract. Trailing edge of wing has broad, dark band, as on typical adult. Tail pattern with broad, pale gray distal band and 1 or 2 thin inner bands is mainly on females. *Note:* This rare variant marking can adorn any age.

58i: Flight silhouette: Flaps wings in even-height up-and-down strokes (pale blue, dark blue demarcation). Wings are held in low to high dihedral when soaring.

a
b
c
d
e
f
g
h
i

Plate 59. GOLDEN EAGLE (*Aquila chrysaetos*) Tails

Ages: Juvenile, 3 immature annual ages, and adult. Molt sequence and data based on Bloom and Clark (2001). Figures simulate birds in flight. Patterns not as reliable as wing molt to age immatures. Figures also show similarities with immature ages of adult Bald Eagles (Plate 55). *Undertail coverts are long (longer than on Bald Eagle) and can conceal tail pattern when tail is closed. Large feet, legs feathered to toes; however, pale feathering is difficult to see in the field. Juvenile.*—Tail is longer than on older ages. Amount of white on base of tail is highly variable. *Immatures.*—Retain remnant juvenile tail feather set(s) through 2-year-old class; sometimes as 3-year-olds. Only a few feathers are molted each year. Molt starts on typical middle feather set (r1), then continues to outermost set (r6) as 1-year-old (immature Bald Eagle molts all 12 feathers as 1-year-old). Molts r2 as 1-year-old or 2-year-old, then r3 and r5 sets as 2-year-old; r4 molted as 2-year-old or sometimes 3-year-old. Amount of white on feather base reduces with each molt. Molt is irregular thereafter. *Adult.*—Pale gray bands are somewhat sexually dimorphic but overlap. Males lack pale gray bands/marbling or have thin to moderate-width pale gray bands; females have moderate to wide, pale gray bands and marbling. Immature traits can be seen past age 4. **Subspecies:** *A. c. canadensis* in N. America; 5 other subspecies in Palearctic of Eurasia. **Color morphs:** None. Rare type, Barthelemyi variant, has variably sized white patch on front of scapular tract; can be found on any age (not shown here). **Size:** L: 27–33" (69–84cm). Females are larger. Juveniles appear longer because of longer tails. **Habits:** Varies from wary to tame in West; shy at nest sites. Often seen along highways on utility poles or feeding on carrion. Pairs remain together year-round; unmated/younger birds are solitary. **Food:** Perch and aerial hunter. Mated pairs hunt cooperatively. Hunts mammals, from size of rabbits to small deer; also large waterfowl and upland game birds, and birds to size of Great Blue Heron. Primarily feeds on carrion in most areas fall–winter, especially on hunter-killed game that was not retrieved and on roadkill. **Flight:** As in all birds, tail is closed when gliding and fanned when banking or soaring. Patterns show more when fanned. **Voice:** Silent.

59a. Juvenile, moderate white, closed (ventral): Undertail coverts conceal much of white basal tail. Outer webs of outer feathers are black (as on immature Bald Eagles; see *55e, f, g*).

59b. Juvenile, moderate white, fanned (ventral): Small spot of white shows next to undertail coverts. Outer webs of outer feathers black.

59c. Juvenile, moderate white, fanned (dorsal): *Inner ½ of tail is white, with clean demarcation. Outer edges are black.*

59d. Juvenile, extensive white, closed (ventral): *Broad, white band on inner tail. Outer webs on outer feathers are white.*

59e. Juvenile, extensive white, fanned (ventral): *Broad, white band on inner tail. Outer webs are white.*

59f. Juvenile, extensive white, fanned (dorsal): *Inner ⅔ of tail is white, with clean demarcation; white outer edges.* Dark band has gray marbling on center feathers.

59g. 1-year-old, extensive white, closed (ventral): Outer set is new (darker and shorter) immature feathers; rest of tail is old juvenile feathers.

59h. 1-year-old, extensive white, fanned (ventral): Outer and middle sets are new (darker and shorter) immature feathers. New feathers are marbled and lack sharp demarcation.

59i. 1-year-old, extensive white, fanned (dorsal): "Split" tail pattern is typical of this age: New center feather (r1) set is darker, with adult-like character, and divides white base of tail. Outermost (r6) set also is new, darker feathers. Rest of tail is retained juvenile feathers.

59j. 2-year-old, fanned (ventral): Outer area may retain 1 or 2 longer, paler juvenile feather set(s); set r4 shown. Rest of tail is immature feathers, with partially barred or marbled adult-like design.

59k. 2- or 3-year-old, fanned (dorsal): Pattern is mix of adult marbling/barring and white-based feathers of younger birds. *Note:* 2-year-olds may retain at least 1 juvenile set. All juvenile feathers are usually replaced on 3-year-olds.

59l. 3-year-old, fanned (dorsal): Adult-like on new center and outermost feather sets; previous year's 2 outer sets are white at base.

59m. Adult, 2-banded, fanned (ventral): 1 complete, irregular, pale gray band and 1 partial, pale gray inner band are visible.

59n. Adult male, unbanded, fanned (dorsal): Dark tail with faint, partial, pale gray bands or marbling on center feathers only.

59o. Adult, 2-banded, fanned (dorsal): Irregular, wide, pale gray distal band and thin, pale gray inner band.

59p. Adult, 3-banded, fanned (dorsal): Irregular, wide distal band and 2 progressively thinner pale gray bands.

59q. Adult female, fanned (dorsal): Very wide, pale gray distal band and 1 or 2 very irregular, pale gray inner bands/marbling.

a
b
c
d
e
f
g
h
i
j
k
l
m
n
o
p
q

GOLDEN EAGLE (*Aquila chrysaetos*)

HABITAT: *Summer/winter.*—Remote open and semi-open, arid, flat or hilly prairie, tundra, and intermountain valleys with adjacent embankments, cliffs, or isolated coniferous and deciduous trees for nest sites. Also inhabits rugged montane regions with cliffs, including alpine tundra. Some habitat areas are in rural locales with light agriculture, especially grass-hay fields. Also inhabits humid, coastal montane areas in Canada and Alaska. *Winter/migration.*—Birds may migrate or move to lower elevations, often to wide-open, remote, arid plain regions, including extensive agricultural lands with wheat fields. Open areas often have tall utility poles and metal power lines, which are favored roosting and hunting perches. Many inhabit semi-open rural locations that host ground squirrels, prairie dogs, hares, and rabbits. Also use semi-open and open remote and rural areas with lakes and rivers hosting large numbers of waterfowl. **STATUS:** Uncommon. Population is stable, but declines have been noted in some breeding areas. U.S. estimated population was 39,000 in 2009 and 40,000 in 2014 (USFWS 2016); no data for Canada. Electrocution, poisoning, shooting, and wind turbines cause mortality. Isolated breeding populations are found in n. and e. mainland areas of Nunavut (Poole 2011), and breeding now occurs in Wapusk Nat'l. Pk., near Churchill, Man. (Asselin et al. 2013). On range map, magenta dashed line in Alaska denotes possible irregular breeding in low-density and/or regular but low-density summer records. Cyan dashed line denotes very low-density irregular winter records. **NESTING:** Begins in Jan. in low-elevation southerly latitudes but in late May–early Jun. in northerly latitudes and higher elevations. Large stick nest is built by both sexes; pairs also have alternate nests. Nests are often reused for decades and become very large, up to 6' (1.8m) in diameter and up to 10' (3m) deep. Nest sites may be on low, accessible embankments along ridgelines, on high cliffs, or in isolated trees or narrow, especially riparian tree stands 10–100' (3–30m) high. Clutch size is 1 or 2 eggs, 3 eggs in some lower-latitude regions. Fratricide is common. **MOVEMENTS:** Adults breeding in U.S. are resident or short-distance migrants. Most U.S.–born juveniles migrate. All ages of Alaskan and Canadian birds migrate. Hundreds a day can be seen at spring hawk watches in Rocky Mts. near Bozeman, Mont. (up to 250 birds/day) and in fall near Calgary, Alta. (up to 600 birds/day). Based on data gathered through telemetry-tracked birds, eagles breeding from N.W.T. east to Lab. may winter in upper Mississippi River valley (M. Martell pers. comm.). *Spring.*—Breeding adults move before younger ages. Movement noted in early Feb. in U.S., in mid-Feb. in s. Canada, and in mid-Mar. in s. Alaska. Peak adult numbers in U.S. occur early–mid-Mar. (as late as 3rd week), mid–late Mar. in s. Canada, and late Mar.–early Apr. in s. Alaska. Younger ages move mainly Mar.–mid-May, with peak in U.S. in early–mid-Apr. and in s. Canada in mid–late Apr. Younger ages arrive in s. Alaska in mid-Apr. *Fall.*—Younger ages begin moving in mid–late Aug. Juveniles are noted on cen. Great Plains by mid-Sep. All ages peak in s. Rocky Mts. of Canada in 2nd–3rd weeks of Oct., but older ages dominate in Nov.–Dec. In n. U.S. and Minn., peak numbers of mostly juveniles and some older birds occur in mid–late Oct., while movement of especially older ages continues into Nov., or into Dec. in Great Plains, cen. Rocky Mts., and Great Basin. **COMPARISON: Turkey Vulture (juvenile).**—Small, gray, bald head *vs.* large, fully feathered head. In flight, wings held in similar dihedral, but vulture shows unstable rocking motion, *vs.* stable flight. **Bald Eagle (immatures).**—Base of bill is gray, brown, or yellow, *vs.* bluish. Throat is a whitish patch, *vs.* dark. In flight, underwing has white axillaries, *vs.* all-dark axillaries. Tail patterns are similar, and both species have white on base of tail. **Rough-legged Hawk (dark morph).**—Head is solid pale tawny or brown *vs.* tawny nape. When perched, wingtips equal or extend beyond tail tip, *vs.* shorter than tail tip. Underside of flight feathers is white, *vs.* gray. Underside of tail is similar to that of young Golden Eagle, with dark band on tip and white basal region.

Figure 1. Jackson Co., Colo. (Jul.) Year-round habitat. Eagles nest on cliffs along this ridge. (Mt. Zirkel Wilderness is across the valley.)

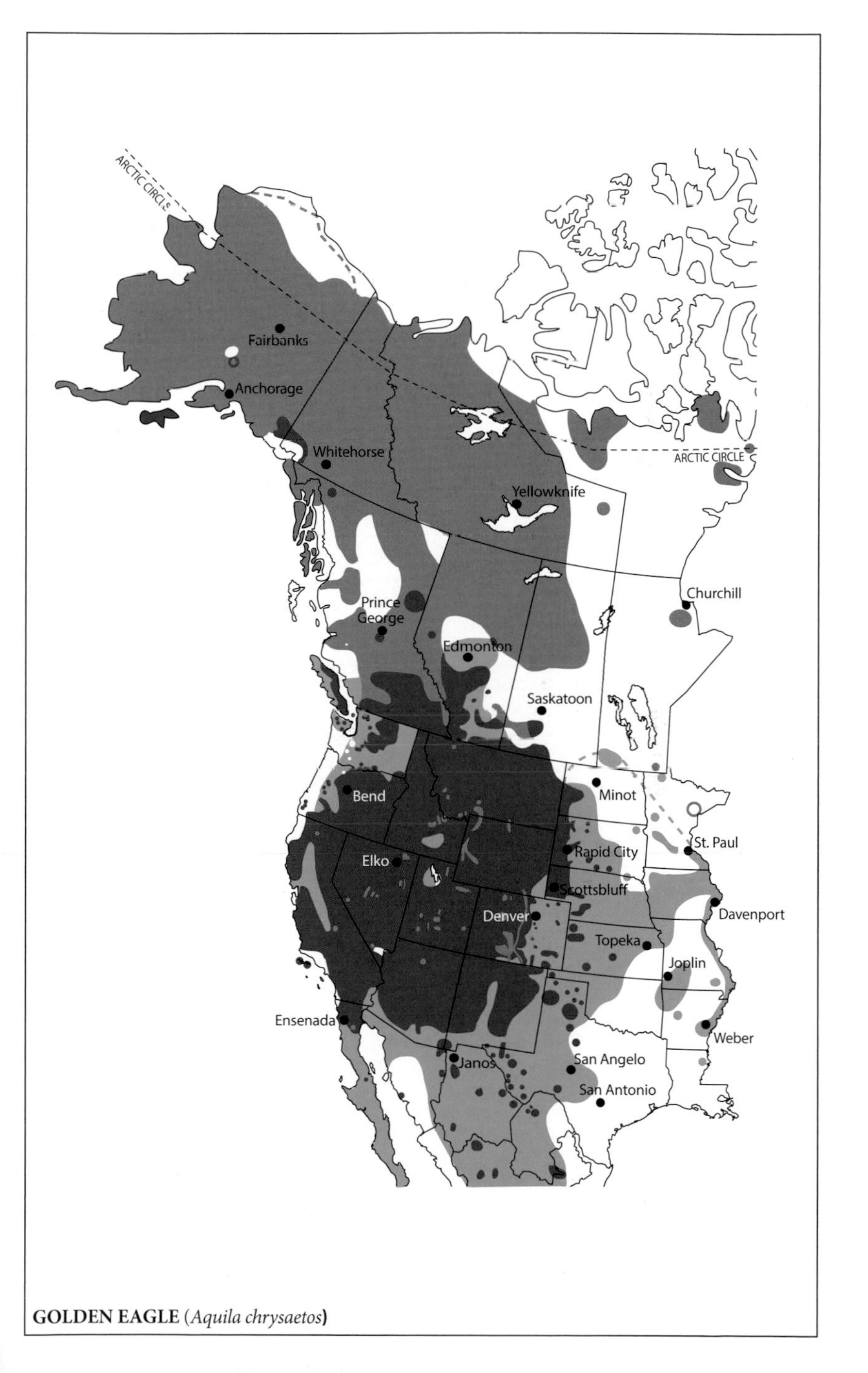

GOLDEN EAGLE (*Aquila chrysaetos*)

Plate 60. AMERICAN KESTREL (*Falco sparverius*) "American" (*F. s. sparverius*) and "Southeastern" (*F. s. paulus*)

Ages: Juvenile and adult. Juveniles have partial 1st-year molt that occurs only on the body in fall and ends by winter; field-visible mostly on males. 1st prebasic molt of body, wings, and tail begins in late spring (Pyle 2005a). *2 black stripes on white face: 1 under eye, 1 on rear of auriculars. Crown is all-blue or all-gray or has small, medium, or large rufous top spot; black streaking may be present.* (Rufous spot more prevalent on birds of northerly latitudes; Smallwood et al. 1999.) Nape is white (mainly on juveniles), white with tawny top and rear areas, or all-tawny. *Large black spot on each side of nape.* Fleshy orbital skin is yellow. Eyes are dark brown. *Underside of primaries has distinct black bar on trailing edge. Males.—Dorsal surface of wings is blue. Rest of upperparts and tail are rufous. There is a row of white spots on trailing edge of underwing.* Flanks have small or large black spots. *Tail is rufous, with white outer feather set and wide black subterminal band; there are many alternate patterns.* Juvenile male's white or tawny underparts are spotted or streaked on breast. Northern birds are fully barred on back, scapulars; birds of mid- and southerly latitudes are barred on distal ½ of scapulars. "Southeastern" is similarly marked or barred only on very distal scapulars. *Adult male's tawny underparts are lightly or moderately spotted on flanks and very lightly spotted or unmarked on belly. Only distal ½ of scapulars is barred. Females.*—In both ages, *3 main patterns of black barring on rufous upperparts.* Underparts have rufous/brown streaks varying from thin to wide. *Rufous tail has thick, moderate-width, or thin black bands.* **Subspecies:** "American" in all of West except se. Tex., where "Southeastern" is found; 15 other subspecies in the Americas. **Color morphs:** None in our area. **Size:** Male L: 8–10" (20–25cm), W: 20–22" (51–56cm). Female L: 9–11" (23–28cm), W: 21–24" (53–61cm). Females average larger. "Southeastern" is smaller. **Habits:** Typically fairly wary but sometimes tame. Perches on exposed objects, especially branch tips and utility wires. Plumage is fluffed in cool weather, making bird appear chunky. **Food:** Perch and aerial hunter of large insects, small rodents, songbirds, reptiles, and amphibians. Insects are often eaten while in flight. **Flight:** Soars with wings held on flat plane; glides with wings bowed slightly, wingtips bent slightly upward. Powered flight is of fluttering wingbeats. Hovers briefly; kites in strong winds. **Voice:** Vocal. Rapid, high-pitched *killy-killy-killy*.

60a. Juvenile male: All white nape. Breast is streaked. *Rufous back is barred. Gray crown is streaked.*

60b. Adult male: *Partially tawny nape. Rufous crown spot. Tawny breast is unmarked.*

60c. Adult male: *Tawny nape. Large rufous crown spot. Tawny breast is unmarked.*

60d. Juvenile male, barred type: *Back and scapulars are barred.* White breast is spotted; large flank spots.

60e. Juvenile male, partially barred type: *Lower ½ of scapulars is barred.* Tawny breast is streaked. *Note*: Feathers are very fluffed in cool weather.

60f. Adult male, moderately marked type: Large flank spots; small belly spots. Lower ½ of scapulars are lightly barred.

60g. Adult male, lightly marked type: Small flank and belly spots. Scapulars are barred on lower ½.

60h. "Southeastern" adult male: Crown is all-blue. Lightly spotted flanks; unmarked belly. Distal scapulars are sparsely barred.

60i. Juvenile male, barred type: *Blue inner wing; black primaries. Back and tail are rufous.*

60j. Adult male: *Row of white spots on trailing edge of wing. Rufous tail has wide, black subterminal band.*

60k. Male tail variations (dorsal): 1. Black-and-white banding on a few outer feathers (common). **2.** Banded, except on center, rufous feather set (fairly common). **3.** Blue and rufous, with 1 blue center feather (uncommon). **4.** Blue on a few center feathers and banded (very uncommon).

60l. Juvenile female: *Crown is streaked. White nape has dark spots.* Ventral streaking is brown.

60m. Female: *Blue crown has rufous top spot. Nape is tawny.* Thin ventral streaks are rufous.

60n. Female, heavily marked type: *Upperparts have thin rufous bars.* Ventral streaks are brown.

60o. Female, moderately marked type: *Upperparts have equal-width rufous and dark bars.*

60p. Female, lightly marked type: *Thin, dark bars on dorsal plumage.* Thin ventral streaks are rufous. *Note*: Feathers are very fluffed in cool weather.

60q. Female: *Pale underwing has moderate-width black bar on trailing edge of primaries.*

60r. Female: *Black flight feathers contrast with rufous upperparts.* Rufous tail has many thin black bands.

a
b
c
d
e
f
g
h
i
j
1
k
2
3
4
l
m
n
o
p
q
r

AMERICAN KESTREL (*Falco sparverius*)
"American" (*F. s. sparverius*) and "Southeastern" (*F. s. paulus*)

HABITAT: "American." *Summer.*—Inhabits wide variety of moderate- and low-elevation semi-open and open habitats within remote regions and in rural, suburban, and urban environments. All areas have natural or human-made holes and crevices in cliffs, live or dead trees, utility poles, buildings, or other structures. Readily adapts to nest boxes placed on poles or tree trunks. Absent from alpine elevations. *Winter.*—Identical habitat as in summer but without regard to cavities. Regularly found on open, harvested agricultural areas and plains that have elevated perches for hunting and roosting. **"Southeastern."** *Year-round.*—Low-elevation areas with semi-open forests of mainly Longleaf Pine (*Pinus palustris*) adjacent to meadows. Wooded areas are mix of live and dead trees, some of which have natural cavities. Small numbers are found in suburban and even urban locales that have natural or human-made cavities, including nest boxes. **STATUS: "American."** Common. Population is stable and thriving and very adaptable to human alteration. Competes with European Starling (*Sturnus vulgaris*) for nest sites. Males winter farther north than females. **"Southeastern."** Very uncommon; occurs sparingly throughout pine woodlands of se. U.S. In West, found only in se. Tex. and s. La.; intergrades with "American" in La. Range (purple) included on "American" map, with approximate northern boundary noted by *white dotted line.* **NESTING:** Nests in natural or human-made cavities at elevated heights, 10–30' (3–9m) or higher. No nest is built; 4–6 eggs are laid in bare cavity. 2 broods are common with early nesting pairs. **"American."** Begins nesting in Feb. in low elevations of southerly and mid-latitudes. Nesting can begin as late as late May in northerly latitudes. Nesting ends Jun.–Sep., depending on elevation, latitude, and number of broods. **"Southeastern."** Similar timing as "American," but often starts as late as Apr.–May. **MOVEMENTS: "American."** Sedentary or short-distance migrant in southerly and mid-latitudes; moderate- to long-distance migrant in northerly latitudes. *Spring.*—Feb.–mid-May, with peak in mid-Apr. Northern population is on migration when southern and mid-latitude population is nesting. Males precede females. *Fall.*—Post-breeding groups form before and during migration in food-rich areas. Both ages move mid-Aug.–early Oct., with peak in mid-Sep. for juveniles and late Sep. for adults. **"Southeastern."** Sedentary. **COMPARISON:** American Kestrels often pump their tails up and down when perched; none of the following do so. **Sharp-shinned Hawk.**—Similar in size. Lack of head markings, *vs.* 2 thin black face stripes. In flight, broad wings with round tips, *vs.* narrow, pointed-tipped wings. Rufous-streaked underparts of some juveniles are similar to those of female kestrel; tail has 3 or 4 distinct black bands, *vs.* rufous with multiple thin, black bands on females. **Aplomado Falcon.**—Single black face stripe, *vs.* 2 black stripes. Black bellyband, *vs.* spotted or streaked belly. Flight feathers are dark ventrally, *vs.* white feathers. **Peregrine Falcon.**—Much larger species. Single black stripe below eyes, *vs.* 2 black stripes. Dorsal parts are gray or brown, *vs.* all-rufous or rufous with blue wings. When perched, wingtips equal tail tip, *vs.* much shorter than tail tip. Wingbeats are slow and steady with long glides, *vs.* very quick beats and short glides. **Prairie Falcon.**—Single black stripe below eye, *vs.* 2 black stripes. Upperparts are brown, *vs.* all-rufous or rufous with blue wings. Black axillaries adorn underside of wing, *vs.* white axillaries. ***American Kestrel female.*** **Merlin.**—Similar in size. Moderately distinct single dark stripe below eye, *vs.* 2 distinct black stripes below and behind eye. Tail is dark with thin, pale bands, *vs.* rufous with many thin black bands. Voice is rapid *ki-ki-ki*, *vs. killy-killy-killy.*

Figure 1. Prowers Co., Colo. (Sep.) Year-round habitat. Often nests in cavities in dead trees.

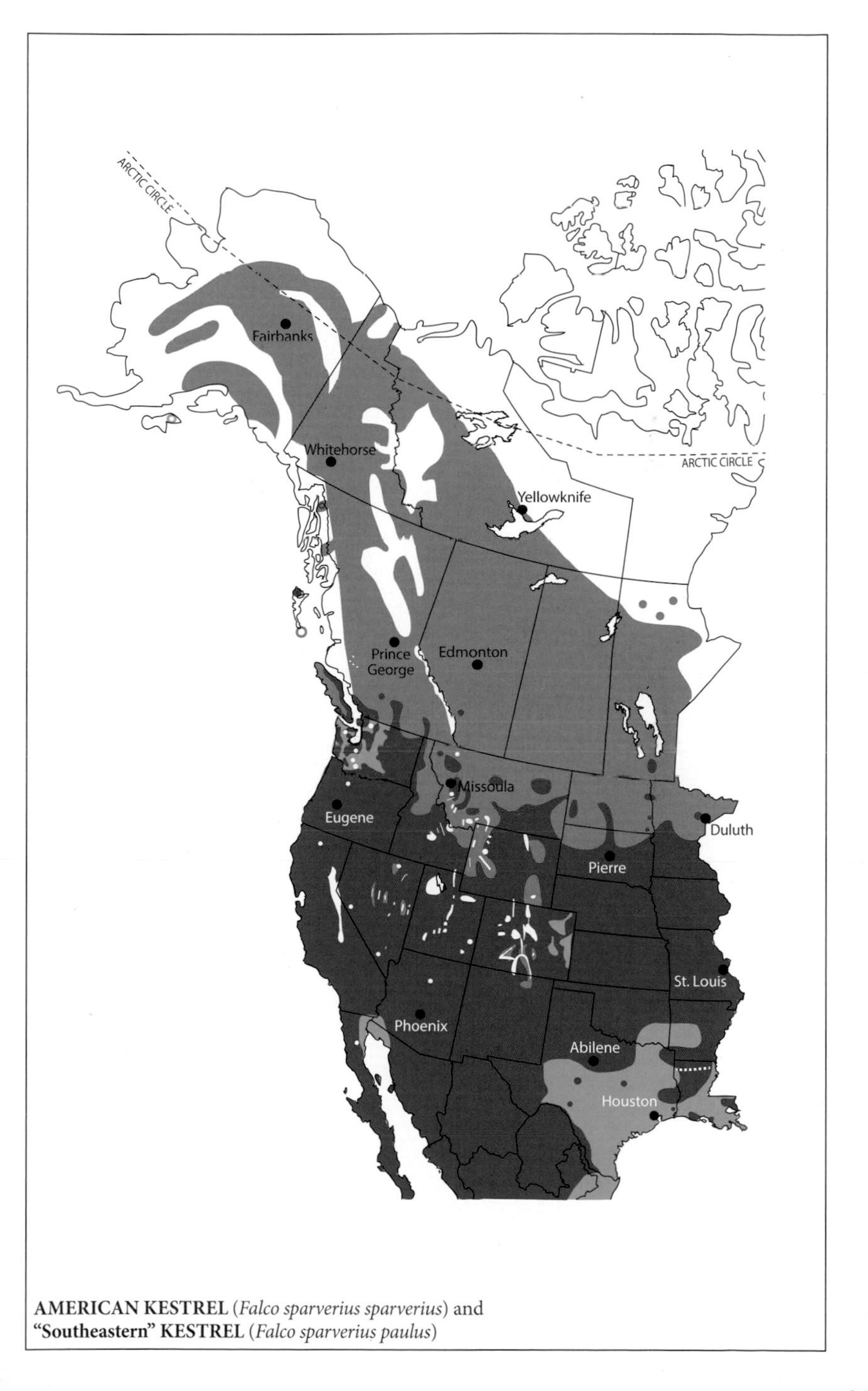

AMERICAN KESTREL (*Falco sparverius sparverius*) and
"Southeastern" KESTREL (*Falco sparverius paulus*)

Plate 61. MERLIN (*Falco columbarius*)
"Taiga" (*F. c. columbarius*)

Ages: Juvenile and adult. Some juveniles begin molt before age 1, in early spring, in partial 1st-year molt or 1st prebasic molt on back, scapulars, and head. Adult plumage attained as 1-year-old by late fall of 2nd year. Orbital skin is yellow. Eyes are dark brown. *There is a thin dark line behind eye.* Pale supercilium is thin and long. Dark malar stripe is faint. *Wings are moderately long, with broad secondaries and pointed wingtips. Uniformly patterned underwing is fairly dark. On flight feather (50q), pale bars/spots cover 50% of dark areas.* Underparts are thinly to thickly streaked with brown. When perched, wingtips are moderately shorter than tail tip. Legs and feet are yellow. *Juveniles (both sexes) and adult female.*—Plumage exhibits individual variation, with some darker and some paler; geographic variation occurs in West. Juvenile males and adult females have grayish cast to dorsal plumage, especially rump and uppertail coverts. Juveniles have tawny crown; adult females have gray crown. Ventrally, flight feathers have pale, tawny spotting. *Dorsal surface of tail is partially banded or has 2 or 3 thin, pale bands and broad, darker subterminal band; juvenile males have gray tail bands, juvenile females have tawny bands, and adult females have gray or tawny bands.* Dorsal tail band patterns are shared; *all have tawny ventral tail banding. Adult male.*—Upperparts are medium blue. *Leg feathers are tawny or pale rufous.* White spots on ventral side of flight feathers. *Black uppertail has 2 or 3 thin, blue bands; white bands on ventral side.* **Subspecies:** 2 other subspecies in N. America, both in West ("Richardson's" may stray eastward; see Plate 70); 6 others in Eurasian Palearctic. Pale type (former "Bendire's") female is similar to juvenile and adult female "Richardson's." **Color morphs:** None. **Size:** Male L: 9–11" (23–28cm), W: 21–23" (53–58cm). Female L: 11–12" (28–30cm), W: 24–27" (61–68cm). Sexually dimorphic. **Habits:** Tame. Perches on exposed objects but may use concealed areas when resting and roosting. In temperament, high-strung and aggressive. **Food:** Aerial hunter of small birds and large insects. **Flight:** Long periods of quick, powerful wingbeats with short glides. Wings held flat when soaring. Hunting flights may undulate. **Voice**: High-pitched, rapid *ki-ki-ki-ki* when agitated.

61a. Juvenile: *Crown is tawny. Supercilium line is thin. Thin, dark eye line; faint malar mark.*

61b. Adult female: *Crown is gray. Supercilium line is thin. Thin, dark eye line; faint malar mark.*

61c. Juvenile male, 3-banded tail (dorsal): *3 thin, gray bands.* Gray rump, uppertail coverts.

61d. Juvenile male, 2-banded tail (dorsal): *2 thin, gray bands.* Gray rump, uppertail coverts.

61e. Juvenile male, partial-banded tail (dorsal): *May have 2 or 3 incomplete, thin bands on center feathers.*

61f. Juvenile female, 3-banded tail (dorsal): *3 thin, tawny bands; may have 2 bands or partial bands.*

61g. Adult female, 3-banded tail (dorsal): *Thin, pale bands are mainly gray but can be tawny; may have 2 bands or, more commonly, partial bands.* Uppertail coverts are very gray and lack tawny tips.

61h. Juvenile/adult female tail (ventral): *1 thin, tawny band shows in full; 2 or 3 other bands are partially hidden by undertail coverts.*

61i. Juvenile male: Dark brown upperparts have grayish cast. *Tail is 3-banded type.*

61j. Juvenile female: Dark brown upperparts. *Tail is partial-banded type (2 partial bands).*

61k. Adult female: Similar to *61i* but larger, often more grayish. *Tail is 3-banded tawny type.*

61l. Adult female, dark type: Darker than *61k.* Supercilium is short. *Tail is 2-banded gray type.*

61m. Adult female, pale type: Medium to pale brown upperparts. Wing coverts may have pale spots. *Tail is partial-banded gray type. Note:* Similar to juveniles/adult female "Richardson's" (*62f, 62g*). Adult male (not shown) is paler blue than "Taiga" male.

61n. Juveniles/adult female: *Uniformly dark underwing has tawny-barred flight feathers and tawny tail band.*

61o. Adult female (Oct.): Molting; shows new and old brown feathers, old outer secondary. Gray uppertail coverts.

61p. Undertail coverts: Thin to moderately thick streak. *Note:* Compare "Richardson's" (*70n*).

61q. Primary flight feather: *Pale barring covers 61% of dark area. Spots/bars are tawny on juveniles and adult females, white on adult males.*

61r. Adult male: *Crown is blue. Pale supercilium is thin; thin, dark eye line and faint malar mark.*

61s. Adult male: *Medium blue upperparts; tawny or rufous leg feathers. 2 thin, blue bands on black tail.*

61t. Adult male: *Medium blue upperparts. Black primaries have small white spots. Tail is 3-banded type.*

61u. Adult male, 2-banded tail (dorsal): *2 moderately thin, blue bands on black tail.*

61v. Adult male tail, fanned (ventral): *1 thin white band; 1 or 2 partially hidden white bands.*

61w. Adult male: *Uniformly dark underwing, with white-barred flight feathers. Tail (closed) shows 1 white band.*

a
b
c
d
e
f
g
h
i
j
k
l
m
n
o
p
q
r
s
t
u
v
w

Plate 62. MERLIN (*Falco columbarius*) "Richardson's" (*F. c. richardsonii*)

Ages: Juvenile and adult. Juvenile plumage is retained for 1 year. Adult plumage attained as 1-year-old, by late fall of 2nd year. Orbital skin is yellow. Eyes are dark brown. *Thin, dark line extends behind eye.* Thin, white supercilium extends from forehead. Malar stripe is faint. *Wings are moderately long, with broad secondaries and pointed wingtips. Uniformly patterned underwing is fairly pale. On flight feather, pale bars/spots occupy more than 50% of dark area.* Underparts are thinly or moderately thickly streaked with rufous or brown. When perched, wingtips are moderately shorter than tail tip. Legs and feet are yellow. Tip of tail has broad, white band. *Juveniles (both sexes) and adult female.*—Juvenile males and some adult females have grayish cast to pale brown upperparts. Adult females also have grayish rump and uppertail coverts. Crown of head is tawny but can be gray on some adult females. Pale barring on dorsal flight feathers is pale tawny. *Uppertail has 3 moderately thin, pale bands and wide, dark subterminal band; tail is rarely partially banded. Juvenile males have pale gray bands; juvenile females have pale tawny bands; and adult females have pale tawny or pale gray bands. All have pale tawny tail bands on ventral side. Adult male.—Upperparts are pale blue. Leg feathers are pale tawny.* Pale spotting/barring on underside of flight feathers is white. *Black uppertail has 2 or 3 moderate-width or wide, pale blue or whitish bands; white bands on ventral side.* **Subspecies:** 3 subspecies in N. America; 6 others in Eurasian Palearctic. **Color morphs:** None. **Size:** Male L: 9–11" (23–28cm), W: 21–23" (53–58cm). Female L: 11–12" (28–30cm), W: 24–27" (61–68cm). Sexually dimorphic. This is largest subspecies. **Habits:** Tame. Perches on exposed objects but may use concealed areas when resting and roosting. In temperament, high-strung and aggressive. **Food:** Aerial hunter of small birds and large insects. **Flight:** Long periods of quick, powerful wingbeats with short glides. Wings are held flat when soaring. Hunting flights may undulate. **Voice**: High-pitched, rapid *ki-ki-ki-ki* when agitated.

62a. Juveniles/adult female: Thin, white supercilium; *thin dark eye line; very faint malar mark.*

62b. Juvenile male/adult female tail (dorsal): *3 moderately thin, gray bands.*

62c. Juvenile female/adult female tail (dorsal): *3 moderately thin, tawny bands. Bands can also be white.*

62d. Juvenile male/adult female, partial-banded tail (dorsal): *3 incomplete, pale gray bands on middle feathers. Note: Uncommon type. Partial bands can be pale tawny on juvenile females and some adult females.*

62e. Juvenile/adult female tail (ventral): *1 or 2 full, pale tawny bands; tawny even on juvenile males.*

62f. Juvenile male (recently fledged): Rich tawny underparts. *This variant has thin brown streaking on underparts. Pale tawny spots on lower scapulars and some upperwing coverts. Tail is as 62b.*

62g. Juvenile female/adult female: White underparts. *This variant has classic rufous streaking indicative of this subspecies. Pale tawny spots on rear scapulars and some upperwing coverts. Tail is as 62c.*

62h. Adult female: *Tawny spotting on wing. 3 pale gray tail bands.* Some have gray uppertail coverts.

62i. Juveniles/adult female: *Uniformly pale, spotted/barred underwings. 2 whitish tail bands.*

62j. Primary flight feathers: *Pale areas are wider than dark areas. Left figure is palest type. Outer feather web (left side of quill) has large, pale tawny spots. Note:* Juvenile/adult female feathers shown, depicting tawny spotting. Adult male has similar, but white spotting.

62k. Adult male: *Pale blue crown; white supercilium; thin dark eye line; faint dark malar mark.*

62l. Adult male: *Pale blue upperparts. Legs are pale tawny.* This variant has thin brown streaking on breast and belly.

62m. Adult male: *Pale blue upperparts. Legs are pale tawny.* This variant has rufous streaking on breast and belly. *Tail has very wide, pale blue bands and very thin black bands.*

62n. Undertail coverts: Unmarked or with thin, dark shaft streak. *Note:* Compare to other subspecies (*61p, 63k*).

62o. Adult male, 2-banded tail (dorsal): *2 pale blue or whitish bands and complete black inner bands.*

62p. Adult male, 3-banded tail (dorsal): *3 pale blue or whitish bands and partial black inner bands.*

62q. Adult male, 3-banded tail (dorsal): *3 pale blue or whitish bands and faint dark inner bands.*

62r. Adult male: *Pale blue upperparts. Black primaries have large white spots. Tail is as 62p.*

62s. Adult male: *Uniformly pale, spotted/barred underwings. Tail has 1 white band.*

62t. Adult male tail (ventral): *1 white inner band shows in full; 1 or 2 white bands visible on widely fanned outer feathers.*

a
b
c
d
e
f
g
h
i
j
k
l
m
n
o
p
q
r
s
t

Plate 63. MERLIN (*Falco columbarius*)
"Black" (*F. c. suckleyi*)

Ages: Juvenile and adult. Juvenile plumage is retained for 1 year. Adult plumage is attained as 1-year-old. Orbital skin is yellow. Eyes are dark brown. Individuals in same brood may be lightly or heavily marked (C. Susie, unpubl. photos, Sitka, Alaska). Supercilium is either small pale patch or is absent. *Wings are moderately long, with broad secondaries and pointed wingtips. Underwing is dark, with pale spots on flight feathers occupying less or much less than 50% of dark areas. Paler types have uniformly spotted flight feathers; darker types have pale spots only on outer primaries and outer secondaries.* Underparts are thickly streaked. When perched, wingtips are moderately shorter than tail tip. Legs and feet are yellow. *Undertail coverts can be barred* (unique to subspecies) or streaked (as on "Taiga"). *Juveniles (both sexes) and adult female.—* Upperparts are uniformly solid very dark brown. Sex and age differences are subtle. Juvenile males and adult females have grayish cast to brown upperparts. Juveniles have faint brownish crown streaking; adult females have faint grayish crown streaking. Spotting on ventral flight feathers is pale tawny. Uppertail may be unmarked or partially banded (partially banded patterns shared by "Taiga"). *If partially banded, juvenile males have gray bands, juvenile females have tawny bands, and adult females have gray or tawny bands. White terminal band is thin.* Ventral surface of tail can be partially banded or fully banded tawny (as banded as on "Taiga"). *Adult male.—*Upperparts are dark blue. *Leg feathers are dark rufous.* Any spotting on ventral flight feathers is white. *Black uppertail has 1 or 2 thin, full or partial blue bands.* Undertail has 1 partial whitish band. **Subspecies:** 3 subspecies in N. America; 6 others in Eurasian Palearctic. **Color morphs:** None. **Size:** Male L: 9–11" (23–28cm), W: 21–23" (53–58cm). Female L: 11–12" (28–30cm), W: 24–27" (61–68cm). Sexually dimorphic. **Habits:** Tame. Favors exposed objects for perches but may use concealed areas when resting and roosting. In temperament, it is high-strung and aggressive. **Food:** Aerial hunter of small birds and dragonflies (including dragonflies in Alaska in summer). **Flight:** Long periods of quick, powerful wingbeats with short glides. Wings are held flat when soaring. Hunting flights may be in undulating fashion. **Voice**: High-pitched, rapid *ki-ki-ki-ki* if agitated.

63a. Juveniles/adult female, lightly marked type: *Very dark head. Pale supercilium is short.* Underparts are heavily streaked. *Note:* Juvenile shown (brownish crown).

63b. Juveniles/adult female, heavily marked type: *Very dark head lacks pale supercilium and appears hooded. Underparts are very heavily streaked. Note:* Adult female shown (grayish crown).

63c. Juveniles/adult female, unmarked tail (dorsal): *Dark, with thin white tip. Subterminal band is darker.*

63d. Juveniles/adult female, partial-banded tail (ventral): Thin, tawny partial band. *Barred undertail coverts.*

63e. Juvenile female, lightly marked type: Upperparts are very dark brown. Underparts are heavily streaked. *Tail is partial-banded type.* Streaked undertail coverts.

63f. Adult female, heavily marked type: Upperparts are very dark brown with a grayish cast. Underparts are very heavily streaked. *Tail is unmarked. Note:* Juvenile male has similar grayish dorsal surface but brownish crown.

63g. Adult female: *All upperparts are very dark brown, including tail. Note:* Juvenile female similar but more brownish.

63h. Juveniles/adult female, lightly marked type: *All flight feathers are lightly spotted or barred.*

63i. Juveniles/adult female, heavily marked type: *Dark flight feathers are faintly barred on outer primaries.*

63j. Primary flight feathers: *Pale barring covers less than 50% of dark area. Note:* Juvenile/adult female feather depicted, showing tawny barring/spotting; adult male has white markings.

63k. Undertail coverts: *Left:* streaked (as on many "Taiga"). *Right: barred on outer web* (unique to subspecies).

63l. Adult male: Short, pale supercilium. Malar mark is dark, and throat is streaked.

63m. Adult male, lightly marked type: *Dark blue upperparts. Underparts are rusty. Tail is as on 63p.*

63n. Adult male, heavily marked type: More heavily streaked than *63m. Tail is as on 63q.*

63o. Adult male: Dark blue upperparts. Primaries are mainly black. *Black tail has 1 thin blue band, as on 63q.*

63p. Adult male, 2-banded tail (dorsal): *Black tail has 2 thin, blue bands.*

63q. Adult male, 1-banded tail (dorsal): *Black tail has 1 thin, blue basal band.*

63r. Adult male tail (ventral): *Black underside has thin whitish band. Barred undertail coverts.*

63s. Adult male, heavily marked type: *Dark flight feathers, with some faint white barring.*

a
b
c
d
e
f
g
h
i
j
k
l
m
n
o
p
q
r
s

MERLIN (*Falco columbarius*)
"Taiga" (*F. c. columbarius*), "Richardson's" (*F. c. richardsonii*), "Black" (*F. c. suckleyi*)

HABITAT: "Taiga." *Summer.*—Low-elevation boreal forest and low-elevation montane regions interspersed with bogs, lakes, and meadows and with moderate-size to tall coniferous trees. Increasingly common in suburban areas of n. states and Canada that have coniferous trees and host populations of American Crow (*Corvus brachyrhynchos*). In northerly latitudes of Alaska and Canada, inhabits taiga zone of semi-open tundra dotted with lakes and ponds and with growth of dwarf spruce and deciduous scrub. Farther north in medium-shrub zone, nests on fairly open tundra among scattered groups of stunted White Spruce and dwarf birch and willow, on gentle slopes adjacent to lakes and ponds. Large open areas, especially bodies of water, are necessary ingredient for most nesting territories (prey is most vulnerable over such areas). *Winter/migration.*—Prefers areas adjacent to bodies of water, whether ocean shores, bays, marshes, or lakes, in order to hunt exposed prey. Found from moderate latitudes to subtropical latitudes. Also acclimates to suburban and urban areas that are not near water. **"Richardson's."** *Summer.*—Semi-open and open, arid, short-grass prairies interspersed with lakes, ponds, or rivers that have riparian growth of single trees or stands of deciduous trees. Also on hilly prairie settings with single or scattered stands of conifers, especially Ponderosa Pine savannas and, to lesser extent, higher-elevation Limber Pine (*Pinus flexilis*) on hillsides. Also uses rural prairie habitat with moderate agricultural fields, often with shelterbelts of deciduous or conifer growth. This subspecies has acclimated to suburban areas with cemeteries and city park and residential locations, especially with conifer trees. All areas have populations of Black-billed Magpie and/or, in rural and suburban locations, American Crow. *Winter.*—Inhabits areas as in summer. Also found on open prairies that have large populations of Horned Lark (*Eremophila alpestris*). These areas may have only fence posts or isolated trees for elevated perches. Fond of inhabited or abandoned prairie homesteads that support tree growth. Readily utilizes prairie-region suburban and urban areas. Absent from high-elevation plains. **"Black."** *Summer.*—Found in semi-open, humid coniferous habitat on islands and near coastal mainland; includes villages and towns. These regions have rivers, inland lakes, and saltwater coasts. On Vancouver Island, B.C., nests in city parks of coastal cities; also on adjacent saltwater coastal areas and on wooded islands of inland lakes. In Alaska, found nesting on spruce-covered hillside overlooking village of Kake on Kupreanof Island; near Sitka on Baranof Island, nested on shoreline of small, inland (but near-coastal), low-elevation lake (C. Susie pers. comm.). *Winter.*—Found in coastal rain-forest, saltwater habitat, as in summer; also in coastal humid and arid regions without regard to vegetation types, including open marshes and seashores. Small numbers regularly inhabit opposite-type habitat in interior arid locales in rural and suburban areas of Great Basin. **STATUS:** Overall stable; suffered during organochlorine (DDT) era, as did other bird-eating birds of prey, but has rebounded on its own. All subspecies adapt well to human disturbance on breeding and winter grounds. **"Taiga."** Uncommon and sparsely distributed during all seasons. Locally common nester in some n. suburban areas where American Crows are present. This subspecies is most common along coastal areas in winter; uncommon in interior regions, where most use riparian and urban areas. A pale variant of this subspecies, which is nearly as pale as respective sexes of "Richardson's," was once labeled as "Bendires" subspecies. It inhabits Kodiak Island and interior Alaska (images from Kodiak Island by R. McIntosh; C. Susie; UAM). *Note:* Adult male was photographed at close range near Sitka, Alaska, in mid-Jun. 2017; unknown whether it was a breeder (C. Susie pers. comm./unpubl. photo). **"Richardson's."** Locally fairly common in its restricted breeding range; common to uncommon in winter areas south of breeding range. Populations of House Sparrows (*Passer domesticus*) in urban areas and Horned Larks in rural and remote areas reflect winter numbers. **"Black."** Uncommon in its breeding range and very sparsely distributed. It is uncommon to very uncommon in wintering areas along Pacific coast and in Utah; rare elsewhere. There are no winter records for Baranof Island (Sitka), Alaska (Webster 2006; C. Susie pers. comm.). **NESTING:** No nest is built. Uses unoccupied nests of large passerines and sometimes small raptors. Clutch is 3–5 but up to 7. Incubation is mainly by female. Youngsters fledge in 28–30 days and remain on natal grounds for at least 2 more weeks. Breeding mainly occurs when 2 years old, but 1-year-olds may breed. **"Taiga."** May begin in early Apr. in southerly latitudes, but most do not begin until later in Apr.–May. A few s. urban pairs may remain on nesting grounds. South of N.W.T. and Nunavut, in range of American Crow, most pairs use old crow nests; old nests of Black-billed Magpie and tops of flattened leafy squirrel nests are also used, as well as high, broken-off trunk stubs. All nest sites are in upper part of tall conifer. North of crow's range, old stick nests, in trees, of Common Raven (*Corvus corax*) or large raptors are used. In tundra region, may nest on ground under a bush (on hillside). **"Richardson's."** May remain on breeding grounds, or migrants arrive Feb.–Apr.; even Canadian birds may arrive as early as late Feb. Adult males arrive 1st. Suburban pairs use old nests of American Crows in conifers; in rural and remote regions, pairs use abandoned nests of Black-billed Magpies in coniferous as well as deciduous trees. Nest sites can be as low as 7' (2m) and average 16' (5m); much higher, up to 50' (15m), in suburban crow nests. **"Black."** Nests only in tall conifer trees, especially in old Northwestern Crow nests (at Kake, Alaska, C. Susie pers. comm.) or in old Steller's Jay (*Cyanocitta stelleri*) nests; also may use tops of tall broken trunks. Arrives on breeding territory early Apr.–early May (C. Susie pers. comm.). **MOVEMENTS: "Taiga."** Mainly a moderate- to long-distance migrant. Males may winter as far south as Cen. America and Gulf of Mexico islands. Females winter farther north. A few s. birds remain on breeding territory. *Spring.*—Most movement is seen during Apr.; continues into May for northern nesting adults and returning 1-year-olds. *Fall.*—Juveniles begin moving in mid–late Aug. and peak in mid-Sep. Adults move later and peak in mid-Oct. in middle latitudes. **"Richardson's."** Can be resident even in n. part of its prairie biome range, including adult male. Both sexes and ages may move short to moderate distances, and a few engage in fairly long-distance movements into interior n. Mexico and s. Baja California. *Spring.*—Adult males begin leaving cen. Great Plains in Feb. Peak numbers are seen in late Feb.–early Mar. Females and 1-year-olds move in Mar.–Apr.; rarely seen after mid-Apr. on cen. plains. *Fall.*—1st juveniles arrive on cen. Great Plains in early Sep., young adult males as early as mid-Sep. A large movement of all ages and sexes is seen on the plains in mid–late Oct. Movement continues later, but this may be due to nomadic nature of this subspecies, especially birds that follow the massive flocks of Horned Larks on the open plains. **"Black."** Little data on this subspecies' movements. Most of Alaskan population probably moves out of breeding territories (shown in year-round purple on map, because published winter data are not determined to subspecies). *Spring.*—Juvenile females noted in s. Calif. in early Apr. Alaska data based on C. Susie (pers. comm.): Birds arrived at nesting territory at Kake on Kupreanof Island in early–mid-Apr. (2008, 2009); adult female photographed at Klukwan (north of Haines) in very late Apr. and very early May (2013); adult male arrived at Sitka, Baranof Island, in early May (2015) and late Apr. (2016). Last sightings of juveniles at natal territory near Sitka were in early Aug.; seen in mid–late Aug. in vicinity of Sitka. *Fall.*—Little data. Based on above data, Baranof Island birds begin moving in mid–late Aug. This subspecies is seen along coastal Wash. in mid-Sep.–Oct. **COMPARISON: Sharp-shinned Hawk.**—Similar in size. Eye is yellow, orange, or red, *vs.* dark brown. Streaked underparts of juveniles are similar. Tail has wide, equal-width light and dark bands, *vs.* all dark brown or dark brown with thin light tawny/white bands. In flight, underwing is pale with thick barring and wingtips are rounded, *vs.* dark underwing and pointed wingtips. **American Kestrel.**—Similar in size. Side of head has 2 black stripes, *vs.* 1 faint dark stripe below eye. Females have similar ventral streaking, but kestrel has reddish upperparts, including tail, *vs.* dark brown, including tail. Similar in flight, but kestrel's wings are pale on underside, and secondaries are narrower, *vs.* dark underside and broad secondaries. **Peregrine Falcon (juvenile).**—Larger species. Distinct black malar stripe extends below eye, *vs.* ill-defined malar mark. Similar very dark brown upperparts, especially "Peale's" and some "Anatum"; "Arctic" has distinct pale edges. When perched, wingtips equal or nearly equal tail tip, *vs.* much shorter than tail tip.

Figure 1. Osgood Church, Weld Co., Colo. (Mar.)
"Richardson's" winter habitat. Trees around old homesteads or buildings provide roost sites on the open prairie.

Figure 2. Thimbleberry Lake, Barnof Island (near Sitka), Alaska (Aug.)
"Black" nesting habitat near lakes and seacoasts. "Taiga" nests in similar habitat but in interior mainland areas. Photo by Chuck Susie

"Taiga" MERLIN (*Falco columbarius columbarius*)

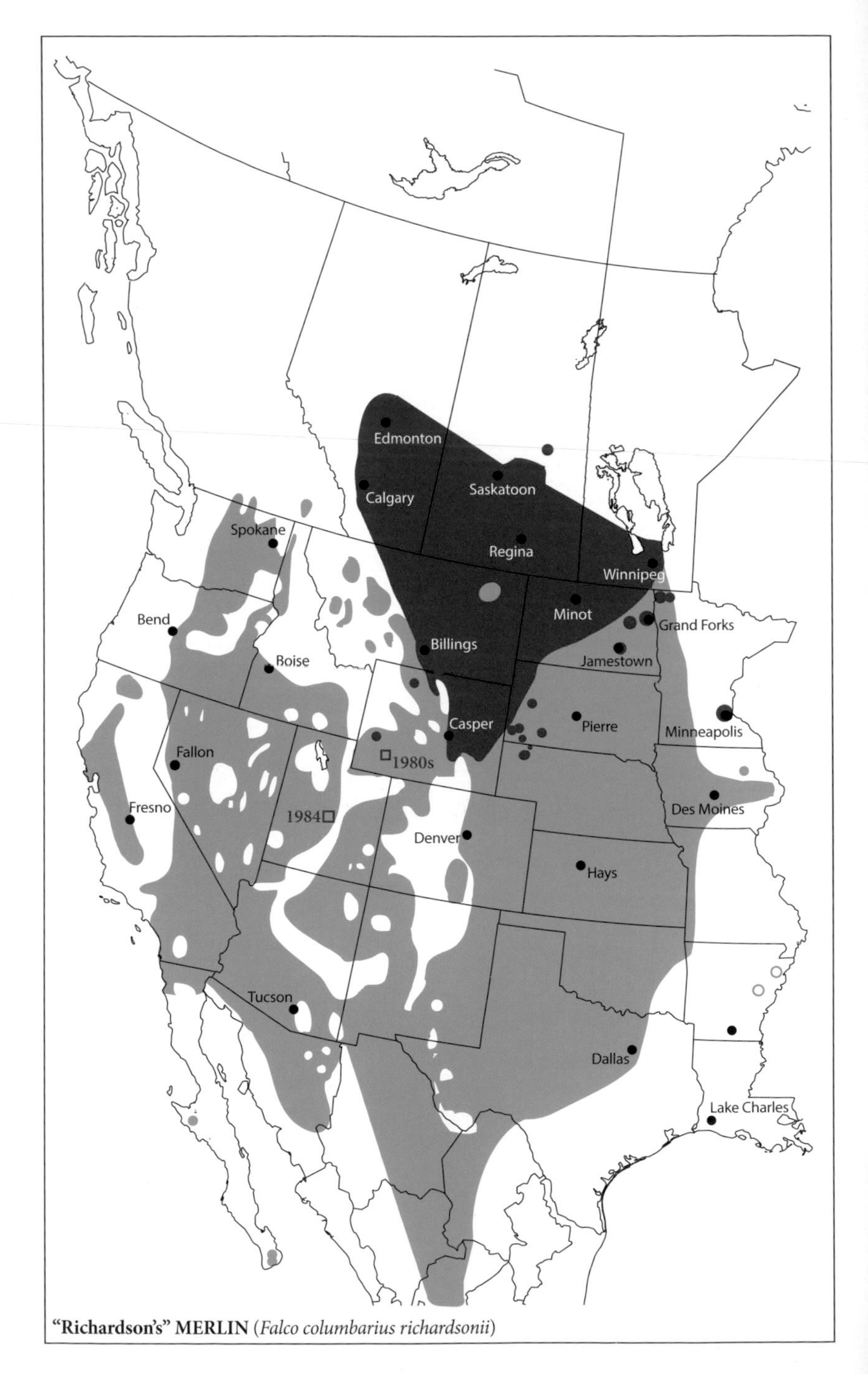

"Richardson's" MERLIN (*Falco columbarius richardsonii*)

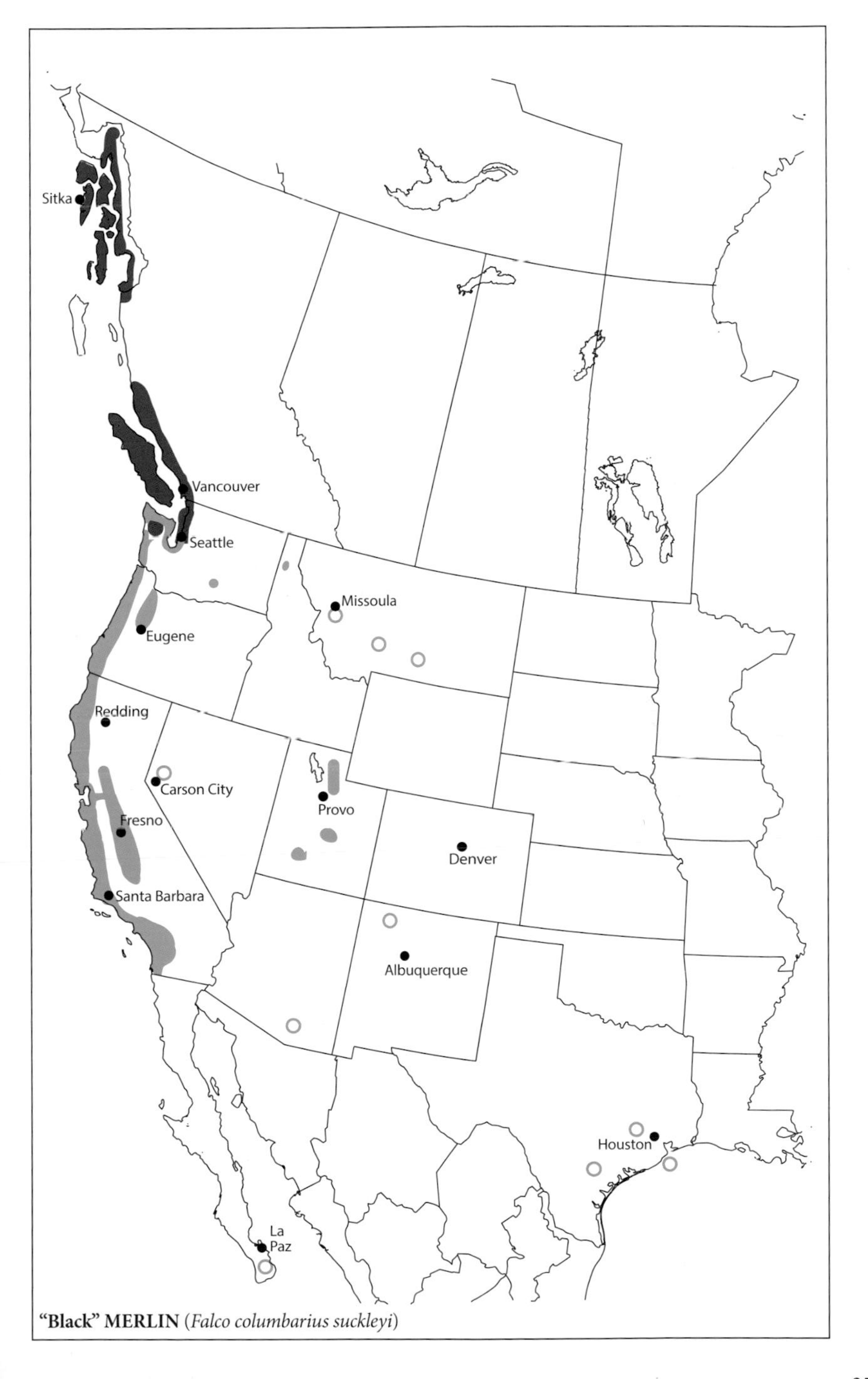

"Black" MERLIN (*Falco columbarius suckleyi*)

Plate 64. GYRFALCON (*Falco rusticolus*) **Juvenile, Perching**

Ages: Juvenile. Nearly all Gyrfalcons seen in s. Canada and n. U.S. are females. Falcons have fleshy orbital skin encircling eye. Eyes are dark brown. Fleshy orbital skin and cere are pale blue. Large pale blue area on basal part of bill. When perched, wingtips are much shorter than tail tip. *Uppertail coverts and dorsal tail surface are uniformly marked.* Feet are pale blue or gray. **Subspecies:** Monotypic. **Color morphs:** Absolute cline from light (white) to dark. There are variations within each color morph. Color morphs are divided into 3 main categories: white (3 variations), intermediate (gray), and dark. 2 main intermediate plumages are shown between main color morph divisions: light intermediate morph and dark intermediate morph. *Note:* Intermediate morph through dark morph can exhibit head markings that are paler than depicted on figures. **Size:** L: 19–24" (48–61cm). Somewhat sexually dimorphic; females average larger. Juveniles are longer than older birds because of their longer tails. **Habits:** Fairly tame to tame. Perches on exposed objects and on ground. **Food:** Perch and aerial hunter of birds up to size of geese and mammals to size of hares. **Voice:** Juveniles are silent on winter grounds. *Note*: Figures on this plate are placed in the same location as the respective-plumage variation of adult on Plate 66.

64a. White morph, lightly marked type: *Nearly all-white head is very lightly streaked on nape and sometimes sides of breast. Note:* There is no dark area encircling orbital skin.

64b. White morph, moderately/heavily marked types: *White head is lightly streaked. Hint of dark malar mark may show. Breast is always lightly spotted or streaked.*

64c. Light intermediate morph: White head is heavily streaked. Breast is moderately spotted or streaked.

64d. Intermediate morph: Heavily streaked head has fairly defined dark malar stripe connecting to dark mask under eye and dark rear auriculars; head can be paler and lack dark mask, as on *64j*. Breast is moderately streaked. *Note:* Head pattern can also adorn dark intermediate morph; rare on dark morph.

64e. Dark intermediate morph: Dark brown head, with pale forehead, short pale supercilium, and light patch on hindneck. White throat is streaked. Breast is heavily streaked. *Note:* This head pattern can also adorn dark morph.

64f. Dark morph: Mainly dark brown head, with small pale forehead and very small pale areas on supercilium and hindneck. White throat often heavily streaked. Breast is very heavily streaked.

64g. White morph, lightly marked type: *White, with lightly streaked upperparts. Small dark tips on flight feathers. Tail and uppertail coverts are white. Breast and flanks may have very sparse dark markings.*

64h. White morph, moderately marked type: *White, with moderate-size dark markings on upperparts. Tail can be partially banded. Uppertail coverts are streaked. Underparts are very lightly spotted.*

64i. White morph, heavily marked type: *White, with large dark markings on upperparts. Tail can be fully or partially banded, with dark bands less than ½ width of white bands. Uppertail coverts are thickly streaked. Primaries are barred. Underparts are lightly spotted.* White undertail coverts.

64j. Light intermediate morph: Dark upperparts, with feathers distinctly fringed with white. Greater wing coverts are barred. Dark primaries are faintly barred. Large dark marks on uppertail coverts. *Tail has equal-width dark and white bands.* White underparts are moderately spotted. Undertail coverts are thinly streaked.

64k. Intermediate morph: Variant with dark mask below eye that connects with dark malar mark. Dark upperparts, with feathers thinly fringed with white. Inner portion of feathers can have pale spots or be partially barred (shown) or can be solid brownish edged with pale fringes, as on *64l*. Greater wing coverts are thinly barred (shown) or solid (*64l*). Uppertail coverts are dark with white feather edges. *Dorsal surface of tail is always banded, with pale bands thinner than dark bands.* Underparts are moderately streaked.

64l. Dark intermediate morph: Dark upperparts may have partial thin, pale feather edges. Uppertail coverts are dark, with narrow white edges. *Tail banded, with faint, thin, pale bands.* Underparts heavily streaked.

64m. Dark morph: *All-dark upperparts, including uppertail coverts and tail.* Underparts heavily streaked.

a
b
c
d
e
f
g
h
i
j
k
l
m

Plate 65. GYRFALCON (*Falco rusticolus*) — Juvenile, Flying

Ages: Juvenile. Nearly all Gyrfalcons seen in s. Canada and n. U.S. are females. Falcons have fleshy orbital skin encircling eye. Eyes are dark brown. Large, pale blue area on basal part of bill. *Uppertail coverts and dorsal tail surface are uniformly marked.* Feet are pale blue or gray. **Subspecies:** Monotypic. **Color morphs:** Absolute cline from light (white) to dark. There are variations within each color morph. Color morphs are divided into 3 main categories: white (3 variations), intermediate (gray), and dark. 2 main intermediate plumages are shown between main color morph divisions: light intermediate morph and dark intermediate morph. **Size:** L: 19–24" (48–61cm), W: 43–51" (109–130cm). Somewhat sexually dimorphic; females average larger. Juveniles have broader secondaries and longer tails than older birds and appear thicker-winged and longer. **Habits:** Fairly tame to tame. **Food:** Perch and aerial hunter of birds up to size of geese and mammals to size of hares. **Flight:** Moderately slow, stiff wingbeats. Glides in irregular sequences with wings held on flat plane; soars with wings on flat plane. Hovers and kites. **Voice:** Gyrfalcons are mainly silent on winter grounds. *Note*: Except for Fig. *65k*, all figures are placed in the same location and in the same pose as respective-plumage variation of adult on Plate 67.

65a. White morph, lightly marked type: *White with sparse, narrow dark streaks on back, scapulars, and lesser upperwing coverts. Uppertail coverts and tail are unmarked. Wingtips are dark.*

65b. White morph, heavily marked type: *White upperparts are heavily marked with brown. Barred flight feathers; dark-tipped primaries. On ventral side, primaries are unmarked except dark tips* (see *65h*). White uppertail coverts have thin dark streaks. *White tail has thin dark bands.*

65c. Light intermediate morph: Slightly darker than *65b*, with barred uppertail coverts. *Tail has light and dark bands of equal width. Tail is similar color as rest of upperparts.* On ventral side, primaries are lightly barred (see *65i*).

65d. Intermediate morph: Dark brown upperparts, with feathers thinly fringed with white. Flight feathers have thin, pale barring. *Tail has pale bands narrower than dark bands.*

65e. Dark intermediate morph: Dark brown upperparts may be solid or have faint pale spotting or feather edges. May have light edges on uppertail coverts. *Tail is always faintly banded, with narrow pale bands.*

65f. Dark morph: Uniformly all-dark upperparts, including tail. *Outermost primaries have very thin, pale barring on ventral side. Note:* Ventral flight position is not depicted. Similar to adult in *67k*, but underparts are darker and thinly streaked with white (not barred as on adult); wing is nearly identical.

65g. White morph, lightly marked type: *All-white underside except dark-tipped primaries. Faint dark subterminal band may show on secondaries.*

65h. White morph, heavily marked type: Very lightly spotted white underparts. *White underwing has dark-tipped primaries, lightly barred secondaries, and lightly streaked coverts.* White tail is thinly banded.

65i. Light intermediate morph: Lightly spotted white underparts. *Secondaries are more thickly barred than primaries, making inner wing appear somewhat darker. Moderately streaked underwing coverts are often more heavily marked on median coverts than on lesser coverts. Tail has equal-width light and dark bands. Note:* Not readily separable from intermediate morph, which has continuous streaking on ventral plumage.

65j. Intermediate morph: Moderately streaked white underparts, with continuous streaks. *Secondaries are much more thickly barred than primaries, making inner wing appear darker. Moderately to heavily streaked and barred underwing coverts; can make coverts appear darker than flight feathers.* Tail has thin, white bands.

65k. Dark intermediate morph: Heavily streaked underparts. *Secondaries are virtually solid gray, with thin, pale barring on a few outermost feathers. Primaries are faintly mottled or barred. Underwing coverts can be quite dark and appear darker than flight feathers.* Tail has faint, thin, pale bands. *Note:* Dark morph (not depicted) is similar but darker on underparts and has faint barring only on outermost primaries; tail is solid gray on ventral side.

a
b
c
d
e
f
g
h
i
j
k

Plate 66. GYRFALCON (*Falco rusticolus*) Adult, Perching

Ages: Adult (includes immatures). Nearly all Gyrfalcons seen in s. Canada and n. U.S. are females. Falcons have fleshy orbital skin encircling eye. Eyes are dark brown. There is a large, pale blue area on basal part of bill. When perched, wingtips are distinctly shorter than tail tip. *Uppertail coverts and dorsal tail surface are uniformly marked. Immatures.*—Interim plumage stage of 1- and 2-year-olds, which are in primarily adult plumage but retain variable amount of worn, faded juvenile feathering, mainly on upperwing coverts and rump (*see* Clum and Cade 1994). Cere and orbital skin are pale green to pale yellow. Feet are pale yellow or pale green. *Adult.*—Full plumage is attained when 2 or 3 years old. Cere, orbital skin, and feet are yellow, brighter on males. **Subspecies:** Monotypic. **Color morphs:** Absolute cline from light (white) to dark. There are variations within each color morph. Color morphs are divided into 3 main categories: white (3 variations), intermediate (gray), and dark. 2 main intermediate plumages are shown between main color morph divisions: light intermediate morph and dark intermediate morph. **Size:** L: 19–24" (48–61cm). Somewhat sexually dimorphic; females average larger. Adults are shorter than juveniles because of their shorter tails. **Habits:** Fairly tame to tame. Perches on exposed objects and on ground. **Food:** Perch and aerial hunter of birds up to size of geese and mammals to size of hares. **Voice:** Gyrfalcons are mainly silent on winter grounds. Nesting pairs are highly vocal when agitated, emitting harsh *cack-cack-cack. Note*: Figures on this plate are placed in the same location as the repective-plumage variation of juvenile on Plate 64.

66a. White morph, lightly marked type: *Nearly all-white head is very lightly streaked on nape and lower hindneck. Note:* There is no dark border encircling orbital skin, as on all darker types.

66b. White morph, moderately/heavily marked type: *White head is lightly streaked. Dark border encircles orbital skin.*

66c. Light intermediate morph: *White head is heavily streaked. Top and rear borders of auriculars may have dark smudge. Narrow, streaked, dark malar is somewhat distinctly marked.*

66d. Intermediate morph: *Distinct, narrow, dark malar mark may merge with dark top and rear borders of auriculars. Forehead is white. Moderate-size white supercilium and large white patch on nape. Note:* Younger adults often have greenish fleshy areas.

66e. Dark intermediate morph: Dark-hooded appearance, with dark malar mark and dark auriculars; white forehead. Partial white supercilium and small white nape patch. White throat is partially streaked.

66f. Dark morph: All-dark head, including forehead. White throat is streaked.

66g. White morph, lightly marked type: *White, with single, short crossbar on each feather on upperparts, including uppertail coverts and some wing coverts. Small dark tips on primaries. Underparts are pure white and unmarked. White tail is unmarked or very partially banded along feather shafts.*

66h. White morph, moderately marked type: *White, with 1 or 2 long, dark crossbars on each feather on upperparts, including upperwing and uppertail coverts. Primaries are barred and have dark tips. White tail is partially or fully banded.*

66i. White morph, heavily marked type: *White, with multiple long, dark crossbars on each feather on upperparts, including wings. Primaries are barred and have dark tips. Flanks and leg feathers are very lightly marked. White tail has complete, narrow, dark bands.*

66j. Light intermediate morph: *Equal-width white and dark crossbars on feathers of upperparts, including wings.* Lightly spotted underparts; streaked undertail coverts. *Whitish tail has broad, mainly equal-width bars.*

66k. Intermediate morph: Broad gray bars on dark brown/gray upperparts. Moderately spotted breast and belly; partially barred flanks and legs. Undertail coverts are barred. Tail is barred with gray.

66l. Dark intermediate morph: Dark upperparts are thinly barred. Underparts are heavily spotted or streaked with blobs; barred flanks and legs. *Dark tail has moderate-width gray bands.*

66m. Dark morph, immature: Dark upperparts are partially barred with thin, gray marks. *Dark tail is thinly banded with gray.* Underparts are very heavily marked with large blobs and appear streaked; flanks and legs are heavily barred. *Note:* Retains faded brown juvenile wing coverts and pale greenish feet.

a
b
c
d
e
f
g
h
i
j
k
l
m

Plate 67. GYRFALCON (*Falco rusticolus*) — Adult, Flying

Ages: Adult (includes immatures). Nearly all Gyrfalcons seen in s. Canada and n. U.S. are females. Falcons have fleshy orbital skin encircling eye. Eyes are dark brown. Large, pale blue area on basal part of bill. *Uppertail coverts and dorsal tail surface are uniformly marked. Immatures.*—Interim plumage stage of 1- and 2-year-olds, which are in primarily adult plumage but retain variable amount of worn, faded juvenile feathering, mainly on upperwing coverts and rump (*see* Clum and Cade 1994). Cere and orbital skin are pale green to pale yellow. Feet are pale yellow or yellow. This age is not depicted on this plate. *Adult.*—Full plumage is attained when 2 or 3 years old. Cere, orbital skin, and feet are yellow, brighter on males. **Subspecies:** Monotypic. **Color morphs:** Absolute cline from light (white) to dark. There are variations within each color morph. Color morphs are divided into 3 main categories: white (3 variations), intermediate (gray), and dark. 2 main intermediate plumages are shown between main color morph divisions: light intermediate morph and dark intermediate morph. **Size:** L: 19–24" (48–61cm), W: 43–51" (109–130cm). Somewhat sexually dimorphic; females average larger. Adults are shorter than juveniles because of their shorter tails. **Habits:** Fairly tame to tame. Perches on exposed objects and on ground. **Food:** Perch and aerial hunter of birds up to size of geese and mammals to size of hares. **Flight:** Moderately slow, stiff wingbeats. Glides in irregular sequences with wings held on flat plane; soars with wings on flat plane. Hovers and kites. **Voice:** Gyrfalcons are mainly silent on winter grounds. Nesting pairs are highly vocal when agitated, emitting harsh *cack-cack-cack*. *Note*: Except for Fig. *67k*, all figures are placed in the same location and in the same pose as respective-plumage variation of juvenile on Plate 65.

67a. White morph, lightly marked type: *White with short, thin, dark crossbars on back, scapulars, and upperwing coverts. Uppertail coverts and tail are all-white. Small dark tips on outer primaries.*

67b. White morph, heavily marked type: *White, with dense, thin crossbars on all dorsal plumage. White tail has thin, dark bands. Dark primary tips.* On underside, outer primaries are partially barred (as on *67h*).

67c. Light intermediate morph: *Equal-width white and dark crossbars on all dorsal plumage. White or pale gray tail has thick, dark bands; all bands may be equal width. Uppertail coverts and tail are uniformly marked.* On underside, primaries are thinly barred (as on *67i*).

67d. Intermediate morph: *Thin, whitish or pale gray crossbars on uniformly marked upperparts;* primaries and greater primary coverts are typically darker than rest of wing. *Uppertail coverts and tail are uniformly marked.* On underside, primaries are thickly barred, with dark tips (as on *67j*).

67e. Dark intermediate morph: Very thin, partial gray crossbars on most of dark brown upperparts. *Uniformly colored uppertail coverts and tail can be paler grayish than rest of upperparts.*

67f. Dark morph: *Grayish, uniformly marked uppertail coverts and thinly banded tail contrast with dark brown back, scapulars, and wings.* Portions of dorsal plumage may have partial, very thin barring.

67g. White morph, lightly marked type: *White, with dark wingtips and trailing edge of flight feathers.*

67h. White morph, heavily marked type: Very lightly marked white underparts, including underwing coverts. *Barred secondaries contrast with unmarked basal area of primaries.*

67i. Light intermediate morph: Lightly marked white underparts, including underwing coverts. *Broadly barred secondaries contrast with more thinly barred primaries and lightly spotted coverts.*

67j. Intermediate morph: Moderately marked white underparts. *Very broadly barred secondaries, which can be nearly solid gray, contrast with more thinly barred primaries. Note: On dark intermediate morph (not shown), heavily marked underwing coverts are darker than flight feathers.*

67k. Dark morph: Very heavily marked underparts, with barred flanks and legs. *Dark underwing coverts contrast with nearly solid gray flight feathers, which have thin pale barring only on outer primaries.*

a
b
c
d
e
f
g
h
i
j
k

GYRFALCON (*Falco rusticolus*)

HABITAT: *Summer.*—Arctic and alpine tundra from sea level to 5,350' (1630m). Breeding pairs inhabit areas with cliffs or embankments, which serve as nest sites and also protective year-round roosts. These areas may be in open tundra or along lakes, rivers, or seacoasts. A few inhabit taiga habitat with stands of larger White Spruce. Nonbreeding birds may be found in any open, high-latitude area without regard to topographic features. *Winter.*—Majority of males and most females remain in breeding latitudes in the Arctic. Many females and some younger males may winter farther south in s. Canada and n. U.S. in open and semi-open areas of prairies, large fields, harbors, marshes, seashores, lakeshores, and airports. Less frequently found in semi-open rural locales. **STATUS:** Uncommon and sparsely distributed in breeding areas. Very uncommon to rare in winter-only areas but fairly common in parts of Dakotas and nw. Wash. Population is affected by cyclic trends of Rock and especially Willow Ptarmigan (*Lagopus muta* and *L. lagopus*, respectively). Most Alaskan and mainland Canadian birds are gray morphs; white morph is on High Arctic Canadian islands, and darker birds occur on Aleutian Islands. **NESTING:** 1st breeds when 2–4 years old. Nesting pairs may remain together year-round. Breeding depends on prey availability; may not nest in years of low prey. Pair formation may begin in Feb.; eggs are laid Mar.–late May, depending on latitude and elevation. Most nests are located on cliff or embankment formation that has protective overhang. Pairs do not build their own nests. Nest sites may be scrapes on soil or vacant stick nests of Golden Eagle, Rough-legged Hawk, or Common Raven placed on cliffs. A few pairs nest in upper portion of White Spruce in vacant Rough-legged Hawk or Common Raven stick nest. Female lays up to 5 eggs and does much of incubation. **MOVEMENT:** Females, especially younger birds, tend to move south to s. Canada and n. U.S., especially nw. Wash. and s.-cen. S.Dak. They very rarely winter south to blue dashed line on range map. *Spring.*—Adult females that winter south of the Arctic may start heading north in late winter. There are sightings of juveniles in n. U.S. in Mar.–Apr. and in s. Canada in May. *Fall.*—South-bound birds are noted in s. Canada as early as late Sep. and in n. U.S. in late Oct. Movements may continue through midwinter. Very few gyrfalcons are seen at fall hawk watch sites. **COMPARISON:** ***Gyrfalcon white morph.*** **Red-tailed Hawk (partial/full albino).**—Plumage is all-white, including wingtips, or has erratically placed darker (normal-color) feathers, *vs.* white plumage with thin, dark barring (adult) or brown with white-fringed dorsal feathers (juvenile), and always with neat, dark wingtips. ***Gyrfalcon intermediate and dark morphs.*** **Northern Goshawk.** *Juveniles.*—Yellow eye (rarely brown), *vs.* dark brown. Thickly barred ventral flight feathers and 3 or 4 equal-width dark tail bands, *vs.* faint barring on underwing and tail unmarked or with multiple thin bands. *Adults.*—Has black auricular stripe and lacks dark malar mark, *vs.* dark auriculars and dark malar mark. Eye is orange or red, *vs.* dark brown. Barred ventral plumage, *vs.* spotted. **Peregrine Falcon.** *Juveniles.*—Can be very similar, especially darker "Peale's." When perched, wingtips reach near tail tip, *vs.* much shorter than tail tip. Underside of all flight feathers is distinctly barred, *vs.* unmarked and gray on secondaries. *Adults.*—Distinct black cap and malar mark, *vs.* black cap and no malar mark. When perched, wingtips equal tail tip, *vs.* shorter than tail tip. Flight feathers are all-barred on underside, *vs.* lack of barring on gray secondaries.

Figure 1. Denali Nat'l. Pk., Alaska (Jun.) Polychrome Pass area, historic year-round habitat.

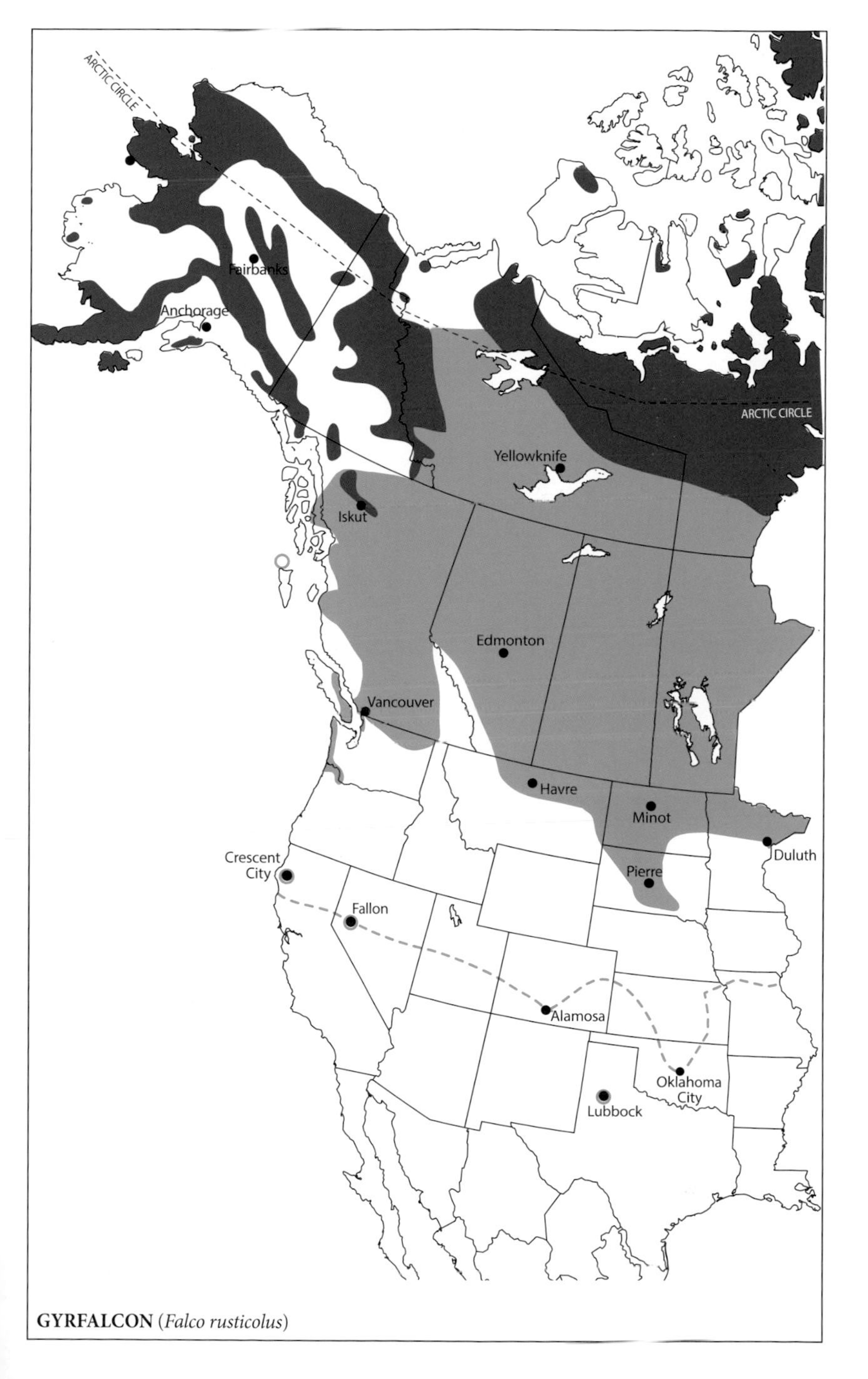

GYRFALCON (*Falco rusticolus*)

Plate 68. PEREGRINE FALCON (*Falco peregrinus*)
"Arctic" (*F. p. tundrius*)

Ages: Juvenile and adult. In flight, wings are long, with pointed tips. Fleshy orbital skin encircles dark brown eye. Dark malar mark is thin to moderately wide. Forehead is pale. *Underside of wings is uniformly patterned. Juvenile.*—Plumage retained for 1st year. Orbital skin is pale blue until late winter–spring, then turns pale yellow. Feet are gray but turn green or yellow by late summer–fall. All variations have pale nape ring. Wings are broader and tail is longer than on adult. When perched, wingtips are a bit shorter than tail tip. *1st prebasic molt stage.*—Continuous full molt in spring–summer; wings and tail molt after much of body is molted. *Adult.*—Attains plumage in winter of 2nd year as a 1-year-old (1st prebasic molt). Some juvenile feathering remains on all body areas during fall. Cere and orbital skin are yellow, brighter on males. When perched, wingtips equal tail tip. Upperparts of males are bluish; those of females are darker blackish or brownish. *Dark tail contrasts with paler bluish upperparts of male, with paler rump of female.* White, lightly marked underparts. Molt starts in summer, is suspended in fall, and is finished during winter. **Subspecies:** 3 native N. American subspecies; 15 others elsewhere. **Color morphs:** None in N. America. **Size:** Male L: 14–16" (36–41cm), W: 37–39" (94–99cm). Female L: 16–18" (41–46cm), W: 40–46" (102–117cm). Sexually dimorphic. **Habits:** Fairly tame. Perches on exposed objects as well as on ground. **Food:** Aerial hunter of birds, from small songbirds and shorebirds to large ducks; some small mammals. Prey is captured by diving or tail chasing. In early part of nesting season in Nunavut, Canada, vital prey species are Snow Bunting (*Plectrophenax nivalis*), male Arctic Ground Squirrrel (*Urocitellus parryii*), and lemmings (Franke 2013). **Flight:** Soars and glides with wings held flat. **Voice:** Silent, except at nest site, where it emits repetitive, harsh *cack-cack-cack*; other calls when courting. *Note:* This subspecies was heavily used in reintroduction programs in e. U.S.

68a. Juvenile, pale-headed "blonde" type: *Pale crown with thin dark eye line. Thin, dark malar stripe is often split at gape.* Full, pale supercilium. Rear auriculars often have rufous tinge.

68b. Juvenile, moderately dark-headed type: *Pale crown with dark eye line that includes upper auriculars. Moderately wide, dark malar stripe may be partially or fully split at gape.* Pale supercilium is partial.

68c. Juvenile, dark-headed type: Moderately dark crown with moderately wide, dark eye line. *Moderately wide, dark malar stripe is usually not split.* Small, pale above-eye spot and pale nape ring.

68d. Juvenile, pale-headed type: *Head as on 68a but here shows more rufous on auriculars.* Underparts are tawny and very thinly streaked. *Dark brown upperparts are tawny-scalloped.*

68e. Juvenile, dark-headed type: *Head as on 68c.* Underparts are tawny and thinly streaked. *Dark brown upperparts are tawny-scalloped.*

68f. Juvenile, all types: *Dark brown upperparts are tawny-scalloped.* Thin, pale bands in tail.

68g. Juvenile, all types: *Uniformly dark gray flight feathers spotted/barred pale tawny; sometimes pale pinkish or whitish.*

68h. Adult male: *Moderate-width black malar stripe. Crown and upper auriculars are bluish gray. White forehead patch is large.* Fleshy areas are bright yellow.

68i. Adult female: *Moderate-width black malar stripe. Crown and upper auriculars are brownish gray. White forehead patch is large.* Fleshy areas are pale yellow. *Note:* May have white mottling on hindneck.

68j. Adult female, West Hudson Bay type: *White crown and supercilium. Dark malar stripe is thin. Thin, dark eye line. Note:* Pale variant found on only *some* adult females (Rankin Inlet, Nunavut).

68k. Adult male: *Upperparts barred bluish gray. White breast; pale, tawny-pink belly; grayish flanks and legs. Belly markings sparse; very thinly barred flanks and legs. Tail is dark.*

68l. Adult female: *Brownish-gray, barred upperparts. White breast; pale, tawny-pink belly; grayish flanks and legs. Belly markings moderate; thinly barred flanks and legs. Tail is dark.*

68m. Female, 1st prebasic molt (Oct.): Retains some patches of brown juvenile feathers on wings and as remnant streaks on belly and flanks; some have pale areas on crown.

68n. Adult female, West Hudson Bay type: White-striped head. White underparts very lightly marked. *Note:* Rankin Inlet, Nunavut.

68o. Adult male: *Dark tail contrasts with paler upperparts. In female, tail contrasts mainly with paler rump region.*

68p. Adult female: *Uniformly patterned underwing.* White breast; pinkish belly and gray flanks are moderately spotted and barred. *Tail is dark.*

68q. Adult female, West Hudson Bay type: *Uniformly patterned underwing is paler than on 68p.* White underparts are very lightly marked and lack pinkish tinge on belly of typically colored females. *Tail is darker and contrasts with body.*

a
c
d
e
b
f
g
h
i
j
m
l
n
o
k
p
q

Plate 69. PEREGRINE FALCON (*Falco peregrinus*) "Anatum" (*F. p. anatum*)

Ages: Juvenile and adult. In flight, wings are long, with pointed tips. Fleshy orbital skin encircles dark brown eye. Dark malar mark is moderately wide to wide. *Underside of wings is uniformly patterned. Juvenile.*—Pale supercilium is short or lacking. Orbital skin is pale blue until late winter–spring, then turns pale yellow. Feet are gray but turn green or yellow by late summer or fall. Pale variations have pale nape ring. Wings are broader and tail is longer than on adult. When perched, wingtips are a bit shorter than tail tip. *1st prebasic molt stage.*—Continuous full molt in spring–summer; wings and tail molt after much of body is molted. Body-only molt *may* start earlier in winter in partial 1st-year molt (P. Pyle pers. comm.; pers. obs.). *Adult.*—Attains plumage as 1-year-old in fall of 2nd year. Forehead is dark or has small pale patch. Cere and orbital skin are yellow, brighter on males. Wide, black malar often extends onto auriculars and can form "helmet" on either sex. When perched, wingtips equal tail tip. Upperparts of males are bluish; those of females are blackish or brownish, with bluish uppertail coverts and rump. *Dark tail contrasts with paler bluish upperparts.* White/rufous underparts are moderately marked. Some females have narrow malar, similar to "Arctic." **Subspecies:** 3 native N. American subspecies; 15 others elsewhere. **Color morphs:** None in N. America. **Size:** Male L: 14–16" (36–41cm), W: 37–39" (94–99cm). Female L: 16–18" (41–46cm), W: 40–46" (102–117cm). Sexually dimorphic. **Habits:** Fairly tame. Perches on exposed objects as well as on ground. **Food:** Aerial hunter of birds, from small songbirds and shorebirds to large ducks. Prey is captured by diving or tail chasing. **Flight:** Soars and glides with wings held flat. **Voice:** Silent, except at nest site, where it emits repetitive, harsh *cack-cack-cack*; other calls when courting. *Note:* Stock from w. U.S. sparingly released in e. U.S. reintroduction programs; w. Canadian stock *only* was used e. Canada reintroduction programs. (See Plate 71.)

69a. Juvenile, pale-headed type: *Wide, black malar stripe and wide, dark eye line.* Crown is mottled with tawny; tawny above-eye patch and nape ring. *Note:* Found throughout breeding range.

69b. Juvenile, dark-headed type: *Wide malar stripe and crown form cap.* Hindneck stripe is pale.

69c. Juvenile, pale-headed type: *Head as on 69a.* Underparts can be moderately streaked (shown) or heavily streaked. In fresh plumage, upperpart feathers are fairly distinctly edged with tawny-rufous.

69d. Juvenile, dark-headed type: *Head as on 69b.* Underparts heavily streaked. Upperparts are dark with little or no pale feather edging.

69e. Juvenile, pale-headed type: Dark brown, with thin tawny-rufous feather edges. Tail is mainly dark.

69f. Juvenile, all types: *Uniformly dark gray flight feathers spotted/barred pinkish tawny; sometimes whitish.* Belly is heavily streaked.

69g. Adult male: *Head appears hooded, with very wide, black malar stripe covering all of auriculars. Note:* Some females are similar.

69h. Adult female, moderate-width malar: Malar has columnar shape; lower auriculars often have spotting.

69i. Adult female, wide malar: *Malar stripe is wide, with hint of columnar shape; auriculars are all-black. Note:* Uncommon type. Found throughout range.

69j. Adult male: *Black-hooded head. Bluish, barred upperparts. Tawny breast and belly; grayish flanks and leg feathers.* Moderate markings on underparts; breast is rarely marked on males. *Tail is dark.*

69k. Adult female: *Head is similar to 69h, except here has dark forehead. Black upperparts have little if any gray barring. Tawny-rufous underparts are heavily marked. Breast is heavily spotted. Flanks and legs are tinged grayish. Tail is dark.*

69l. Adult female, whitish type: *Head pattern is similar to that on 69h.* Underparts are white, except tawny-pink wash on belly. Black dorsal plumage is barred and more bluish than on most females. *Note:* Found throughout range, especially w. Canada.

69m. Adult female, narrow malar: Narrow black malar mark occurs on only some females (width as on "Arctic"). *Black upperparts are barred.* Rufous ventral plumage; breast is unmarked. *Note:* Nesting bird, Mesa Co., Colo. (Jul.).

69n. Female, 1st prebasic molt (Jul.): Mainly adult body contrasts with faded brown juvenile flight feathers and coverts; tail is juvenile but with 2 new adult deck feathers. *Pale uppertail coverts contrast with dark tail. Note:* Wing molt of incoming (blackish) feathers can be seen on mid-primaries and mid-secondaries.

69o. Adult female: *Uniformly marked underwing. Flight feathers barred tawny-pink. Tail is dark.*

69p. Adult female: *Pale bluish rump and uppertail coverts contrast with dark tail and blackish upperparts.*

a
b
c
d
e
f
g
h
i
j
k
l
m
n
o
p

Plate 70. PEREGRINE FALCON (*Falco peregrinus*) "Peale's" (*F. p. pealei*)

Ages: Juvenile and adult. In flight, wings are long, with pointed tips. Fleshy orbital skin encircles dark brown eye. Dark malar mark is narrow. Forehead is pale. *Underside of wings is uniformly patterned.* There is much plumage variation (data assisted by C. M. White): birds from Haida Gwaii (Queen Charlotte Islands), B.C., to Gulf of Alaska average lightly to moderately marked, but heavily marked on some juveniles; those from outer Aleutian Islands, Alaska are consistently the heaviest marked. *Note:* Captive stock of this subspecies was used extensively in e. U.S. reintroduction efforts. *Juvenile.*—Plumage is held for 1st year. Orbital skin is pale blue until late winter–spring, then turns pale yellow. Feet are gray but turn green or yellow by late summer–fall. All variations have pale nape ring. Wings are broader and tail is longer than in adult. When perched, wingtips are a bit shorter than tail tip. *1st prebasic molt stage.*—Molt starts in spring and extends into fall in this subspecies. Much of body is molted into adult plumage before wings and tail molt, creating contrast of adult body and juvenile wings and tail. *Adult.*—Plumage attained during fall–winter of 2nd year. Cere and orbital skin are yellow, brighter on males. When perched, wingtips equal tail tip. Upperparts of males are bluish; those of females are darker blackish or brownish, with bluish uppertail coverts and rump. *Dark tail contrasts with paler bluish upperparts.* Lightly to heavily marked white underparts may have a yellowish tinge and not the rufous-pink wash found on belly of other subspecies. Flank and leg feathers are often more grayish than those of breast and belly. Molts in summer through fall. **Subspecies**: 3 native N. American subspecies; 15 others elsewhere. **Color morphs:** None in N. America. **Size:** Male L: 16.5" (42cm), W: 36" (92cm). Female L: 19" (48cm), W: 44" (111cm). Largest subspecies. **Habits:** Fairly tame. Perches on exposed objects as well as on ground. **Food:** Aerial hunter of birds, from small alcids and shorebirds to large ducks. This subspecies captures prey by tail chasing. **Flight:** Soars and glides with wings held flat. **Voice:** Silent, except at nest site, where it emits repetitive, harsh *cack-cack-cack*; other calls when courting.

70a. Juvenile, lightly marked type (recently fledged): *Moderate-width black malar stripe and black eye line.* Partial, tawny supercilium wraps around nape and connects to pale hindneck stripe. Crown has pale mottling. Underparts are thinly streaked. *Note:* From Haida Gwaii, B.C.

70b. Juvenile, moderately marked type: *Wide black malar stripe and black eye line.* Lower auriculars are streaked. Partial, white supercilium wraps around nape but is isolated from white hindneck stripe.

70c. Juvenile, heavily marked type (spring): *Wide, black malar stripe. Head is dark, with small white patches. Auriculars are heavily streaked.* Fleshy areas are pale yellow. Underparts are heavily streaked. *Note*: Found in all regions of range.

70d. Juvenile, heavily marked type: *Uniformly dark gray flight feathers spotted/barred whitish or pinkish-tawny.* Underparts heavily streaked.

70e. Juvenile, lightly marked type: *Head as on 70a but faded.* Underparts thinly streaked. Thinly rufous-edged feather tips on upperparts (wear off by late fall). *Note:* Mainly Haida Gwaii, B.C.

70f. Juvenile, moderately marked type (recently fledged): *Head has traits of 70a and 70b.* Shown in rich tawny fresh plumage. Has thin, rufous-edged feather tips, which wear off by fall. Underparts are heavily streaked.

70g. Juvenile, heavily marked type: *Head as 70c but lacks pale spot over eyes.* Underparts are very heavily marked; dark upperparts. Dorsal tail lacks pale banding. *Note:* From Haida Gwaii, B.C.

70h. Adult female, lightly marked type: *Moderate-width black malar stripe. White forehead patch is small. Lower auriculars and breast are white and unmarked. Note*: From Haida Gwaii, B.C.

70i. Adult female, moderately marked type: *Similar to 70h, except spotted lower auriculars and breast. Note:* From Haida Gwaii, B.C.

70j. Adult female, heavily marked type: *Very heavily spotted auriculars and breast. Note:* From Aleutian Islands, Alaska.

70k. Adult male, lightly marked type: *Head as 70h, but malar stripe is narrower.* Flanks and belly are heavily barred. *Note:* Similar to adult "Arctic" (*68k*), except belly is whitish and more heavily barred. *Note:* From Haida Gwaii, B.C.

70l. Adult male, moderately marked type: *Head as 70i. Moderately spotted white breast; heavily barred white flanks and belly. Tail is dark. Note:* From Haida Gwaii, B.C.

70m. Adult male, heavily marked type: *Heavily spotted white breast; barred flanks and belly. Note:* From outer Aleutian Islands, Alaska.

70n. Adult female, heavily marked type: *Black upperparts; heavily streaked and barred underparts. Note:* From outer Aleutian Islands Alaska.

70o. Adult female, moderately marked type: *Uniformly colored underwing. Flanks and belly are barred. Note:* From Haida Gwaii, B.C.

a
b
c
d
e
f
g
h
i
j
k
l
m
n
o

Plate 71. PEREGRINE FALCON (*Falco peregrinus*) Non-Native Subspecies

Eastern native "Anatum" Peregrine (*F. p. anatum*) was extirpated east of the Great Plains by DDT in the 1960s. To restore Peregrine Falcons back into this region, U.S groups released birds of 4 captive-bred non-native subspecies, along with much larger numbers of the 3 N. American subspecies, "Peale's" (*F. p. pealei*), "Arctic" (*F. p. tundrius*), and w. "Anatum." This massive captive-release program, conducted by nonprofit groups and state agencies, began in the mid-1970s and ended in the 1990s to the early 2000s. Canada had its own captive breeding program for e. provinces during the same period but released only native w. "Anatum." Birds of these origins are found in e. part of West region (as well as throughout East). **Ages:** All plumage data on Plates 68–70 also pertain to these foreign subspecies. The juvenile figures on this plate reflect either sex; adult figures depict only females, which, as on our native subspecies, are more brownish on dorsal areas than males. **Subspecies:** 3 native N. American subspecies; 15 others elsewhere. *F. p. peregrinus* of Europe (Scotland), *F. p. brookei* of the Mediterranean (Spain), *F. p. macropus* of Australia, and *F. p. cassini* of s. S. America (Argentina) were part of the e. U.S. restoration program. Australian and S. American subspecies were the least commonly used, mainly in the Midwest, and most failed to survive early releases. Spanish *brookei* has done well and flourishes, especially in the eastern part of the West along Mississippi River. *Note:* With 7 subspecies interbreeding, subspecies designation is not logical for peregrines breeding in the cen. U.S. and s.-cen. Canada (some U.S. stock emigrated to s.-cen. Canada), where it is virtually impossible to label most individuals to origin. Some Peregrine Falcon mixes appear similar to the former eastern *F. p. anatum* (C. M. White pers. comm.). It may take hundreds of years for a recognizable subspecies to evolve in the East. The term "Eastern" Peregrine is coined herein to address the population created by these interbreeding stock. **Color morphs:** None in N. America. **Size:** Male L: 13–16" (33–41cm). Female L: 15–18" (38–46cm). Wingspans are smaller or equal to those of N. American subspecies. Sexually dimorphic. Scottish and S. American subspecies are similar to N. American "Arctic" and w. "Anatum," but Spanish and Australian birds are much smaller. **Habits, Food, Flight, and Voice:** Information is identical to that on Plates 68–70.

71a. Juvenile *peregrinus*: Moderate-width black malar stripe. Head is similar to darker juvenile "Arctic" (*68c*).

71b. Juvenile *brookei*: Similar to *71a*, but head often darker; rufous on pale nape areas.

71c. Juvenile *macropus*: Black head and very wide black malar create hooded appearance.

71d. Juvenile *cassini*: Inverted triangle-shaped black malar mark. Head has small tawny nape spots.

71e. Juvenile *peregrinus*: Moderate-width to wide ventral ***streaking***. In fresh plumage, feathers of upperparts are distinctly pale-edged. Uppertail is distinctly banded (more banded than on most "Arctic" juveniles).

71f. Juvenile *brookei*: Similar to *71e* but smaller. Pale nape and hindneck areas are often rufous.

71g. Juvenile *macropus*: Dark head. Rufous underparts are thinly streaked. Upperparts are dark.

71h. Juvenile *cassini*: Moderately thin streaking on underparts. Upperparts are solid dark brown.

71i. Adult *peregrinus* (fresh plumage): Wide black malar stripe; often grayish and spotted on auriculars, and can be somewhat hooded. Breast is typically lightly spotted but unmarked on some males.

71j. Adult *brookei*: Head is often identical to that of *peregrinus*. Rufous breast is unmarked or spotted.

71k. Adult *macropus*: Black head and very wide black malar create hooded appearance. Rufous breast is always unmarked.

71l. Adult *cassini*: Black head with inverted triangle-shaped malar mark. Breast is spotted.

71m. Adult female *peregrinus* (fresh plumage): Rufous underparts are lightly or moderately spotted on breast and thickly barred on flanks and belly. *Note:* Similar to "Anatum" (*69k*) in this plumage.

71n. Adult female *peregrinus*: As on *71m*, but underparts fade to mainly white by winter.

71o. Adult female *brookei*: Similar to *71m*, but rufous underparts are more thinly barred. Breast may be partially or fully spotted, or unmarked.

71p. Adult female *macropus*: Thin, gray barring on black upperparts. Rufous underparts are unmarked on breast and thinly barred on flanks and belly.

71q. Adult female *cassini*: Rufous or tawny edges on many dorsal feathers. Rufous underparts are spotted on breast and heavily barred on flanks and belly.

a
b
c
d
e
f
g
h
i
j
k
l
m
n
o
p
q

PEREGRINE FALCON (*Falco peregrinus*)

"Arctic" (*F. p. tundrius*), "Anatum" (*F. p. anatum*), "Peale's" (*F. p. pealei*), "Eastern" (no subspecies designation)

HABITAT: "Arctic." *Summer.*—Canadian and Alaskan high-latitude, low-elevation mainland and island tundra regions with lakes, rivers, or seacoasts with low or high cliffs or embankments. *Winter/migration.*—Mainly found along the Pacific coast, without regard to cliff formations, from Wash. and southward. Majority winters in riparian and coastal areas in S. America. During spring migration, utilizes Tex. coastal areas, and interior lakes and marshes of Great Plains. **"Anatum."** *Summer.*—Low- or high-elevation areas with tall cliffs that are adjacent to lakes, marshes, and rivers. Also found along coastal mainland and islands with cliffs from Wash. south to Baja California, Mexico. *Winter/migration.*—Occupies similar areas as nesting habitat but without regard to cliff formations. Spring migrants share Tex. coastal and interior plains areas with "Arctic" birds. **"Peale's."** *Summer.*—Moist, sea-level, coastal mainland and islands with low to tall cliff formations or embankments. All of Canadian and s. Alaskan range is in coastal rain-forest biome; Aleutian Island region is in damp, barren coastal cliff areas. *Winter/migration.*—Resident in much of coastal, cliff-type habitat noted above. Many move southward along open beach locales that lack cliff formations. **"Eastern."** *Summer.*—Middle and upper portions of Mississippi River that have cliff formations, bridges, and tall smokestacks, such as those at power companies. Also, found in interior urban centers with tall buildings. *Winter/migration.*—May remain in above-noted habitat or access marshes, lakes, rivers, and Gulf coast areas, without need for tall structures, including such areas south of U.S. **STATUS: "Arctic."** Population was greatly reduced by organochlorine (DDT) use in 1940s–70s. Numbers are now stable and increasing on their own accord since DDT was banned in 1972. Rankin Inlet region of Nunavut, Canada, has highest breeding density of this subspecies (G. Court pers. comm.). **"Anatum."** This subspecies was extirpated, by DDT, west of Rocky Mts. in U.S. and s. Canada by 1967 (by 1962 in e. U.S.). U.S. and Canadian groups released w. "Anatum" raised in captive-breeding facilities or taken from healthy populations of w. Canada and reestablished the subspecies in areas that suffered heavy losses. Releases occurred from mid-1970s until early 1990s. This subspecies is now healthy and increasing. "Anatum" occurs naturally from w. Tex. north throughout Canada and Alaska; most of range depicted in Sask. and Man. is from reintroduction programs. **"Peale's."** This subspecies was not affected by organochlorine use, thanks to its diet of marine birdlife. Captive stock of "Peale's" was used extensively to augment decimated Peregrine population in e. U.S. (but not e. Canada). **"Eastern."** Conglomerate of multiple subspecies raised and released mainly from falconry stock through private, state, and federal programs in order to reestablish Peregrine Falcon back into e. U.S. Large-scale releases of a few thousand Peregrines occurred from mid-1970s to mid-1990s in parts of West. U.S. programs released large numbers of captive-raised "Peale's" and "Arctic" subspecies. A fairly large number of birds of foreign subspecies were also used to bolster populations. European (from Scotland) *F. p. peregrinus* and Mediterranean (from Spain) *F. p. brookei* adapted quite well. Small numbers of captive-bred Australian *F. p. macropus* and South American *F. p. cassini* were also released, but at least in Midwest and Mississippi River region they did not survive their initial introduction and were not used thereafter. (Midwest programs tracked their released birds.) All captive-released birds have been interbreeding for several decades, and have emigrated to and from s. Canada and into e. parts of West. Some birds are virtually identical to the original dark and richly colored e. "Anatum" subspecies. Other mixes appear more whitish underneath and similar to heavily or lightly marked "Peale's" or similar to very lightly marked "Arctic." *Note:* "Eastern" breeding areas depicted on map are east of Mont. south to w. Tex.; major population inhabits areas along Mississippi River and larger cities of Midwest. Some s. Canada (Man.) breeders are emigrants from U.S. **NESTING:** No nest is built. Nest sites are shallow scrape in soil, whether natural or, in the case of many released birds, artificially supplied soil. All Peregrines nest on tall natural cliffs or other rock structures or, especially with released stock, on cliff-like artificial urban sites, such as ledges and gravel areas on protected roofs of tall buildings, in covered or open wooden boxes placed on buildings, on bridge girders and concrete supports, and on ledges of tall smokestacks. Some 1-year-old females nest with older males, but most females do not breed until 2 years old, males until 3–5 years old. **"Arctic."** Begins nesting activities in late May–Jun. (much later than other subspecies, because of latitude). May nest on easy-to-access, low soil or rock embankments; often uses old nests of Common Raven or Rough-legged Hawk located on cliffs or embankments. Youngsters fledge mid-Aug.–late Sep. **"Anatum."** Begins in Mar.–Apr., but as late as May along Mackenzie River, N.W.T. Youngsters fledge Jun.–Aug. **"Peale's."** Also begins Mar.–Apr.; fledges Jun.–Jul. **"Eastern."** Begins Feb.–Apr.; fledges May–Jul.

MOVEMENTS: "Arctic." This is a highly migratory subspecies and longest-distance migrant bird of prey. Round-trip from Arctic breeding grounds to S. American wintering grounds—as far south as cen. Chile and se. Brazil—can exceed 15,000 mi. (24,000km). Migration data herein are based on Court (2008), Franke (2013), Raptor Research Center (2017), Seegar et al. (2015), and St. John (2014). *Spring.*—Adults move prior to returning 1-year-olds. Bulk of population passes through and stages along ancestral path on Gulf coast of Tex. (especially Padre Island) in early–mid-Apr. From coastal Tex., migrants radiate northward to Arctic regions spanning Alaska to Greenland. Island Girl, who nested annually on s. Baffin Island, Nunavut (e. Canada), was telemetry-tracked for 5 years (2009–13). She wintered in same location in cen. coastal Chile, and used this historic coastal Tex. pathway each year. From there she headed due north to w. side of Hudson Bay (Man. or Nunavut), then veered northeast or east across or above Hudson Bay to her nesting grounds further east. Adults peak on Great Plains in late Apr.–early May. Birds arrive on nesting grounds in nw. Hudson Bay in 3rd week of May, in High Arctic regions by early Jun. Returning juveniles move mainly late Apr.–early Jun. *Fall.*—Small numbers move along Pacific coast, where Yakutat, Alaska, notes 1st birds in mid–late Aug. (C. Susie pers. comm.); small numbers also seen Sep.–Oct. along Wash. coast (D. Varland pers. comm.). However, both ages from most of w. and cen. Arctic migrate diagonally south and east into U.S., using 3 main routes to S. America: (1) Great Plains to coastal Tex., land-based route through Cen. America, (2) Midwest to mid-Gulf region and trans-Gulf route, and (3) Atlantic coast and/or Fla. and trans-Caribbean route. Falcons of w. Hudson Bay and farther west may head south to coastal Tex. and mid-Gulf route; some head directly to Fla., while others aim farther north to mid-Atlantic before angling south to Fla. and across Caribbean. "Island Girl" used mid-Gulf route 2 of her 5 southbound trips and angled farther west 1 year to take coastal Tex. route (other 2 years used Atlantic coast route). Concentrations occur from Cape Cod, Mass., and southward, especially at Cape May Co., N.J.; Assateague Island of Md. and Va.; and at Little Crawl Key, Fla. Adults may leave nesting grounds in mid–late Aug. Juveniles may migrate soon after fledging. None of the se. Tex. hawk watches tally large numbers of Peregrines. On Fla. Keys, peak is in mid-Oct. (Highest tallies in North America are at Curry Hammock St. Pk., Little Crawl Key; 2015 had record fall total of 4,559 birds, 1,506 on Oct. 10). **"Anatum."** Birds can be resident in southerly and low-elevation middle latitudes but are short- or long-distance migrants from high-elevation and northerly latitudes. Canadian juveniles have been tracked by telemetry to wintering grounds in s. Cen. America and n. S. America. *Spring.*—Large concentrations of migrants are seen along with "Arctic" in coastal Tex. and on Great Plains within same time span. *Fall.*—Birds move in more of a due-southerly direction from breeding/natal origins than "Arctic," including across interior of Canada and U.S. Peak dates are similar to those of "Arctic." **"Peale's."** Mainly sedentary, especially on Aleutian Islands. Based on banding data, movement of especially B.C. birds is noted in spring and fall along beaches of Pacific coast of Wash. (D. Varland pers. comm.). *Spring.*—In coastal Wash., movement occurs late Feb.–Apr., with peak in Mar. Adults move earlier than juveniles. *Fall.*—Movement along sw. Wash. coast in early Sep., with larger numbers seen in late Sep.–Oct., which aligns with movement of other subspecies; smaller numbers move in Nov. and later. **"Eastern."** These birds are sedentary or moderate- to long-distance migrants, with spring and fall dates that correspond to those of above subspecies. **COMPARISON: Gyrfalcon (intermediate morph, both ages).**—Head features, including malar mark, are diffused, *vs.* distinct markings with sharply defined dark malar mark. When perched, wingtips are much shorter than tail tip, *vs.* equal or nearly equal to tail tip. In ventral flight view, flight feathers are paler than coverts, *vs.* same colored. On juvenile, dorsal feathers have white edges, *vs.* tawny edges. **Gyrfalcon (juvenile dark morph).**—Similar to juvenile Peregrine when perched, but wingtips much shorter than tail tip, *vs.* nearly equal or equal to tail tip. Underside of flight feathers is solid pale gray with paler primaries and darker coverts, *vs.* uniformly dark with barred flight feathers. Dorsal body plumage can be similar solid brown on juveniles. **Mississippi Kite (all ages).**—Similar shape when gliding overhead; kite has very short outermost primary, *vs.* outer primary nearly equal to wingtip. Tail is notched, *vs.* square- or round-tipped. **Northern Harrier (adult female and juvenile).**—Similar shape when gliding overhead, but harrier's wing has longer distance from body to bend at wrist, and wide black bar along trailing edge of secondaries, *vs.* uniformly pale-marked (adult) or dark-marked (juvenile) secondaries. **Merlin (female).**—Similar brown upperparts and streaked underparts pattern as juvenile Peregrine, but smaller bird. When perched, wingtips are distinctly shorter than tail tip, *vs.* equal or nearly equal to tail tip. Underwing pattern is similar to that of juvenile Peregrine. Wingbeats are rapid, with short glide sequences, *vs.* slow, powerful wingbeats, with long glide sequences.

Figure 1. Meliadine River, Nunavut (Jul.)
Tundra breeding habitat of "Arctic" at Rankin Inlet. Photo by Gordon Court

Figure 2. Char River, Nunavut (Jul.)
"Arctic" birds nest on low granite cliffs near rivers at Rankin Inlet. Photo by Gordon Court

Figure 3. Grand Canyon Nat'l. Pk. (South Rim), Coconino Co., Ariz. (Jun.)
Year-round "Anatum" habitat. Birds nest on high cliffs.

Figure 4. Matushka Island, Gulf of Alaska (Jul.)
Year-round "Peale's" habitat. Birds nest on island bluffs and cliffs.

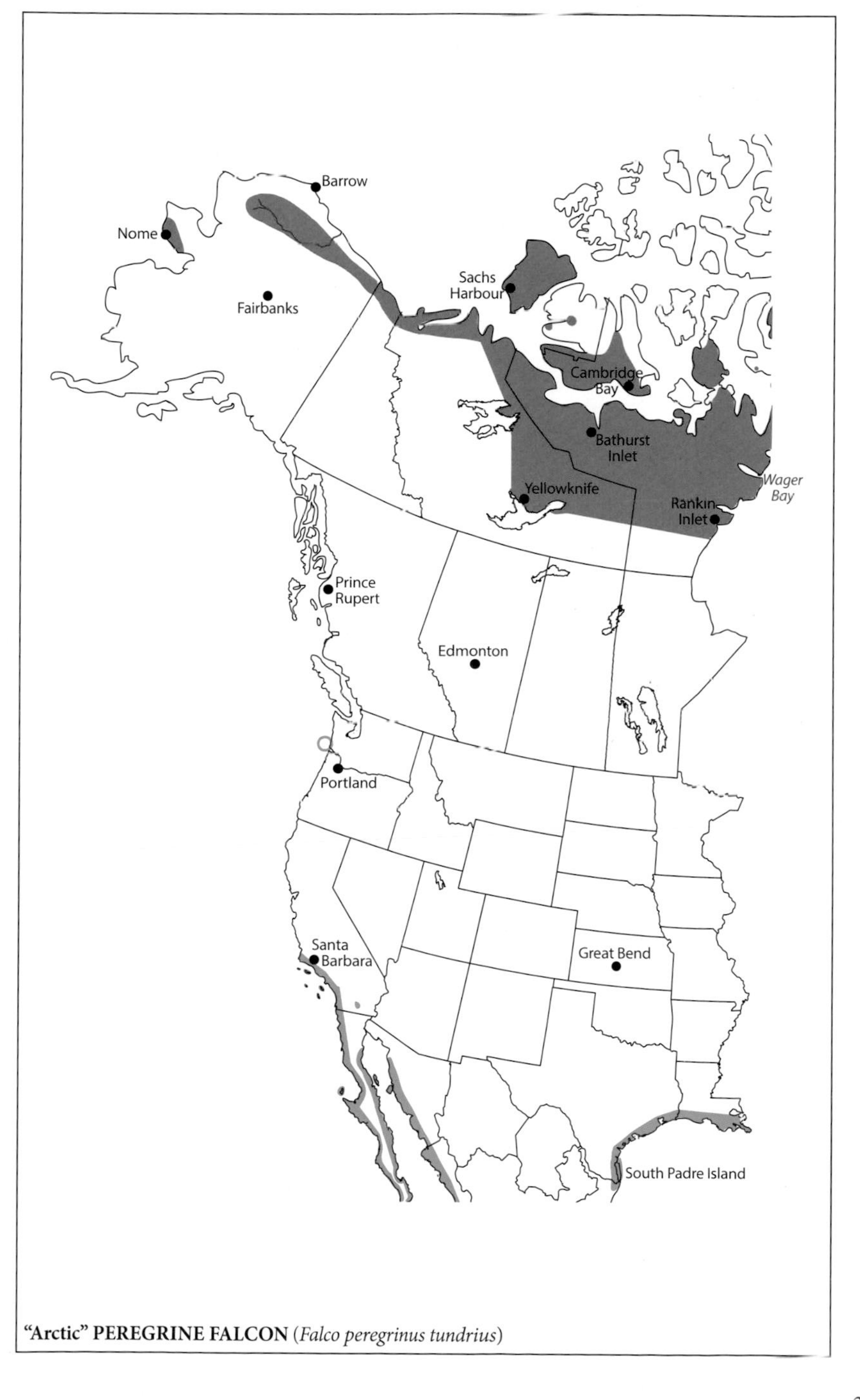

"Arctic" **PEREGRINE FALCON** (*Falco peregrinus tundrius*)

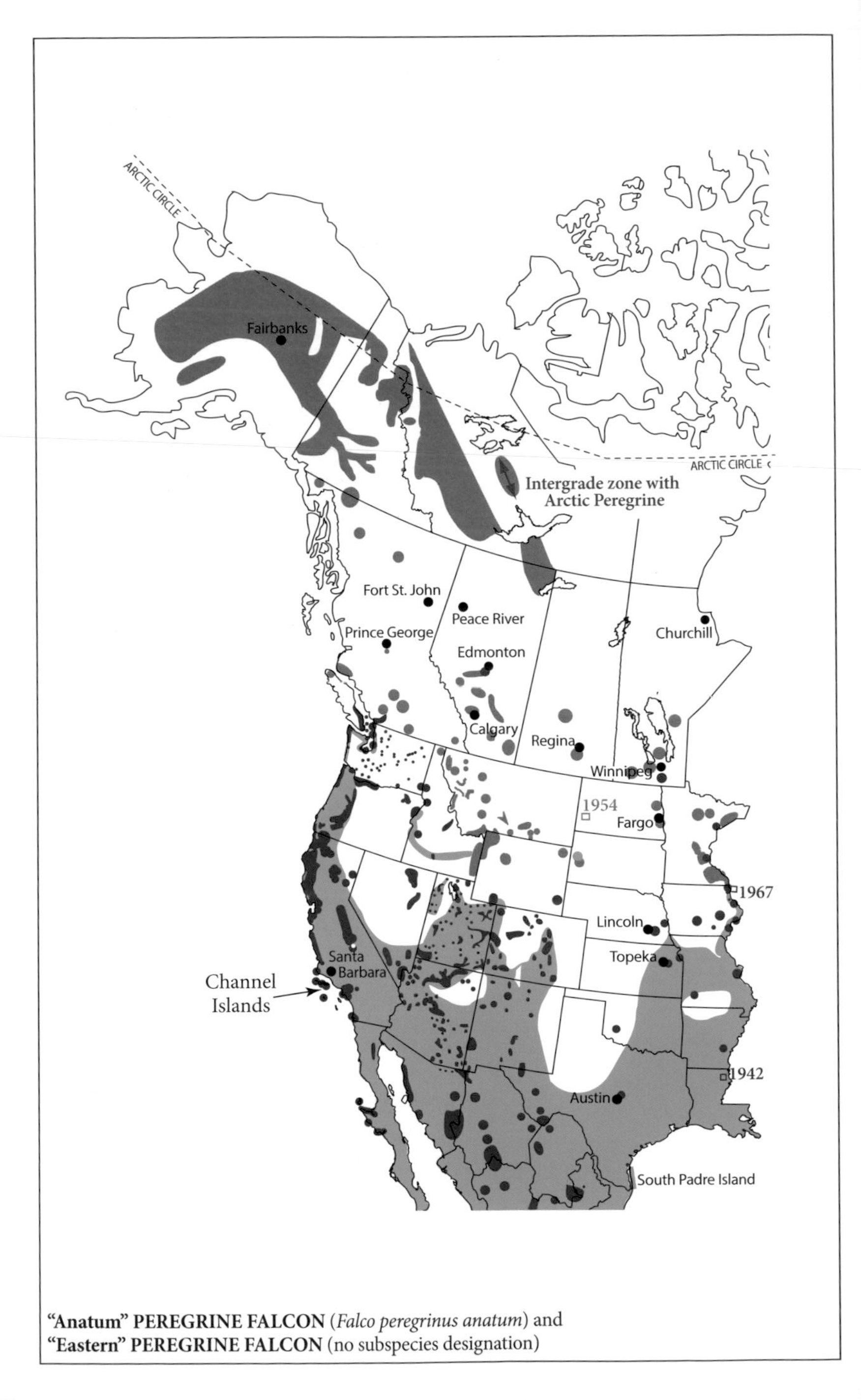

"Anatum" PEREGRINE FALCON (*Falco peregrinus anatum*) and
"Eastern" PEREGRINE FALCON (no subspecies designation)

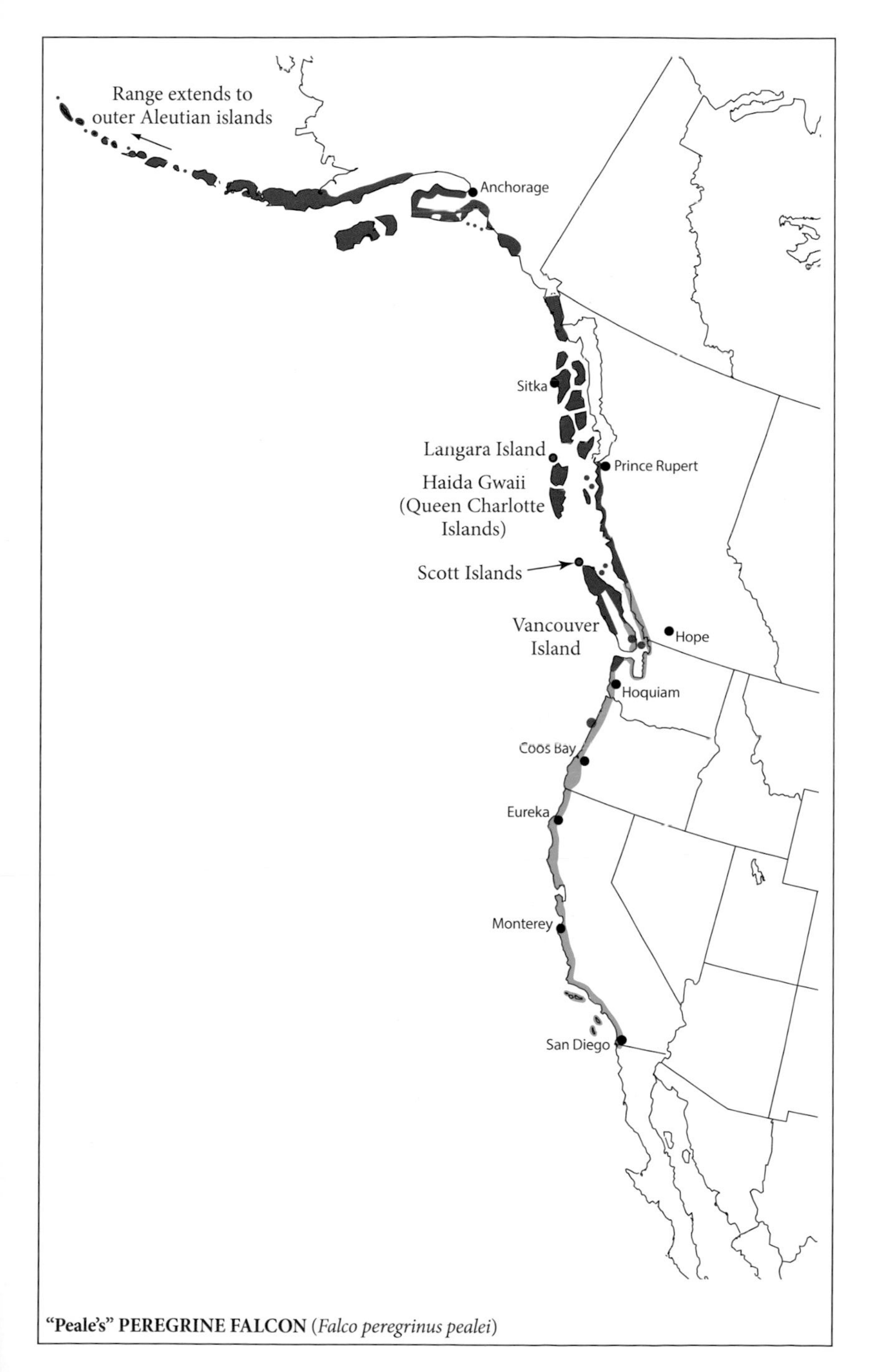

"Peale's" PEREGRINE FALCON (*Falco peregrinus pealei*)

Plate 72. PRAIRIE FALCON (*Falco mexicanus*)

Ages: Juvenile and adult. Juvenile plumage is retained for 1st year. Attains adult plumage as 1-year-old by end of 2nd year. Falcons have fleshy orbital skin encircling eye. Eyes are dark brown. *Narrow white stripe on cheek extends up behind eye and is bordered by brown patch on auriculars.* Dark malar stripe is thin. White forehead; full or partial white supercilium. Females tend to have darker heads than males. When perched, wingtips are shorter or much shorter than tail tip. *On ventral surface of wing, axillaries are solid black, and flight feathers are white, with thin, uniform barring. Juvenile.*—Orbital skin and cere are pale blue until winter, then turn pale yellow. Legs and feet are grayish when fledged but turn yellow by late summer or fall. Underparts are variably streaked; *flanks are nearly solid dark brown.* Feathers on legs are streaked. Plumage fades by fall and winter. Variable patterns on median underwing coverts in both sexes. *Adult.*—Orbital skin, cere, and legs and feet are yellow, often brighter on males. White underparts are spotted. Leg feathers are spotted or partially barred. *Uppertail is always paler than rest of upperparts. Underwing coverts:* Males do not exhibit variance in white spotting on dark median underwing coverts, and coverts are always spotted. In females, coverts are variably marked, typically darker than on males, and can be all-dark; very small number of females have white-spotted coverts, similar to those of males. **Subspecies:** Monotypic. **Color morphs:** None. **Size:** Male L: 14–16" (35–41cm), W: 36–38" (91–96cm). Female L: 16–18" (41–46cm), W: 41–44" (104–112cm). Sexually dimorphic; females average larger. **Habits:** Can be fairly tame during nonbreeding season. Solitary. Perches on open objects, such as outer branches and tops of poles and other human-made objects, and on ground. **Food:** Perch and aerial hunter. Feeds mainly on small to large passerines and small game birds during nonbreeding season. May also capture mammals up to size of rabbits. Eats carrion. **Flight:** Mainly low-altitude, in direct, rapid manner, with shallow, stiff wingbeats, especially when hunting. Soars at times, with wings held on flat plane. **Voice:** Silent away from nesting grounds.

72a. Juvenile (fall): *White behind eye; dark auriculars; thin, dark malar stripe.* Head is often tawny until winter, then wears to more brownish. *Note:* Pale-headed type with full white supercilium.

72b. Juvenile (winter–spring): *White behind eyes; dark auriculars; thin, dark malar stripe.* Head may be brown or gray in all seasons. *Note:* Dark-headed type with partial supercilium.

72c. Juvenile (recently fledged): Tawny underparts, with moderate-width dark streaking and *darker flanks.* Dark brown feathers of upperparts are edged with tawny. Tail is same color as rest of upperparts at this age.

72d. Juvenile, barred type (fall): Underparts fade to white; ventral streaking thin; *dark flanks.* Brown upperparts are barred and similar to adult's. *Tail is paler than rest of upperparts.*

72e. Juvenile (winter–spring): Upperparts fade by winter to medium brown with thin, pale feather edges. *Tail is paler than rest of upperparts.*

72f. Juvenile (recently fledged): Dark brown upperparts. Tail is dark at this age.

72g. Juvenile (fall–spring): Medium brown upperparts. *Tail and often uppertail coverts are paler than rest of upperparts.*

72h. Juvenile (fall–spring): *Black axillaries merge with all-black median underwing coverts. Flight feathers are uniformly barred. Note:* All-black median coverts occur on either sex on juveniles.

72i. Juvenile underwing: *Black axillaries; median coverts lightly speckled; underwing often paler, as on 72p and 72o, on either sex.*

72j. Adult male: *White behind eye; dark auriculars; thin, dark malar stripe.* Head is often grayish. White supercilium is complete. Fleshy areas are bright yellow. *Note:* Some adult females have identical pattern.

72k. Adult female: *White behind eye; dark auriculars; thin, dark malar stripe.* Head is dark brown or grayish. White supercilium is incomplete. *Note:* Some adult males have identical pattern.

72l. Adult male: Head as on *72j* but brownish. Upperparts are barred, often with tawny. Lightly spotted underparts. *Tail is paler than upperparts.* Tail can be marked with thin, dark bands (shown) or be unmarked. *Note:* Some adult females have similar upperparts, but lack dorsal tail banding.

72m. Adult female, dark-backed type: Head as on *72k.* Upperparts are barred only on distal scapulars and greater wing coverts. Heavily spotted underparts; partially barred leg feathers. *Tail is pale, unmarked.*

72n. Adult male/female: *Tail and often uppertail coverts are paler than rest of upperparts in all seasons.*

72o. Adult male: *Black axillaries; heavily spotted median coverts are typically paler than on most females.* Flight feathers are uniformly barred.

72p. Adult female underwing: *Black axillaries; lightly spotted median coverts. Note:* Can also be less spotted as on *72i.*

72q. Adult female: *Black axillaries; solid dark median coverts.* Flight feathers as on *72o.*

a
b
d
e
f
c
g
h
i
j
k
n
m
l
p
q
o

PRAIRIE FALCON (*Falco mexicanus*)

HABITAT: *Summer.*—Breeds in mainly open short-grass prairies and steppe regions interspersed with low to high embankments and cliffs, at least 20' (6m) high, which are used for nest sites. Small numbers nest on cliffs within wooded areas, foraging in open habitat, sometimes several miles/kilometers from nest sites. Most territories are in undisturbed locales, though this falcon may inhabit regions of light agriculture and/or near human settlements. Primarily nests below timberline, but some territories are far above tree line, at 12,000' (3700m) in cen. Rocky Mts. *Winter/migration.*—Mid–late summer dispersal takes many recently fledged birds into semi-open montane parks and alpine meadows far above tree line, up to 14,000' (4300m). By late summer–early fall, high-elevation populations move to lower, open, moderate-elevation intermountain valleys. All birds move to lower-elevation, open plains, deserts, and agricultural areas for winter. These open regions support ample avian prey base and have elevated perching and roost sites, such as utility poles and fence posts. Small numbers inhabit semi-open rural and even urban regions. On Pacific coast, they may access open salt flats, marshes, and coastal pastures. In far e. part of range, they inhabit open areas of airports as well as expanses of tall-grass prairie and fields. **STATUS:** Uncommon. An estimated 4,300–6,000 pairs live in w. North America, where numbers have always been rather low. Breeding density is sporadically distributed wherever ample nest sites and prey base exist. Snake River Birds of Prey National Conservation Area in sw. Idaho has largest and most densely concentrated breeding population, supporting about 200 pairs. Population is overall stable. Accessible nest locations provide easy access for legal falconry take. Since this species is primarily a mammalian predator during breeding season, it was minimally affected by organochlorine use in mid-20th century. Winter density is very low in all regions. **NESTING:** Occurs Dec.–Jul., with s. low-elevation pairs on territory by Dec., though most territories are not attended until Feb.–Mar. Nesting cycle and success in many areas of cen. and n. West are timed according to life cycles of *Spermophilus* ground squirrels, which emerge from hibernation in Jan.–Feb. at lower elevations at southerly and middle latitudes. Nest sites are in shallow caves, crevices, or potholes in soft earth or rock embankments or cliffs. All sites have protected overhangs that provide shelter during severe weather in early part of nesting season. No nest is built. Nest locations can be as low as 15' (4.5m) on low embankments or low cliffs, or hundreds of feet/meters high on sheer rock faces of high cliffs. In prime areas, pairs may nest and defend territories 200' (60m) apart. Pairs also commonly use abandoned stick nests of Red-tailed Hawk, Golden Eagle, and Common Raven placed in sheltered areas on embankments and cliffs. Nest sites may be used for many decades and often show white excrement stains below them. Female attends to nest duties, and male hunts. Clutch is 4 or 5 eggs. **MOVEMENTS:** *Spring.*—Mid-Jan.–May. Adults move 1st, males preceding females. Adults are on nesting grounds by mid-Feb.–mid-Mar. Northbound juveniles move in Mar.–Apr., stragglers into May. *Fall.*—Post-breeding dispersal of mostly juveniles and adult females in quest of ample food occurs mid-Jun–Aug. (Breeding-season diet is mainly of adult ground squirrels, which engage in summer estivation in May–Jul. that extends into fall–winter hibernation.) Summer dispersal is generally a northeasterly movement, but encompasses areas north, east, or southeast of breeding/natal areas. Most move to higher elevations or onto Great Plains. Actual fall migration is an extension of this summer dispersal and sometimes extends in more easterly rather than southerly direction. Birds that go to alpine elevations above tree line in mid–late summer depart for lower elevations of plains and deserts. On cen. Great Plains, juveniles and adult females move 1st, peaking mid-Sep.–Oct. Adult males mainly move Oct.–Nov. This species is often nomadic in winter, following avian food sources. **COMPARISON: "Taiga" and "Richardson's" Merlin (juveniles and adult females).**—Head pattern is nondescript, *vs.* distinct dark malar and white patch behind eyes. Merlin tail usually has 3 thin, pale bands; when it is similar solid brown on dorsal surface, it is same color as rest of dorsum, *vs.* paler than rest of dorsum in Prairie Falcon. In flight, underwing is uniformly marked, *vs.* pale with black axillaries and/or median coverts. "Richardson's" female can be deceptively similar because of its large size and pale brown dorsum, but its upperparts are uniformly brown, *vs.* brown with paler tail; its tail has 3 or 4 thin bands, *vs.* unbanded or indistinct bands; and its underwing is uniformly pale, *vs.* dark axillaries and median coverts. **Gyrfalcon (gray morph, both ages).**—Nondescript head pattern, *vs.* thin, dark malar and distinct white patch behind eye. Adult is barred with pale gray or whitish, and juvenile has scalloped whitish edges on upperparts, *vs.* tawny barring (adult male) or scalloping (juvenile). Underwing is uniformly pale, *vs.* black axillaries/median coverts. All dorsal areas same color, tail marked with thin whitish bands, *vs.* dorsal side of tail unmarked and paler than rest of upperparts. **Peregrine Falcon (juveniles).**—"Arctic" juvenile can be very similar in often having thin, dark malar mark, but lacks dark ear patch of Prairie Falcon. All dorsal areas are uniformly brown, *vs.* tail that is paler than rest of upperparts. Underwing is uniformly dark, *vs.* pale with black axillaries and/or median coverts.

Figure 1. Albany Co., Wyo. (Jul.)
A 30' (9m)-high nest cliff at 7,000' (2100m) elevation on arid short-grass prairie. Note excrement "whitewash" below aerie (left of pine tree in foreground midway on cliff face).

Figure 2. Jackson Co., Colo. (Jul.)
A 200' (61m)-high nest cliff at 8,200' (2500m) elevation overlooking a montane valley.

Figure 3. Albany Co., Wyo. (Jun.)
American Badger (*Taxidea taxus*) shares short-grass prairie habitat and preys on same ground-squirrel species as Prairie Falcon.

Figure 4. Albany Co., Wyo. (Aug.)
Pronghorn (*Antilocapra americana*) buck. The terrestrial speedster of the plains. The Pronghorn also shares short-grass prairie with Prairie Falcon.

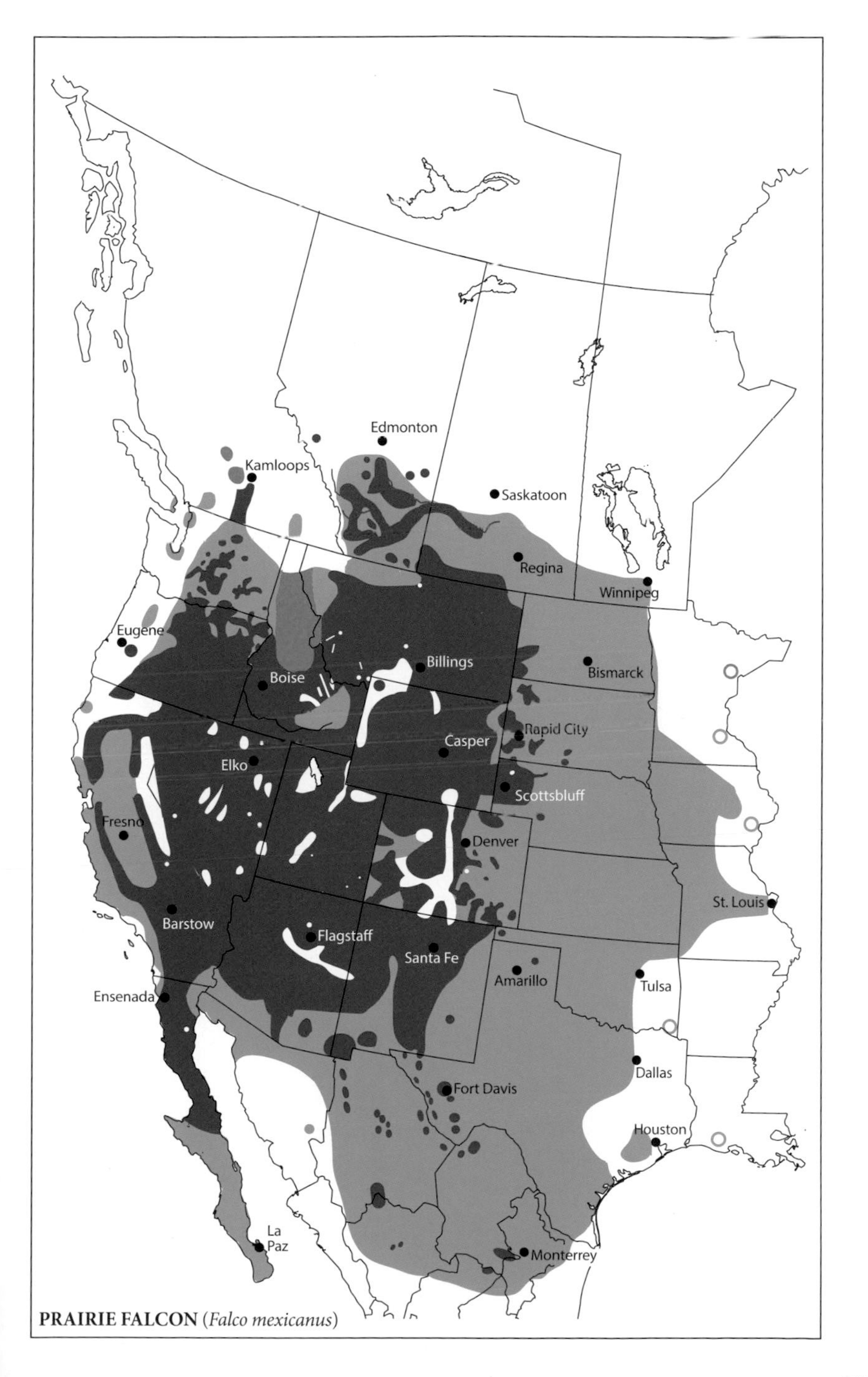

PRAIRIE FALCON (*Falco mexicanus*)

Southwestern Specialty Species

Ten subtropical and/or tropical raptor species are either resident or breed mainly in s. and sw. states of La., Tex., Ariz., and N.Mex. One of these species, Swallow-tailed Kite, has very limited breeding area in West in s. La. and se. Tex. that is contiguous with its broad range in se. U.S. during summer months. However, historically this kite nested in old-growth riparian bottomlands as far north as n.-cen. Minn. (Becker Co.).

Two species, Crested Caracara and Short-tailed Hawk, also are found in East (represented by same subspecies) but only in Fla. (Short-tailed Hawk of this w. region, which is of Mexican origin, has discernibly different tail markings from birds of Fla. population.) The very small numbers of Short-tailed Hawks in West region are either resident or depart for more southerly latitudes for winter.

The remaining 7 species extend the n. part of their breeding range from Mexico into Ariz., N.Mex., and Tex. Aplomado Falcon, Harris's Hawk, Hook-billed Kite, and White-tailed Hawk are resident, although some kites may migrate south for the winter. Common Black Hawk and Zone-tailed Hawk typically head south to extreme s. Tex. or mostly south of U.S. during winter. Most Gray Hawks also depart breeding areas for winter quarters in Mexico, although population of s. Tex. is resident.

Crested Caracara, Harris's Hawk, and Swallow-tailed Kite are known for rather far-flung northward dispersal movements extending throughout much of Great Plains. Records exist as far north as Minn. and s. Canada for caracara and kite. Caution has to be noted for some Harris's Hawk sightings, because it is a very common falconry species; wild stock has nested north of typical range in Okla. and Kans. Common Black Hawk and Zone-tailed Hawk will at times venture somewhat north of their typical breeding ranges, into Colo., and Common Black Hawk has nested in Sonoma Co., Calif. (Zone-tailed Hawk has been photo-recorded in Fla. Keys, on U.S. East Coast, and in Maritimes, Canada.)

Figure 1. Hook-billed Kite, adult female, Hidalgo Co., Tex. (May)
In the United States, this species is found only in lower Rio Grande valley of Tex. in thornbush tracts near the river.

Figure 2. Harris's Hawk, juvenile (lightly marked type), Webb Co., Tex. (Dec.) A widespread and adaptable species of the Southwest.

Plate 73. HOOK-BILLED KITE (*Chondrohierax uncinatus*) Juvenile

Ages: Retention period of juvenile plumage is unknown. Probable partial 1st-year molt begins partway through 1st year. 1st prebasic molt of wings and tail occurs as 1-year-old in late spring–summer. Molt begins on head and neck, then progresses downward on body; most of body is molted into respective adult plumage before wings and tail molt. *Note:* Mexico/U.S. birds show adult-like plumage on head and forward part of body in spring; museum specimens show 1st signs of molt on head/neck as early as midwinter. *Very large, hooked bill. Head has fleshy, yellow supraorbital teardrop-shaped patch and yellow cere; lores are greenish.* Eyes are tan. Nape feathers can be elevated in cool weather. Short legs and feet are yellow. When perched, wingtips extend halfway down tail. In flight, wings are broad, with very rounded wingtips. There is much plumage variation, including tail variations, within light morph and some in dark morph. In light morph, lightest are males, moderately marked are either sex, and heavily marked are females (museum examples; Clark and Schmitt 2017). **Subspecies:** 2 currently accepted; based on Johnson et al. (2007), *C. u. uncinatus* occurs from s. Tex. to at least s. Cen. America, but DNA differs in S. America birds. *C. u. mirus* is in Grenada. **Color morphs:** Light and dark, with no intergrades. **Size:** L: 16–20" (41–51cm), W: 34–38" (86–97cm). Females are larger. **Habits:** Tame to moderately tame but secretive. Perches on concealed and exposed branches. **Food:** Perch hunter of mainly Whitewashed Rabdotus (*Rabdotus dealbatus*) snails. **Flight:** Moderately slow, floppy wingbeats. Soars with wings held flat and wingtips flexed upward; glides with wings held in arch. Often flies at treetop level, but soars to high altitudes. **Voice:** Staccato *keh-keh-eh-eh-eh-eh* (call is similar to that of Golden-fronted Woodpecker, *Melanerpes aurifrons*).

73a. Juvenile light morph, lightly marked, pale-cheeked type: *Black crown and nape form cap; narrow black strip extends under eye. Supraorbital patch is yellow. Tawny collar wraps around hindneck.* Ventral parts are unmarked or partially barred; tawny when recently fledged. *Note:* Male with Whitewashed Rabdotus snail in its bill.

73b. Juvenile light morph, moderately marked, partial gray-cheeked type: *Gray extends under eye to cheek and jaw. Supraorbital patch is yellow. Hindneck collar is white.* Underparts are moderately barred.

73c. Juvenile light morph, moderately marked, gray-cheeked type: *All-gray cheeks and auriculars. White hindneck collar. Supraorbital patch is yellow.* Underparts are moderately barred.

73d. Juvenile light morph, heavily marked type: All of head is dark brown, with darker cap. *Supraorbital patch is yellow.* Hindneck has faint white collar. Underparts are broadly barred with brown. *Note:* Female.

73e. Juvenile dark morph: Head is dark brown or blackish, often with darker black cap. Throat and breast may be barred with white. *Supraorbital patch is yellow. Note:* Male (per museum data).

73f. Juvenile light morph, lightly marked type: Head as on *73b*. White underparts have sparse, thin barring. Typical tail has 3 equal-width brown and black bands on dorsal surface.

73g. Juvenile light morph, heavily marked type: Head as on *73d*. Hindneck collar is partial. White underparts are broadly barred. Alternate light morph tail: *2 pale bands (brown to white) and 3 equal-width black bands. Note:* Heavily marked types are females.

73h. Juvenile light morph, moderately marked type: Moderately barred underparts, often with reduced or with no belly markings. *Broad, round-tipped wings are narrow at body junction; all flight feathers are distinctly barred.*

73i. Juvenile dark morph: Dark brown body, with rufous edges on most dorsal body feathers and upperwing coverts when in fresh plumage. Tail is found *only* on dark morph: *2 broad white bands and 2 equal-width black bands. Note:* Plumage is more brownish than on either sex of adult dark morph.

73j. Juvenile dark morph: Dark brown body and underwing coverts. *Broad, round-tipped wings are narrower at body junction. Underside of wings is black and distinctly barred or spotted with white (white markings may be absent on outer primary).* Tail has broad black and white bands, 2 of each color.

73k. Male in partial 1st-year molt/1st prebasic molt (spring): *White eye, as on adult.* Gray adult male feathers on head, breast, back form "hood"; rest of ventral plumage and tail are fully juvenile, lightly marked type (*73f*). Dorsal mainly juvenile-like brown with some incoming gray adult feathers, depending on how far along molt is.

73l. Female in partial 1st-year molt/1st prebasic molt (spring–early summer): *White eye, as on adult.* Head, breast, much of flanks have adult female feathers; rest of ventral plumage, leg feathers and wings, and tail are juvenile, moderately marked type.

73m. Flight silhouette: *Soaring.* Wings are held on flat plane, with wingtips flexed upward.

73n. Flight silhouette: *Gliding.* Wings are held in arched position.

a
b
c
d
e
f
g
h
i
j
k
l
m
n

Plate 74. HOOK-BILLED KITE (*Chondrohierax uncinatus*) Adult

Ages: Adult plumage is attained as 1-year-old by 2nd fall. *Very large, hooked bill. Head has fleshy, yellow outer lores and cere; supraorbital teardrop-shaped patch is orange; inner lores and eyelids are green.* Eyes are white. Legs and feet are orange-yellow; short legs. When perched, wingtips extend halfway down tail. In flight, wings are broad, with bowed secondaries that narrow at body; very rounded wingtips. Nape feathers can be elevated in cool weather. Considerable plumage variation occurs within both sexes of light morph; only sexual differences in dark morph. **Subspecies:** 2 currently accepted; based on Johnson et al. (2007), *C. u. uncinatus* occurs from s. Tex. to at least s. Cen. America, but DNA differs in S. America birds. *C. u. mirus* is in Grenada. **Color morphs:** Light and dark, with no intergrades. **Size:** L: 16–20" (41–51cm), W: 34–38" (86–97cm). Females are larger. **Habits:** Tame to moderately tame. Secretive species; perches on concealed and exposed branches. **Food:** Perch hunter of mainly Whitewashed Rabdotus snails. **Flight:** Moderately slow, floppy wingbeats. Soars with wings held on flat plane, with wingtips flexed upward; glides with wings held in arched position. Often flies at treetop level but may soar to high altitudes. **Voice:** Staccato *keh-keh-eh-eh-eh-eh* when agitated (similar to that of Golden-fronted Woodpecker).

74a. Adult male light morph, all-barred type: *Gray head, white eye, yellow/green facial skin and cere; orange supraorbital patch.* Underparts are barred with white and gray. *Note:* Holds Whitewashed Rabdotus snail in its bill.

74b. Adult female light morph, rufous-barred type: *White eye.* Yellow/green facial skin and cere. *Crown and cheek are gray; nape is black. Auriculars and hindneck are rufous.* Underparts are barred with rufous.

74c. Adult female light morph, brown-barred type: *White eye.* Yellow/green facial skin and cere. *Supraorbital patch is orange.* Gray cheeks and auriculars; black nape. *Hindneck is rufous or dark rufous.* Underparts are barred with dark brown. *Note:* Head and nape feathers are fluffed, making head large and rufous nape pronounced.

74d. Adult male dark morph: *White eye. Yellow/green facial skin and cere; orange supraorbital patch. Head and body are uniformly dark gray.*

74e. Adult female dark morph: *White eye. Yellow/green facial skin and cere; orange supraorbital patch.* Black body has slight brownish cast. Crown and nape may be darker black.

74f. Adult male light morph, all-barred type: Solid gray head and upperparts. Gray underparts are thinly or broadly barred with white. *Tail has 2 broad black bands and 1 wide gray band.*

74g. Adult male light morph, rufous-barred type: Solid gray head and upperparts. Hindneck may have partial rufous collar. Rufous-tinged underparts are thinly or broadly barred with white. *Tail has 2 broad black bands.*

74h. Adult female light morph, rufous-barred type: *Black crown and nape. Hindneck collar is rufous.* Underparts are barred with rufous. Tail has 2 broad black bands with partial white edges.

74i. Adult female light morph, brown-barred type: *Black crown and nape; gray cheeks and auriculars. Hindneck collar is rufous.* Underparts are marked with broad, brown bars. Tail is similar to juvenile's, with equal-width dark bands. *Note:* Underwing coverts may be dark brown rather than typical rufous-barred.

74j. Adult male light morph, bibbed type: Gray underparts are thinly or broadly barred, except breast, which is solid gray and forms bib. *Broad, round-tipped wings are narrow at body junction. Black flight feathers are barred or spotted with white on primaries; secondaries are solid black.* Tail is broadly banded with black and white.

74k. Adult female light morph, rufous-barred type: *Rufous-barred underparts. Broad, round-tipped wings are narrow at body junction. Flight feathers are pale and distinctly barred.*

74l. Adult male dark morph: *Uniformly dark gray body. Dorsal side of tail has broad white band.*

74m. Adult female dark morph: *Uniformly black body. Broad, white mid-tail band is grayish on distal half.*

74n. Adult male dark morph: *Dark gray body. Flight feathers are all-black. Mid-tail has broad, white band.*

74o. Adult female dark morph: Uniformly black body. *Flight feathers are uniformly black.* Mid-tail has broad white band; may show partial thin white band on basal area when tail is fanned.

74p. Adult female dark morph underwing, spot-winged type: *White spotting on black outer primaries.*

a
b
c
d
e
f
g
h
i
j
k
l
m
n
o
p

HOOK-BILLED KITE (*Chondrohierax uncinatus*)

HABITAT: *Year-round.*—Densely wooded tracts or semi-open patches of medium-size to large thornbush tracts along Rio Grande in s. Tex. **STATUS:** Uncommon and local. There are possibly up to 20 pairs in U.S.; population fluctuates. They inhabit patchily distributed thornbush tracts in s. Hidalgo and sw. Starr Cos. in locations depicted on range map. *Color morph status.*—Virtually all of U.S. population is of light morph. Dark morph was 1st noted in late 1998, and documented nesting in 2002 (light–dark pair). **NESTING:** Mar.–Aug. Nests are placed in dominant-size tree in wooded tract. See-through, flimsy nest is made of thin sticks and is 12" (30cm) in diameter. Nests are placed in middle of horizontal branch in lower canopy, about 20' (6m) high. Egg-laying dates span Apr.–Jun. **MOVEMENT:** Sedentary or short-distance migrant. Birds can be seen all year in inhabited areas; however, notable movement is seen in e. Mexico in autumn. Dispersal occurs, and occasional birds show up in fall at e. Tex. hawk watch sites north of established range. *Note:* Tex. population originated from northward-dispersing Mexican birds. 1st record was in 1964, but the species was not seen again until 1975; thereafter population steadily grew. **COMPARISON:** ***Hook-billed Kite adult male light morph.*** **Gray Hawk (adult).**—Dark eye, *vs.* white eye. Long legs, *vs.* short legs. White bands on dorsal tail surface, *vs.* gray bands. In flight, whitish underwing, *vs.* blackish underwing with white spots. ***Hook-billed adult female light morph.*** **Cooper's Hawk (1-year-olds, adult females).**—Orange/red eye, *vs.* white eye. Long legs, *vs.* short legs. Similar size, rufous nape, rufous-barred underparts, and tail length and pattern. In flight, Cooper's has rapid, stiff-winged flap, *vs.* slow and floppy flap. **Red-shouldered Hawk (adult).**—Eye is dark, *vs.* white. Long legs, *vs.* short legs. Tail has thin, white bands (dorsally), *vs.* gray bands. **Broad-winged Hawk (adult).**—Orange-brown eye, *vs.* white eye. Underparts are similarly rufous-barred; tail similarly gray-banded on dorsal surface. In flight, Broad-wing's wingtips are pointed, *vs.* rounded. ***Hook-billed juvenile light morph.*** Juveniles of above-noted species all have streaked underparts, *vs.* brown-barred underparts. ***Hook-billed dark morph (all ages).*** No other dark raptor has white eye. In flight view, no other black-bodied raptor has round-tipped wings that are all-black or have white spots on primaries (adults) or have thick black-and-white barring on all flight feathers (juvenile). Adults of Broad-winged Hawk, Common Black Hawk, and Zone-tailed Hawk share white bands on undertail, visible in flight. **Harris's Hawk (all ages, flight).**—Body is dark brown *vs.* black. Has similar round-tipped wings and all-black flight feathers, but underwing coverts are rufous *vs.* black. Tail has white base *vs.* white mid-tail band(s). Legs are long *vs.* very short.

Figure 1. Hidalgo Co., Tex. (Apr.)
Whitewashed Rabdotus snail, Hook-billed Kite's primary prey item in Tex.

Figure 2. Hidalgo Co., Tex. (May)
Nest tree in thornbush habitat (nest is halfway up right side of tree).

Figure 3. Hidalgo Co., Tex. (May)
Two nestlings in typical flimsy, see-through nest.

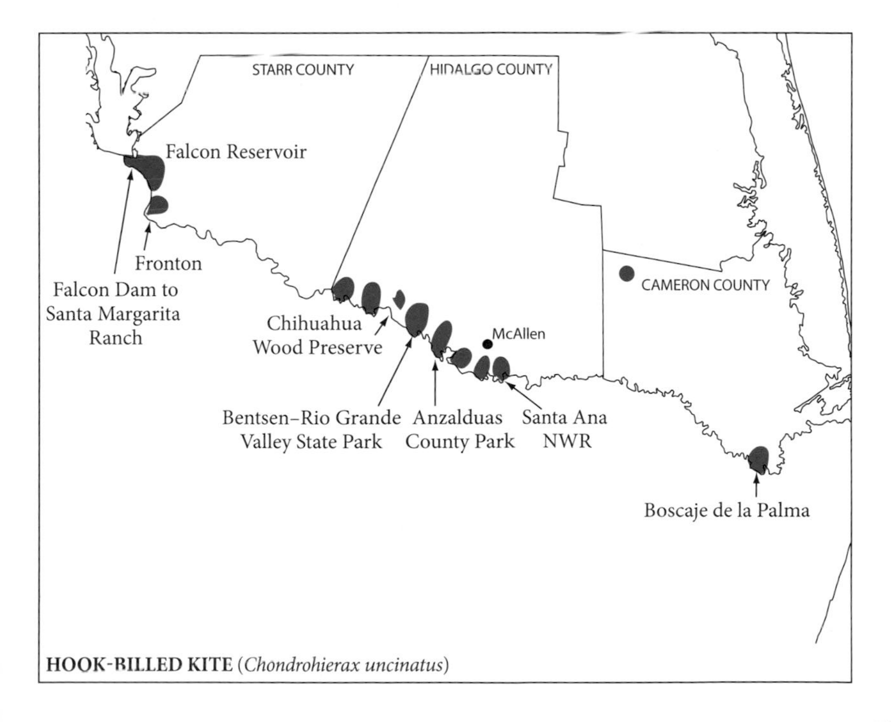

Plate 75. SWALLOW-TAILED KITE (*Elanoides forficatus*)

Ages: Juvenile, 1-year-old (early 1st prebasic molt stage), and adult. *Black tail is very long and deeply forked. Back, forward scapulars, and upperwing coverts are black and contrast with paler iridescence of rest of blackish upperparts. Large white patch shows on inner tertials and greater secondary coverts when perched. Juvenile.*—Body plumage is held, as in all kites, for a few months, with molt starting while on winter grounds south of U.S. Tawny-rusty wash on crown and underparts wears off quickly once fledged. Upperparts are thinly fringed with white, which also rapidly wears off. When perched, wingtips equal tail tip. Iridescent upperparts are greenish or purplish. Black tail is long and deeply forked. *1-year-old (1st prebasic molt).*—This stage encompasses wing and tail molt from juvenile to adult feathering. Retained juvenile wing (including most coverts) and tail feathers are molted when birds return to U.S. in late spring. *Adult.*—Upperparts are iridescent purple-blue. Black tail is very long and deeply forked. **Subspecies:** Nominate (*E. f. forficatus*) in U.S.; 1 other subspecies in S. America. **Color morphs:** None. **Size:** L: 20–25" (51–64cm), W: 47–54" (119–137cm). Females are marginally larger. Adults are longer than juveniles because their very long tails are at least 2"/5cm longer. **Habits:** Perches in trees; does not use utility wires, poles, or posts. **Food:** Aerial hunter. Flying insects; arboreal frogs, lizards, and snakes; nestling birds are captured while in flight. **Flight:** Elegant, flowing maneuvers with constant tail adjustments. Wings are held on flat plane with wingtips bent slightly upward when soaring. Glides with wings on flat plane or bowed slightly downward. Flight may be at high altitude or at treetop or ground-skimming height. **Voice:** High-pitched clear *klee, klee, klee*; also high-pitched *eeeep* and rapid, high-pitched chitter notes.

75a. Old nestling/fledgling: White head with tawny wash on crown, nape, and underparts.

75b. Fledgling/recently fledged juvenile: White head with thin, brown feather-shaft streaks that are visible at very close range (sometimes found on older birds); pale tawny wash on breast. *Note:* Tawny areas fade quickly during late nestling/early fledgling periods; this bird is slightly older than *75a*.

75c. Adult: White head; typically lacks fine head streaking.

75d. Old nestling/fledgling: White head; tawny on crown, nape, and underparts. *Lower scapulars and wings are iridescent greenish. When perched, wingtips equal moderate-length forked tail; primaries and tail are ¾ grown.*

75e. Fledgling/recently fledged juvenile: White head and underparts but may have slight tawny wash on breast, belly, and flanks, which fades to all-white by mid–late summer. *Lower scapulars and wing are iridescent greenish but sometimes bluish or purplish (similar to color of adult).* Many have thin white tips on wing coverts and primaries. When perched, wingtips equal long, deeply forked tail (primaries and tail are fully grown for age class).

75f. Adult: White head and underparts. *Lower scapulars and wing are iridescent purplish or bluish. Very long, deeply forked tail extends far beyond wingtips.*

75g. Juvenile (mid–late summer): *Black flight feathers and white coverts on underside of long, pointed wings. Black tail is long and deeply forked.* Wings are broader and tail shorter than on adult.

75h. Juvenile (mid–late summer): *Greenish iridescence on upperparts. Long tail is deeply forked.*

75i. 1-year-old (late spring–early summer): As on *75g*, but has begun molt of inner primaries: Inner 2 primaries (p1, p2) are new, darker feathers; p3 and p4 are dropped and not yet regrown; rest of wing is old juvenile feathering. White translucent patches on underwing coverts are due to molting (missing) coverts. 2 middle tail feathers are newly molted and darker; others are old juvenile feathers. *Note:* Bird is eating.

75j. 1-year-old: Purple iridescent variation of upperparts is similar to adult. White spots on wing coverts, back, and scapulars due to molting, missing feathers. Inner 2 primaries (p1, p2) are new, p3 is dropped and not yet regrown; rest of wing is old juvenile feathering. Tail molt as on *75i*.

75k. Adult: *Purplish or bluish iridescent upperparts. Black tail is very long and deeply forked. Note:* Non-molting of wings and tail is typical in early spring for nonbreeding birds, which begin molting in Apr.–May, and in spring through midsummer for breeding adults, which begin molting in Jun.–Jul. (latter in males).

75l. Adult: *Black flight feathers and white coverts on long, narrow, pointed wings. Black tail is very long and deeply forked.*

75m. Flight silhouette: *Soaring.* Wings are held on flat plane with wingtips bent slightly upward.

75n. Flight silhouette: *Gliding.* Wings are often slightly bowed downward.

a
b
c
d
e
f
g
h
i
j
k
l
m
n

SWALLOW-TAILED KITE (*Elanoides forficatus*)

HABITAT: *Summer.*—Humid, lowland, semi-open regions with small or large tracts of tall deciduous and/or pine trees. Most areas are within or adjacent to marshes, swamps, rivers, or lakes. Meadows, agricultural fields, or small suburban areas may be nearby and are often utilized for foraging. **STATUS:** Very uncommon and highly local in its limited summer haunts in West. Tex., where nesting 1st occurred in 1994, has about 30 nesting pairs. There are about 100 pairs in La., and a small colony in se. Ark. (Total U.S. estimated population is about 4,000 birds). **NESTING:** Mar.–Jul. Pairs nest singly or in small, loosely grouped colonies. 1 or 2 single birds may accompany a breeding pair but rarely help with nest duties. Medium-size, often oblong-shaped nest is made of thin sticks, and also lichens and moss, and is placed at least 100' (30m) high in top of dominant-size tree of a stand. Both pair members build nest. Female lays single clutch of 1–3 but usually 2 eggs. **MOVEMENTS:** Highly migratory, moves from U.S. to and from winter quarters in w.-cen. and sw. Brazil. Migrates along Tex. coast, or makes over-water trans–Gulf of Mexico flights, especially to and from Yucatán Peninsula, Mexico. Kites may remain over water for up to 4 days and nights (Powell 2017). *Spring.*—Adults move prior to returning 1-year-olds, arriving on breeding grounds in early–mid-Mar., but may be as early as late Feb. Younger birds and nonbreeding adults arrive in U.S. in mid-Apr.–mid-May. *Fall.*—This is a very early-season migrant. Pre-migration assembly of over 100 birds has been seen in s. La. along Pearl River in late Jul. (groups of 1,500 form in Fla.). Adults move before young birds. Movement may begin in mid-Jul. and extend through Sep.; most birds move in Jul.–Aug. Winter quarters in S. America are reached in Oct. *Extralimital dispersal.*—This species engages in annual spring and late summer–early fall dispersal activities. Regular extralimital records extend north to Iowa and s. Colo., but birds have been noted as far north as Minn. (several), S.Dak. (Sep. 2008), and s. Man. (May 2014, 1st record in 122 years, Nicholson 2014). **COMPARISON:** No other bird of prey has deeply forked tail. **Swainson's Hawk (light morph).**—Somewhat similar light–dark underwing pattern, but has pale, square-tipped tail, *vs.* forked tail. **Magnificent Frigatebird (*Fregata magnificens*).**—Exhibits similar long, black, fork-shaped tail, but underwing is all-black, *vs.* white-black underwing pattern.

Figure 1. Liberty Co., Tex. (Jun.)
Semi-open riparian breeding habitat with tall deciduous trees. Photo by John Economidy

Figure 2. Liberty Co., Tex. (Jun.)
Three kites flying above dense riparian tract of older-growth trees. Photo by John Economidy

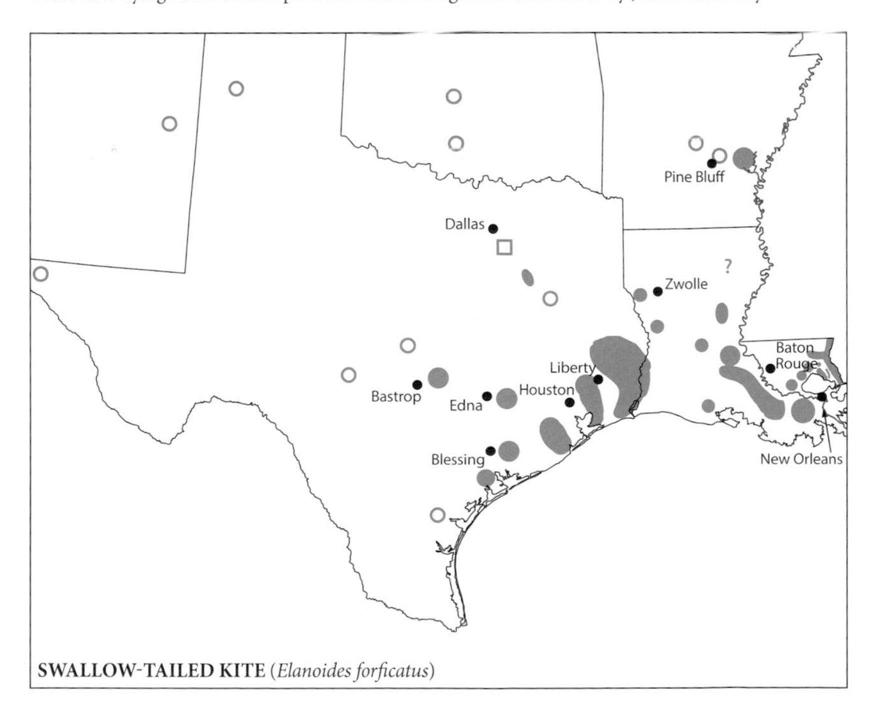

Plate 76. COMMON BLACK HAWK (*Buteogallus anthracinus*)

Ages: Juvenile, 1-year-old, and adult. Juvenile plumage is retained for 1 year. Full adult plumage is attained as 2-year-old by late fall. *In flight, wings are broad, with blunt tips.* Yellow legs are long and extend beyond undertail coverts in flight. *Juvenile.—Basal ½ of bill is pale blue, and cere is dull yellow or greenish yellow. Dark eye line is thick. Pale supercilium is short and wide. Broad, dark malar mark extends onto sides of neck. Flanks are black.* Wings are narrower than on adult but still broad. *Inner primaries have pale panel/window. Mainly white tail has variable patterns; often irregularly banded with thin or thick V-shaped bands; subterminal band always widest. Tail is moderately long, and longer than adult's. 1-year-old.*—Adult plumage, but retains some juvenile feathering, mainly on outer primaries and some underwing coverts. *Adult.—Basal ½ of bill is yellow, brighter on male. Cere and lores are yellow, brighter on male. Some females have white patch that extends from lores under eyes and onto auriculars. Body plumage is grayish black on male and brownish black on female. Small white patch adorns underside of outer 2 or 3 primaries. Ventral flight feathers are pale gray, pale rufous, or a mix; often more rufous on female. Tail always shows broad white band on dorsal and ventral sides.* **Subspecies:** *B. a. anthracinus* inhabits sw. U.S. south to n. S. America; 2 other subspecies, 1 in Cuba, 1 in n. S. America. **Color morphs:** None. **Size:** L: 20–22" (51–56cm), W: 40–50" (102–127cm). Females average larger. **Habits:** Fairly tame. Perches for long periods on concealed or open branches, rocks, or on ground, usually near water. Does not perch on utility poles or wires along roadways in U.S. **Food:** Perch hunter of fish, amphibians, and reptiles; also hunts small mammals, birds, and large insects. **Flight:** Powered flight is of slow wingbeats, often at low altitudes; regularly soars to high altitudes. May briefly hover over water when hunting or kite briefly along cliffs. **Voice:** Quite vocal, with loud, haunting, high-pitched, metallic *kree, kree, kree, he-he-he*; sometimes only *kree, kree, kree*.

76a. Juvenile: Basal ½ of bill is pale blue, and cere is dull yellow. *Dark eye line is broad. Dark malar mark extends onto sides of neck.* Supercilium is short and thick.

76b. Juvenile tail, wide-banded (dorsal): *3 or 4 thick, V-shaped inner bands.*

76c. Juvenile tail, thin-banded (dorsal): *3 or 4 thin, offset bands; may be thicker on basal region.*

76d. Juvenile tail, thin-banded (dorsal): *Numerous thin inner bands; thicker on basal region.*

76e. Juvenile: *Broad wings. Pale window on inner primaries; secondaries are reddish. Underparts are streaked; black flanks. Tail is white and irregularly banded.*

76f. Juvenile: *Thick, dark eye line and malar mark. Flanks are black. Broad secondaries and greater coverts are thinly barred. Tail is similar to 76b.*

76g. Juvenile (fresh plumage): *Tawny on pale areas of head. Wings are broad, with pale tawny panel on inner primaries. Tail is mix of 76c and 76d patterns.*

76h. Adult male: *Basal ½ of bill and cere are bright yellow; lores are pale yellow. Head is black.*

76i. Adult female: *Basal ½ of bill, cere, and often lores are yellow. Whitish extends under eye (some lack whitish area). Black head has brownish cast on crown and nape. Note:* Head/neck feathers are raised.

76j. Adult: *Very broad wings, with small white patch on outer primaries. Flight feathers are thinly barred with pale gray, pale rufous, or mix of gray and rufous. Long legs place feet into white tail band.*

76k. 1-year-old underwing: Adult-like; outer primaries and some coverts are retained juvenile feathers.

76l. Adult male: *Blackish body and very broad wings; 1 broad, white tail band.*

76m. Adult tail (dorsal): *1 broad, white mid-tail band, visible with tail closed or fanned. White-tipped coverts.*

76n. Adult female: *Blackish dorsal plumage on body and wings has more brownish cast than on male. White tail band is broad. Legs are long and yellow.*

76o. Flight silhouette: Glides and soars with wings held on flat plane.

76p. Adult tail (ventral): *1 broad white tail band; feet extend into white band. White-tipped coverts.*

76q. Adult tail (ventral): 1 or 2 thin, white inner tail bands may show when tail is fanned; more so on female.

a
b
c
d
e
f
g
h
i
j
k
l
m
n
o
p
q

COMMON BLACK HAWK (*Buteogallus anthracinus*)

HABITAT: Hot, arid lowland and rugged lower-elevation montane areas with permanent, shallow, clear-watered creeks, rivers, and ponds. These water sources are bordered by isolated tall, old-growth Fremont and Narrowleaf Cottonwoods and Arizona Sycamore (*Populus fremontii*, *P. angustifolia*, and *Platanus wrightii*, respectively), or small or large groves or gallery forests of tall, old-growth trees of these species. Breeding habitat ranges from 3,000' to 5,000' (900–1500m) in primary range in Ariz. and N.Mex.; lower elevations in Tex. **STATUS:** Very uncommon to uncommon. This species has historically suffered from habitat loss. It is locally distributed and found in low density in its restrictive habitat, with about 220–250 pairs in Ariz. and N.Mex.; a very small number breed in w. Tex. and s. Utah. Range is expanding slightly in N.Mex., with recent nesting in San Miguel Co.; state population may be up to 100 pairs. Population now appears stable in all breeding regions. *Note on Calif. status:* Adult female was 1st documented at Stockton (San Joaquin Co.) in Oct. 2004 and early 2005; probable same bird has been summer resident, then resident near Santa Rosa (Sonoma Co.) since May 2005. Successfully nested with male "California" Red-shouldered Hawk in 2012 (S. Moore pers. comm.; Hug 2013); interspecies nesting occurred until at least 2015. Female still at Santa Rosa as of Dec. 2016 (D. Hofmann 2016, photos). (The subspecies found in U.S. is fairly common to common southward; range extends south through n. fringe of S. America.) **NESTING:** Mar.–mid-Sep. The large stick nest lined with greenery is built in primary or secondary crotch in mature, tall, live tree, 30–90' (9–27m) high, near or adjacent to water source. Tree may be isolated or in grove. (Nest tree may be close to Zone-tailed Hawk nest trees, with no interaction.) New nest may be built, or old nest refurbished. Clutch is 1 or 2 eggs; fratricide is common, and often only 1 chick survives. Fledglings remain in natal territory until fall migration. Breeding does not occur until age 2 or 3. **MOVEMENTS:** Short-distance migrant. Bulk of U.S. population moves south for winter into at least nw. Mexico and southward; a few winter in extreme se. Tex. *Spring.*—Adults arrive on breeding grounds Mar.–early Apr. Recent spring migration route noted in se. Ariz. along Santa Cruz River in Santa Cruz Co. near Tubac, in 2nd–3rd weeks of Mar. Adults are seen singly or in flocks of over a dozen birds. Returning juveniles arrive in U.S. in May–early Jun. *Fall.*—All ages move southward during Oct. *Dispersal.*—Fairly regular extralimital movements noted for s. Calif. (9 Mar.–May, 1 Sep. [excludes Santa Rosa, Stockton records]; Calif. Bird Records Committee 2017), s. and w. Colo. (10 Apr.–May, 1 Jun., spanning 1983–2013; Colo. Bird Records Committee 2017), n. Tex and ne. Tex. **COMPARISON:** ***Black Hawk juvenile.*** **Gray Hawk (juvenile).**—Smaller but has similar head markings, barred leg feathers, and long legs. Flanks are streaked, *vs.* all-black. **Red-tailed Hawk.**—Eye typically much paler yellow, gray, or tan, *vs.* medium brown. Dorsal tail surface is covered with thin, neat bands, *vs.* mainly whitish with irregularly shaped wide black bands. In flight, wings are narrower and lack banding on inner primaries, *vs.* very broad wings with strongly barred pattern on all feathers. ***Black Hawk adult.*** **Black Vulture.**—Similar shape and size in flight, with very broad wings and short tail. Distinct white patch on primaries, and otherwise uniform black underwing and all-black tail, *vs.* wide black band on trailing edge of wing and broad white mid-tail band. **Zone-tailed Hawk.**—Shares similar habitat and similar in size. Bill is mainly all-black, *vs.* black on distal ½ and yellow on basal ½. Adult has gray band(s) on dorsal tail surface, *vs.* white band. When perched, wingtips equal or extend past tail, *vs.* slightly shorter than tail. In flight, moderate-width wings held in dihedral, *vs.* very broad wings held flat. In ventral flight view, short legs and tips of toes are within black undertail coverts, *vs.* toes extending well past coverts to edge of broad, white band. **Rough-legged Hawk (dark morph).**—Range overlap Mar., Oct. Similar size and color. Forehead is white, *vs.* black forehead. Legs are short and feathered, *vs.* long and bare. Primaries project far beyond secondaries, *vs.* short projection. Moderate-width wings with whitish underside of flight feathers, *vs.* broad wings with gray/rufous underside. In ventral flight view, small yellow feet are well within undertail coverts, *vs.* extending well past black coverts. Share similar white band on undertail. **Golden Eagle (juvenile, younger immatures).**—Tawny nape, *vs.* black nape. In flight, feet extend within long undertail coverts, *vs.* extending well past coverts. White mid-tail band is very similar.

Figure 1. Gila River, Grant Co., N.Mex. (Jul.)
Prime breeding habitat along clear, shallow water source with tall cottonwood or sycamore trees.

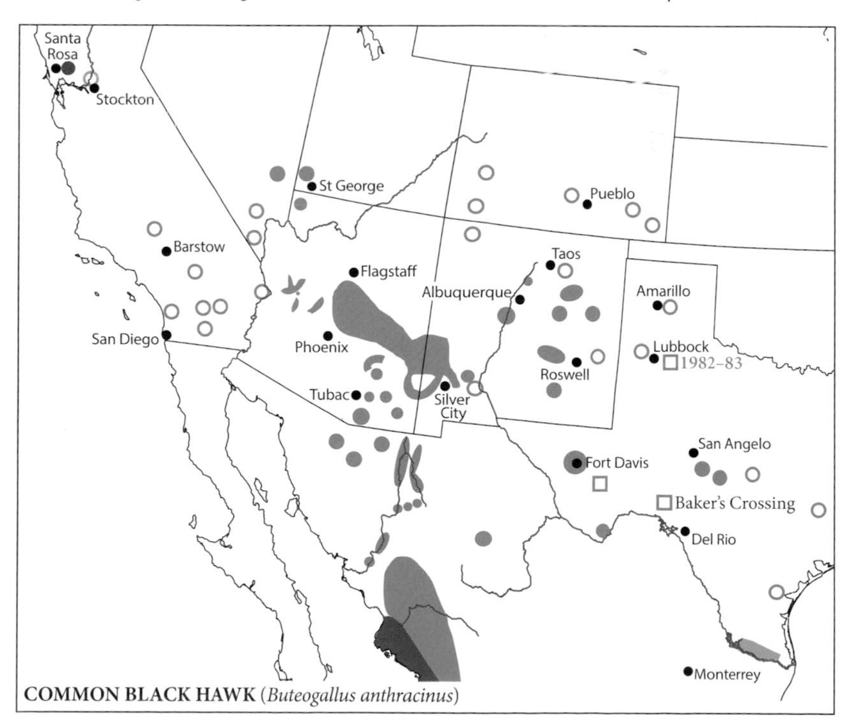

COMMON BLACK HAWK (*Buteogallus anthracinus*)

Plate 77. HARRIS'S HAWK (*Parabuteo unicinctus*)

Ages: Juvenile and adult. Juvenile plumage is held for 1 year. Juvenile-plumaged (and molting) birds can be encountered in any month. *Pale blue bill is short and thick, with small dark tip. Yellow facial region of supraorbital ridge, eye ring, lores, and cere. Lesser and median upperwing coverts form large rufous "shoulder" patch. Lesser and median underwing coverts and axillaries are rufous. In flight, wings are moderate length and broad, with rounded wingtips. Tail is long. When perched, wingtips reach halfway down tail. Uppertail and undertail coverts are white. Bases of outer tail feathers are white and blend with white coverts. Juvenile.*—Underparts have variable amount of white markings, often with darker bib pattern. *Upperwing greater coverts have pale barring. White panel on underside of primaries contrasts with medium-gray secondaries. Wings are narrower than adult's. Leg feathers are variably barred or are solid rufous.* Thin, pale terminal tail band, which wears off quickly. *Note:* Birds in 1st prebasic molt have 1 or more adult (gray) inner primaries. *Adult.*—Body is solid dark brown. *Underside of flight feathers is solid dark gray. Legs feathers are rufous.* Black tail has broad, white terminal band. *Note:* Clark and Schmitt (2017) note that ventral surface of flight feathers of 1-year-old is paler gray than on older birds; may retain at least juvenile outer primary. **Subspecies:** *P. u. harrisi* extends from s. U.S. to Costa Rica; *P. u. unicinctus* in S. America. **Color morphs:** None. **Size:** L: 18–23" (46–58cm), W: 40–47" (102–119cm). Average wing chord (based on Palmer 1988): adult male (221 birds): 324mm; adult female (176 birds): 361mm. Sexually dimorphic; females visibly larger. **Habits:** Tame. Groups of related/unrelated helpers may nest and hunt cooperatively. They use exposed objects, especially utility poles and cacti, for perches. Often perch side by side in pairs or groups. **Food:** Perch hunter. Prey can be as large as jackrabbits, but mainly rabbits, quail, songbirds, and lizards. **Flight:** Series of moderate-speed flaps with long glides, typically at low altitude. May soar to high altitude. *Glides with wings bowed downward*; soars with wings bowed or on flat plane. Sometimes hovers at low altitude but does not kite. **Voice:** Vocalizes when disturbed with drawn-out, grating *irrrr*.

77a. Juvenile, lightly marked type: *Short, thick, blue bill with dark tip. Facial skin and supraorbital ridge skin are yellow or greenish.* Head, throat, and breast are streaked.

77b. Juvenile, moderately/heavily marked type: *Short, thick, blue bill. Facial skin and supraorbital ridge skin are yellow or greenish.* Dark brown head and breast may be lightly streaked.

77c. Juvenile, lightly marked type: *Large rufous shoulder patch. Greater wing coverts have distinct tawny barring.* Underparts are streaked with brown and white; breast streaking is often darker and forms bib. *Legs feathers are white with thin rufous barring but can be darker, as on 77d. Note:* Plumage type found on either sex.

77d. Juvenile, moderately marked type: *Large rufous shoulder patch. Dark breast forms bib above brown-and-white streaked/speckled belly and flanks. Legs are rufous with some white barring but can be lighter, as on 77c, or darker, as on 77e. Note:* Common plumage type for either sex.

77e. Juvenile, heavily marked type: *Large rufous shoulder patch. Dark breast forms bib above lightly speckled dark belly and flanks. Legs are solid rufous or can be lighter, as on 77d. Note:* Common plumage for either sex.

77f. Juvenile, moderately marked type: *Dark bib on breast and streaked belly. Underwing shows rufous coverts and axillaries, white panel on primaries, and medium-gray secondaries. Gray tail has fine barring and white basal area. Note:* Underwing and tail patterns similar for all plumage types.

77g. Juvenile, all types: Rufous shoulder; barred greater coverts. White uppertail coverts and tail base.

77h. 1st prebasic molt, lightly marked type: Streaked belly. Inner primaries are white; secondaries are gray. 2 new, dark gray adult-like inner primaries (p1, p2) have grown in. *Note:* Plumage is juvenile.

77i. Adult: *Short, thick, blue bill. Facial skin and supraorbital ridge skin are yellow or greenish.*

77j. Adult: *Dark brown body. Rufous shoulders and leg feathers. Long, black tail has broad white tip.*

77k. Adult: *Solid dark gray flight feathers; rufous underwing coverts and axillaries. Black tail has white base and broad white tip.*

77l. Adult: *Rufous shoulders. Tail coverts and base of tail are white; broad white tip on tail.*

77m. Flight silhouettes: *Gliding (top):* Wings bowed when gliding. *Soaring (bottom):* Wings flat or bowed when soaring.

a
b
c
d
e
f
g
h
i
j
k
l
m

HARRIS'S HAWK (*Parabuteo unicinctus*)

HABITAT: All areas have some water source, such as ground seepage, creek, river, lake, or livestock water tank. **Arizona.** Hot, arid flatlands or gentle rolling hills covered with thorn-scrub of palo verde (*Parkinsonia* spp.) and Saguaro, or palo verde and other cacti. Also found in semi-open thorn-scrub savannas with mesquite and hackberry and in thorn-scrub areas with scattered tall cottonwood trees along riparian stretches. **New Mexico.** Arid, seasonally hot desert savannas covered with mesquite-cactus and mesquite-oak thorn-scrub. **Texas.** Interior regions encompass seasonally hot or seasonally very hot, arid, semi-open and dense mesquite and mesquite-cactus regions, including mesquite savannas in n. area of range. Coastal areas are hot and humid, with semi-open lush mesquite-cactus thorn-scrub savannas, including areas of tall live-oak patches. **STATUS:** Common and widespread in Tex.; uncommon to common but local in Ariz. and N.Mex.; very local, with very small population in s. Calif. Nesting occurred in Kans. in 1963. Population is stable overall but has suffered massive historic habitat loss due to agriculture and current-day eradication of mesquite and fragmentation of suitable-size mesquite thorn-scrub tracts to sustain breeding pairs/groups. **NESTING:** Pairs remain together as long as mates survive. Breeding takes place in any month, and double or triple annual clutches are produced if there is ample prey. Breeding may not occur if there is inadequate prey. Most breeding activity takes place in winter, early spring, and late summer. Ariz. breeders typically engage in polyandry (2 or more males with 1 female); polygyny (multiple females with 1 male) also occurs. N.Mex. and Tex. birds mainly practice monogamy, with limited polyandry. Breeding pairs often have helpers, which form social group that assists in hunting and fending off predators. Large stick nest is placed 5' (1.5m) to 25' (8m) high in Saguaro cactus, tree, or utility pole; sometimes other human-made structures are used. Nests are lined with greenery. Clutch size is 1–5, most commonly 3 or 4. Fledglings may assist in group hunting within 2 months of fledging, and may stay with parents and social group as helpers for up to 3 years. **MOVEMENTS:** Sedentary. However, both ages are prone to irregular northward dispersal, especially in fall and winter. Irregular movements have been noted in Colo., Kans., Mo., Nebr., and Okla. **COMPARISON: Hook-billed Kite.**—Similar in shape and flight mannerism, with very rounded wings and slow, floppy wingbeats, but kite has dark uppertail coverts and lacks rufous underwing coverts, *vs.* white uppertail coverts and rufous underwing coverts. ***Buteo* (dark species and morphs).**—These *Buteo*s lack rufous upperwing coverts. Adult dark morph Swainson's Hawk has similar rufous underwing coverts but has long, pointed wings, *vs.* blunt, rounded wingtips. **Red-shouldered Hawk (juvenile).**—Similar rufous upperwing coverts, streaked underparts, and long legs as on lightly marked juvenile Harris's, but has dark uppertail coverts, *vs.* white.

Figure 1. Pima Co., Ariz. (Apr.)
Year-round Saguaro cactus habitat.

Figure 2. Duvall Co., Tex. (Aug.)
Year-round mesquite-cactus habitat.

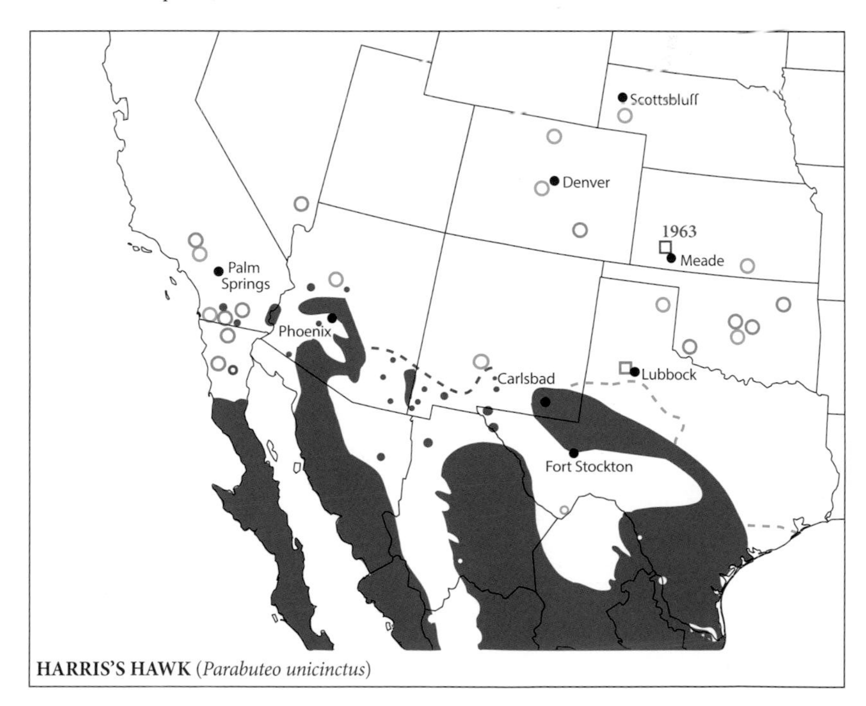

HARRIS'S HAWK (*Parabuteo unicinctus*)

Plate 78. WHITE-TAILED HAWK (*Geranoaetus albicaudatus*) Juvenile

Ages: Juvenile plumage is held for 1st year. Eyes are medium brown (can be reddish brown at certain light angles). *Basal ½ of bill is pale blue. Cere is bluish green.* Plumages are highly variable, with cline from lighter to darker types (appear color-morph-like), which are divided into 3 categories. Forehead and lores are white. *Upperwing has pale tawny patch on lesser coverts (shoulder).* Legs are long and yellow. Moderately long tail is finely banded and longer than on older age classes. *When perched, wingtips extend just past tail tip. In flight, wings are long and moderately wide, but narrower than those of older age classes. Rear edge of secondaries is distinctly bowed and narrows considerably at body junction; wingtips are pointed. Underside of flight feathers is moderately gray and often darker on inner primaries.* **Subspecies:** *G. a. hypospodius* in s. Tex. to S. America; 2 more subspecies in S. America. **Color morphs:** None in this subspecies (dark morph in other 2 subspecies). Extreme plumage variations within juveniles (and 1-year-olds) appear as polymorphic. **Size:** L: 18–22" (46–56cm), W: 49–53" (124–135cm). Females average larger. Tail is 1–1.5" (2.5–3.7cm) longer than that of birds of older age classes and makes juvenile longer and appear larger. **Habits:** Tame. This is a gregarious species. It sits on exposed perches, including utility poles and wires; also on ground. **Food:** Perch and aerial hunter of amphibians, reptiles, large insects, small mammals, and terrestrial birds up to moderate size. Eats carrion. **Flight:** Soars in dihedral; kites, hovers. **Voice:** Vocalizes when agitated, especially when harassed by Crested Caracara. Emits soft, short, nasal, high-pitched *kair* or *kee-aah*.

78a. Lightly marked type: *Broad white patch encompassing rear of supercilium, auriculars, and side of neck. White edges border dark throat.* Breast is all-white.

78b. Moderately marked type: *White patches of supercilium, auriculars, and side of neck are isolated into 2 or 3 groups. White edges border dark throat.* Breast is all-white.

78c. Heavily marked type: *Faint tawny or white patches on supercilium, auriculars, and side of neck. Throat is all-dark. Small white patch adorns mainly dark breast.*

78d. Lightly marked type: *Large white head patch* (as on *78a*). *Throat is dark. Wings have small, tawny shoulder patch.* White underparts are moderately marked on belly, flanks, and leg feathers. *Wingtips extend beyond tail tip. Tail is finely banded on middle and distal parts* (as on *78h*).

78e. Moderately marked type: *2 white head patches* (as on *78b*), *but can also have 1 large patch* (as on *78a*). *Wings have small, tawny shoulder patch.* Breast is white; dark and speckled belly, flanks, and leg feathers. *White undertail coverts. Wingtips extend beyond tail tip. Tail is fully banded* (as on *78g*).

78f. Heavily marked type: *Faint pale head patches* (as on *78c*). *Body is dark, except small, vertical white streak on breast. Undertail coverts are white and barred. Wingtips extend beyond tail tip. Note:* Rarely, breast can be all-dark (W. S. Clark pers. comm., photos).

78g. Tail, banded type (dorsal): Numerous, equal-width bands. *White uppertail coverts.*

78h. Tail, finely banded type (dorsal): *Numerous, thin, V-shaped bands that become obscure on basal region. White uppertail coverts.*

78i. Lightly marked type: *Flight feathers are gray; trailing edge of secondaries is very bowed. Wingtips are pointed. Axillaries are mainly dark.* Forward edge of wing is heavily marked; can appear as patagial mark. *Pale head patch and dark throat.*

78j. Moderately marked type: *Flight feathers are gray; trailing edge of secondaries is very bowed. Wingtips are pointed. Underwing coverts are darker than on 78i. Breast is white, and belly and flanks are dark.*

78k. Heavily marked type: *Flight feathers are gray; trailing edge of secondaries is very bowed. Wingtips are pointed. Underparts are dark, except white patch on breast and white undertail coverts.*

78l. All types: *Dark upperparts with tawny patch on forward edge of wing and white uppertail coverts.* Flight feathers are uniformly dark; secondaries are barred.

78m. Flight silhouette: Wings are held in dihedral when soaring and kiting.

a
b
c
d
e
f
g
h
i
j
k
l
m

Plate 79. WHITE-TAILED HAWK (*Geranoaetus albicaudatus*) 1-Year-Old

Ages: 1st of 2 immature annual stages between juvenile and adult. Body plumage has 3 types of color-morph-like variations similar to those found in juveniles. Head is always black, including throat (paler types have small white chin patch); forehead and lores are white. Eyes are medium brown. *Basal ½ of bill is pale blue. Cere is bluish green. Upperwing has small, rufous patch only on shoulder.* Legs are long and yellow. *When perched, wingtips extend far past tail tip; tail is adult length and much shorter than juvenile's. In flight, wings are long, with pointed wingtips and broad secondaries* (width of adult's). *Trailing edge of secondaries is distinctly bowed and narrows considerably at body junction. Underside of flight feathers is gray, with inner primaries darker; wide black band on trailing edge. Note:* Wing molt is identical to that of other large raptors (variances depicted by W. S. Clark, unpubl. data). May retain outermost 1–4 juvenile primaries (usually outer 1 or 2 [p9, p10]), which are brownish and worn, but many complete the primary molt. May retain several juvenile secondaries, *which are shorter and faded brown, in 1 or 2 groups. White uppertail coverts. Short, adult-length tail is gray, variably banded with dusky subterminal band. Note:* New data on *unique* "1-year-old partial tail molt" into adult-like white feathers with wide black subterminal band that occurs mainly in autumn (pers. obs./photos; statistics based on unpubl. independent data by W. S. Clark): About 75% molted into at least 1 adult-like feather; many molted 2 or more feathers and not always in sequential sets. None completely molted all of tail. About 25% did not start tail molt until typical 1st prebasic molt in spring. **Subspecies:** *G. a. hypospodius* in s. Tex. to S. America; 2 more subspecies in S. America. **Color morphs:** None in this subspecies (dark morph occurs in other 2 subspecies). **Size:** L: 18–22" (46–56cm), W: 49–53" (124–135cm). Females average larger. **Habits:** Tame. This is a gregarious species. Uses exposed perches, including utility poles, wires, and ground. **Food:** Perch and aerial hunter of amphibians, reptiles, large insects, small mammals, and terrestrial birds up to moderate size. Eats carrion readily. **Flight:** Soars in dihedral; kites, hovers. **Voice:** Vocalizes when agitated, especially if harassed by Crested Caracara. Emits soft, short, nasal, high-pitched *kair* or *kee-aah.*

79a. Lightly/moderately marked type: *Black head and throat, sometimes with retained juvenile feathering on supercilium and cheeks. Cere is always green. Lores are white. Breast is white.*

79b. Heavily marked type: *Black head and throat. Cere is green. Lores are white.*

79c. Tail, banded (dorsal): *Pale gray, with partial or nearly full, thin dark bands; dusky subterminal band.*

79d. Tail, 1-year-old partial molt (dorsal): *Pale gray with few or no thin bands; dusky subterminal band. 2 central feathers depict partial tail molt that attains adult-like gray/white pattern with fairly wide black band.*

79e. Lightly marked type: *White underparts have rufous and black barring and mottling on belly, flanks, and leg feathers. Rufous shoulder patch. Scapulars are black. Wingtips extend far past tail tip.*

79f. Moderately marked type: *White breast; belly, flanks, and leg feathers are variably marked with rufous and black. Rufous shoulder patch. Scapulars are black. Wingtips extend far past tail. Tail is banded type.*

79g. Heavily marked type: Black body except for thin strip of white on center of breast. *Rufous shoulder patch. Wingtips extend far past tail tip. Note: Partial molt into adult-like tail feathers with wide black band (79d).*

79h. Lightly marked type: *Dark gray flight feathers; very bowed trailing edge.* Outer 3 primaries (p8–10) and 4 secondaries (s3, s4, s8, s9) are retained juvenile feathers. *Patagial area is rufous.* Belly is mottled.

79i. All types: *Black, with rufous shoulder patch and white uppertail coverts.* A secondary (s4) is a retained brownish and shorter juvenile feather.

79j. Moderately marked, rufous type: White breast. *Patagial area is rufous. Flight feathers are gray.* Outer 2 primaries (p9, p10) and 2 middle secondaries (s4, s9) are retained juvenile feathers.

79k. Moderately marked type: *Rufous on patagial area. Gray flight feathers are darker on inner primaries.* Outermost primary (p10) and a secondary (s8) are retained juvenile feather. Mid-tail has adult-like wide, black band of "1-year-old partial tail molt".

79l. Heavily marked type: *Body is black, with small white breast patch. Patagial area is rufous.* All juvenile flight feathers are replaced (uncommon). Tail has some adult-like white feathers with black band of "1-year-old partial tail molt".

a
b
c
d
e
f
g
h
i
j
k
l

Plate 80. WHITE-TAILED HAWK (*Geranoaetus albicaudatus*) 2-Year-Old and Adult

Ages: *2-year-old and adult.* 2-year-old is 2nd of 2 immature annual stages between juvenile and adult. More or less attains adult plumage as 3-year-old. Eyes are medium brown (appear reddish brown at certain angles). *Basal ½ of bill is pale blue. Cere is bluish green.* Forehead and lores are white. *Upperwing has rufous patch on shoulder.* Legs are long and yellow. *Tail is short, with wide black band. When perched, wingtips extend far past tail tip. In flight, wings are long and wide* (same width in both ages). *Trailing edge of broad secondaries is distinctly bowed and narrows considerably at body junction; wingtips are pointed. Underside of flight feathers is gray, with inner primaries distinctly darker; wide black band on trailing edge. 2-year-old.—* Upperparts are black (as on 1-year-old); *may have rufous on scapulars as on adults.* Belly and flanks vary from barred to nearly solid black. Tail is mainly whitish; black band not as wide as adult's. Some 1-year-old body and wing covert feathers may be retained. *Adult.—Gray upperparts; scapulars are mainly rufous.* Females are darker gray dorsally, have darker upperwing coverts, and are often more heavily barred on underparts than males. *White tail has wide, black subterminal band.* Younger birds (3-year-olds?) may have gray throat. **Subspecies:** *G. a. hypospodius* in s. Tex. to S. America; 2 more subspecies in S. America. **Color morphs:** None in this subspecies (dark morph in other 2 subspecies). **Size:** L: 18–22" (46–56cm), W: 49–53" (124–135cm). Females average larger. **Habits:** Tame. This is a gregarious species. Uses exposed perches, including utility poles and wires; also perches on ground. **Food:** Perch and aerial hunter of amphibians, reptiles, large insects, small mammals, and terrestrial birds up to moderate size. Eats carrion regularly. **Flight:** Soars in dihedral; kites, hovers. **Voice:** Vocalizes only when agitated, especially when harassed by Crested Caracara. Emits soft, short, nasal, high-pitched *kair* or *kee-aah.*

80a. 2-year-old: Black head, throat, and side of neck. White spot adorns chin. Breast is white.

80b. Young adult: *Gray head and throat; white chin spot.* Cere is bluish green.

80c. Adult: *Gray head and white throat.* Cere can be dull yellow on males.

80d. 2-year-old, lightly/moderately marked type: *Black upperparts with rufous shoulder patch; rufous on scapulars.* White underparts are moderately barred on belly, flanks, and leg feathers. *Whitish tail has moderately wide black band. Wingtips extend far past tail tip.*

80e. 2-year-old, heavily marked type: Black upperparts, including scapulars. *Rufous patch adorns only shoulder on this bird.* Belly and flanks are mainly black. *Note:* Leg feathers are also barred.

80f. Adult female: *Gray upperparts with rufous on shoulder and scapulars. Belly, flanks, and leg feathers are barred. White tail has very wide, black subterminal band.* Upperwing coverts are darker than on male.

80g. Adult male: *Gray upperparts with rufous on shoulder and scapulars. Underparts are white, barred only on flanks. White tail has very wide, black subterminal band. Note: Some older females may be similarly marked on ventral areas.*

80h. 2-year-old, lightly marked type: *Black head. Inner primaries are darker than secondaries; black band on trailing edge. White tail has moderately wide black band. Note:* Has a few 1-year-old underwing coverts (dark blotches).

80i. 2-year-old, all types: *Black upperparts with rufous on shoulder and scapulars.* Lower back is speckled or barred black and white. Uppertail coverts are white. *Whitish tail has moderately wide black band.*

80j. Adult tail: *White tail with thin inner bands and very wide, black subterminal band.*

82k. 2-year-old tail: Many feathers are dusky, not white, and often not as neatly banded as on adult.

80l. 2-year-old tail: *Adult-like whitish, but black subterminal band can be a bit narrower.*

80m. Adult: *Gray upperparts with rufous shoulder and scapular patches. Lower back, uppertail coverts, and tail are white; wide, black subterminal band on tail.*

80n. Adult: *Gray head. Inner primaries are darker than secondaries; black band on trailing edge. White tail has wide, black band. Note:* Belly is barred as on *80f;* typical of some heavily marked females.

a
b
c
d
e
f
g
h
i
j
k
l
m
n

WHITE-TAILED HAWK (*Geranoaetus albicaudatus*)

HABITAT: Semi-open and open, grazed grasslands and savannas with isolated or small tracts of mesquite thornbush or small to large trees. Coastal areas are seasonally hot and humid; interior areas are seasonally hot and arid. In fall through winter all ages, but especially younger birds, are found in large groups in agricultural locales that are being control-burned, harvested, or plowed. Does not breed in extensive agricultural or wooded regions found in portions of some coastal Tex. counties. **STATUS:** Fairly common in its restricted range in Tex., where it is listed as threatened species. There are 200–400 pairs in the state; population is currently deemed stable. This hawk has undergone severe habitat loss since the 1920s caused by massive agriculture conversion in s. Tex. and urban sprawl in n. area of range. The spread of dense mesquite tracts from late 1800s to early 1900s in grassland areas that were not converted to agriculture also reduced habitat, but programs since then to eradicate mesquite and regain grassland habitat have assisted the species. Decline was also noted during organochlorine era of 1940s–70s. The species (including 2 additional subspecies) is fairly common south of U.S. in proper habitat, often in rather fragmented areas, southward throughout S. America. **NESTING:** Pairs begin nesting activities in Jan.–Feb. Both members build large nest, lined with fresh greenery, which is placed in middle to upper part of live or dead bush or tree. Nests can be as low as 1.5' (0.5m) or as high as 41' (12.5m), but average 9' (2.7m) high. Clutch has usually 2 eggs but can have 1–4. Female does most of incubation. Fledglings become independent by mid–late summer. **MOVEMENTS:** Mainly sedentary. There is considerable seasonal shifting fall–spring within and adjacent to breeding habitat. Younger ages and some adults move somewhat north, west, or south of breeding areas. There does not appear to be a true migration. Younger birds tend to have irregular extralimital dispersal northeast and northwest of typical range mainly within Texas but now occurs regularly northeast to sw. La. **Louisiana.** Dispersing young birds have been seen fairly regularly in sw. Cameron, Calcasieu, and Jefferson Davis Cos. The first record (specimen) was in 1888. Sightings did not occur again until 1995. Documented records have been regular since then. Majority of sightings are of juveniles and immatures, but also a few adults. Most occur Sep.-May, with most spanning Oct.-Mar. There are 21 accepted records through Aug. 2017 (losbird.org 2017). **Texas.** Species has been found as far north as Hopkins Co., northwest to Lubbock Co. (1 record), and as far west as Jeff Davis Co. (in early Jul.; M. Lockwood pers. comm.). **Arizona.** Sight records in Cochise, Pima, and Maricopa Cos., but none documented with photographs. **COMPARISON:** ***White-tailed Hawk juvenile/immatures.*** **Swainson's Hawk (juveniles and 1-year-olds, all color morphs).**—Dihedral wing attitude, pointed wingtips, and grayish underside of flight feathers are similar; but rear edge of wing is more straight-edged on Swainson's, *vs.* bowed. Cere is usually yellow, but can be greenish, *vs.* bluish green. When perched, wingtips are either slightly shorter than or equal to tail tip, *vs.* extending well beyond tail tip. Throat is either all-white/all-tawny or has thin, dark mid-throat streak, *vs.* large, black mid-throat patch or all-black throat on darker plumage types. **Red-tailed Hawk (juveniles, all subspecies, color morphs).**—Greenish cere, *vs.* usually bluish green. Dark bellyband on lighter types and white breast markings on darker types, especially "Harlan's," are similar to this species: use caution and use features described below to separate. Eye is pale yellow, gray, or tan, *vs.* dark brown. When perched, wingtips are distinctly shorter than tail tip, *vs.* extending far beyond tail tip. Undertail coverts on darker birds are dark, *vs.* mostly white or white with dark blotches. Wings are held on flat plane when soaring, *vs.* held in dihedral. **Ferruginous Hawk (rufous/intermediate and dark morphs, both ages).**—Eye is pale yellow to medium brown, *vs.* dark brown. Cere is always bright yellow, *vs.* bluish green. Legs are fully feathered, *vs.* long, bare legs. When perched, wingtips are somewhat shorter than tail tip, *vs.* extending far beyond tail tip. Dihedral wing attitude and pointed wingtips in flight are similar, but underside of flight feathers is unmarked and white, *vs.* dark gray and thinly barred. Tail is white or gray and if banded, as on juvenile, is coarsely marked, *vs.* thin bands or wide, dark subterminal band. ***White-tailed Hawk immatures/adult.*** **"Krider's" Red-tailed Hawk (adult).**—Has similar white tail, but subterminal band is thin, *vs.* wide, black. **"Harlan's" Red-tailed Hawk (adult darker morphs).**—White breast on black plumage is virtually identical, as is tail pattern of 1-year-old White-tailed Hawk: Use caution! When perched, Harlan's wingtips are shorter than tail tip, *vs.* extending far past tail tip. Wings held on flat plane, and wingtips rounded, *vs.* dihedral and pointed wingtips.

Figure 1. North Padre Island, Kenedy Co., Tex. (Sep.)
Year-round coastal savanna habitat.

Figure 2. Cameron Co., Tex. (Jun.)
Unoccupied old nest in a small Honey Mesquite (*Prosopis glandulosa*) on coastal savanna.

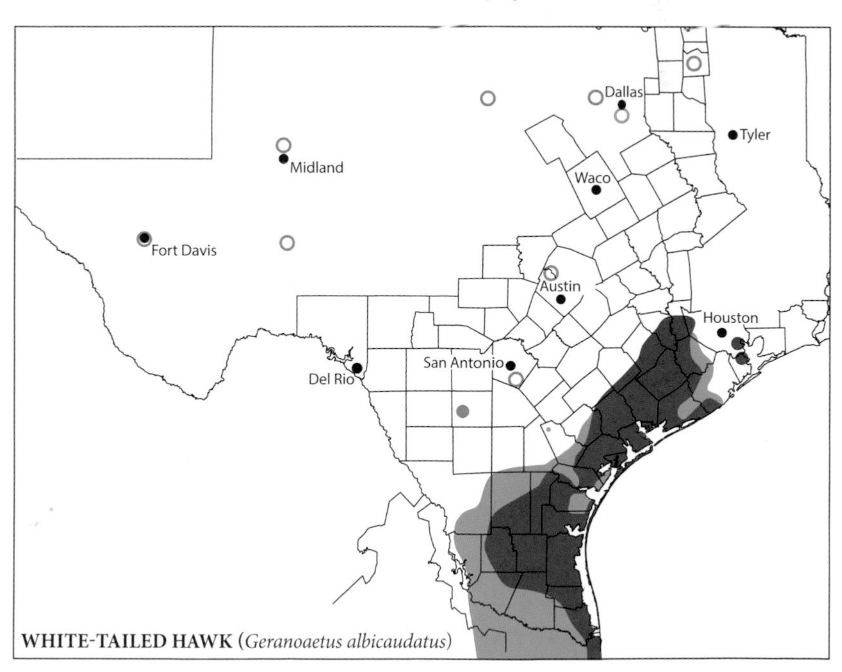

WHITE-TAILED HAWK (*Geranoaetus albicaudatus*)

Plate 81. GRAY HAWK (*Buteo plagiatus*)

Ages: Juvenile and adult. Juvenile plumage is retained for 1st year. *In flight, wings are relatively short, with rounded wingtips, and are held rigidly perpendicular to body. Black bill is short and deep, with large yellow cere.* When perched, wingtips are far shorter than tail tip. *White uppertail coverts.* Legs are long and thick. *Juvenile.*—Facial pattern is striped: *White supercilium and auriculars contrast sharply with dark eye line and thin, dark malar mark.* Eyes are medium brown. Typically has dark center-throat streak (shared by some juvenile Broad-winged [*22a*] and Red-shouldered Hawks [*19a, b*]). Streaked underparts have heavier cluster of streaking on side of neck. Leg feathers are crossed by thin bars. *On dorsal side, pale brown secondaries and their greater coverts are crossed by thin black bars. All flight feathers are finely barred on pale underwing.* Tail is much longer than adult's. *Tail has 7 or 8 dark, V-shaped bands that get progressively thinner toward base of tail. Dark spot adorns each white uppertail covert. Adult.—Underparts are finely barred gray and white.* Females typically have darker heads and more brownish upperparts than males. *Black tail has 2 white bands: thin basal band and thick mid-tail band. Uppertail coverts are plain white.* **Subspecies:** Monotypic. **Color morphs:** None. **Size:** L: 14–18" (36–46cm), W: 32–38" (81–97cm). Females average larger. **Habits:** Fairly tame to tame (tamer south of U.S). Perches on exposed or concealed branches; especially fond of utility poles and wires. **Food:** Perch hunter. Feeds on lizards but also eats small birds, rodents, and large insects. **Flight:** Quick, snappy accipiter-like wingbeats. Hunting occurs at very low altitudes. Wings are held on flat plane when soaring. Does not hover or kite. **Voice:** Loud, piercing *kah-lee-oh*; also, repeated *kah-lee*. *Note*: Scientific name was changed from *Asturina nitidia plagiatus* to *Buteo plagiatus* in 2012 (AOU 2012).

81a. Juvenile: *White supercilium and white auriculars, with distinct, thick dark eye line and dark malar mark. Bill is black, with large yellow cere.* Compare Broad-winged Hawk juvenile (*22a, b*).

81b. Juvenile secondary feather: *Pale brown with distinct, thin black bars.* Compare juvenile Red-shouldered Hawks ("Eastern"/"Southern," *19e*; "California," *21e*); juvenile Broad-winged Hawk (*22h*).

81c. Juvenile: *Striped head. Pale brown secondaries and greater coverts are distinctly barred. Legs are long and thick. Long, brown tail has V-shaped bands that get progressively thinner toward base; some bands are edged with thin, pale lines. Note:* Upperwing greater coverts are also pale brown and barred.

81d. Juvenile: *Pale, short, rounded wings are held perpendicular to body. All flight feathers are thinly barred. Long tail has black bands that get progressively thinner toward tail base.*

81e. Juvenile tail (dorsal): *White uppertail coverts have small, dark spots. Uppertail is brown, with 7 or 8 V-shaped black bands that get progressively thinner toward base of tail.* Compare juvenile Red-shouldered Hawks ("Eastern"/"Southern," *19g, 19i*; "California," *21d*); juvenile Broad-winged Hawk (*22d, 22e*).

81f. Juvenile: *Distinctly barred flight feathers and upperwing greater coverts. White uppertail coverts have small, dark spots. Long tail has V-shaped black bands that get narrower toward tail base.*

81g. Flight silhouette: *Soaring.* Wings held on flat plane.

81h. Adult: *Gray head and dark eye. Bill is black, with large, bright yellow cere. Underparts are finely barred with gray and white.*

81i. Adult male: *Gray head and dark eye.* Upperparts are grayish brown; often more cross-barred than on female. *Gray and white underparts are finely barred. Tail is black, with 1 thin and 1 wide white band.*

81j. Adult female: *Gray head and dark eye.* Upperparts are brownish, especially on scapulars and back; often less cross-barred than on male. *Gray and white underparts are finely barred. Tail is black, with 1 wide and 1 thin white band.*

81k. Adult: *Pale, short, rounded wings with front edge held perpendicular to body. Underparts are finely barred with gray and white. Short, black-and-white-banded tail. Note:* Only 1 wide, white distal band shows when tail is closed.

81l. Adult: Grayish upperparts. *White uppertail coverts. Black tail has 1 wide, 1 thin white band.*

81m. Adult tail (dorsal): *Uppertail coverts are white. Black tail has 2 white bands, 1 thin, 1 thick.*

a
b
c
d
e
f
g
h
i
j
k
l
m

GRAY HAWK (*Buteo plagiatus*)

HABITAT: Arizona. *Summer.*—Lowland arid regions of mesquite-hackberry thornbush tracts adjacent to streams, rivers, or dry washes with large, solitary cottonwood or Emory Oak (*Quercus emoryi*) or gallery forests of cottonwood. (Does not need water other than for growth of large trees for nest sites.) Mainly found in undisturbed areas but may occupy rural areas, even near occupied homesteads, but does not reside in suburban locales. All breeding areas are between 2,000' and 4,000' (600–1220m) elevation. There is recent expansion into lower-elevation montane canyon areas north of Phoenix that have tall cottonwood growth. **New Mexico and West Texas.** *Summer.*—Habitat is similar to that of Ariz. Pairs in sw. N.Mex. and Jeff Davis Co., Tex., are found in lower-elevation montane riparian canyons with tall oaks and cottonwoods. Birds in s.-cen. N.Mex. and breeders in Eddy Co., N.Mex., occupy low-elevation thornbush areas adjacent to riparian stretches with tall cottonwoods. **South Texas.** *Year-round.*—Humid and seasonally hot region with tracts of thornbush interspersed with single or tracts of tall Emory or live-oaks or large mesquite. **STATUS:** This species is very common south of U.S.; found from sw. U.S. to nw. Costa Rica (AOU 2012). Gray Hawk was given full species status in 2012 (AOU 2012). It was previously considered the northern counterpart of the-now Gray-lined Hawk (*Asturina nitidia*), the southern counterpart that ranges south of Costa Rica. **Arizona.** Very local and uncommon summer resident numbering 80-plus pairs. There is recent expansion north and west of historic range. One-third of population breeds along Santa Cruz River in Santa Cruz Co. A few pairs reside in the Patagonia area and along Sonoita Creek in Santa Cruz Co. A few pairs nest in Buenos Aires N.W.R. and near Arivaca in Pima Co. Numerous pairs nest along San Pedro River south of Gila River in Graham and Cochise Cos. **New Mexico.** This species has recently expanded its range into this state. Data that follow are based on Williams and Krueper (2008): Regular in Grant and Hidalgo Cos. since 1992; seen annually since 2000. 1st state nesting was in Guadalupe Canyon, Hidalgo Co., in 2004. A pair nested in 2007 along Pecos River, Eddy Co. Recent records, but no documented breeding, along Rio Grande in Doña Ana, Sierra, and Socorro Cos. **Texas.** This is a very uncommon and local species in the state. Status is unknown. Breeds in Jeff Davis Co. 1 or 2 pairs breed in Big Bend Nat'l. Pk. in Brewster Co. and Big Bend Ranch S.P. in Presidio Co. Many pairs are resident in isolated tracts along Rio Grande from w. Starr Co. to e. Cameron Co. Found in isolated patches north of the river in e. Hidalgo and w. Willacy Cos. **Extralimital.** 1 record of an adult in Clay Co., Kans. (1st in state), in 1993. Juvenile photographed in Santa Barbara Co., Calif. (1st state record) late Nov. 2012–mid-Mar. 2013 (Pike et al. 2014). Minor northward dispersal noted at Corpus Christi, Tex., with 1st-ever juvenile (photographed) in 3rd week of Aug. 2017, at Hazel Bazemore Co. Pk., Nueces Co. **NESTING:** Begins in late Mar.; ends late Jul. Breeding occurs at age 2, when in full adult plumage. Nests are located 40–100' (12–30m) high in dominant live tree. Clutch of 2 or 3 eggs is incubated by female. Nesting activities will be abandoned with prolonged direct human disturbance. Fledglings may remain in natal territory until fall migration in Ariz. and w. Tex.; undetermined time in s. Tex. **MOVEMENTS:** Ariz., N.Mex., and w. Tex. populations are migratory. Those in s. Tex. are sedentary or migratory. *Fall.*—Both ages migrate together in Sep.–Oct. *Spring.*—Adults return to breeding territory in Mar. Returning 1-year-olds arrive in natal areas in May. Some temporarily occupy adult breeding territories at this time. **COMPARISON:** ***Gray Hawk juvenile.*** **Cooper's Hawk (juvenile).**—Pale gray or yellow eye, *vs.* dark brown eye. Nondescript brown head, *vs.* strongly marked dark malar mark and dark eye line. Dark uppertail coverts, *vs.* white coverts. Dorsal side of long tail has 3 or 4 wide, equal-width light and dark bands, *vs.* 6 or more thin dark bands that get progressively thinner on basal part of the likewise long tail. Wing shape is similar, but Cooper's has thick dark bars on outer primaries, *vs.* thin, dark bars. **Red-shouldered Hawk (juvenile).**—Head pattern lacks sharply defined white and dark marks of Gray Hawk; eyes similar medium brown, can be paler. Ventral markings, barred leg feathers, and long legs are similar. Pale bars on dark dorsal side of secondaries, *vs.* dark bars on pale brown secondaries. Dark uppertail coverts are edged in white, *vs.* white coverts. Pale tawny crescent-shaped patch on base of outer primaries in flight, *vs.* uniformly brown outer primaries. Thin tail bands are equal-width or wider on basal area, *vs.* dark bands progressively thinner toward basal area. **Broad-winged Hawk (juvenile).**—Shares similar dark mid-throat streak, dark malar, and often thinly barred leg feathers. Lacks dark line behind eye, *vs.* pronounced dark line. Wingtips are pointed in flight, *vs.* rounded. Uppertail coverts are white-edged, *vs.* white. Dorsal tail has wide subterminal band and nearly equal-width thin inner bands, *vs.* wide subterminal band and thin inner bands that get progressively thinner towards base of the tail. ***Gray Hawk adult.*** **Hook-billed Kite (Tex. only).**—White eye, *vs.* dark brown. Yellow teardrop-shaped fleshy mark on lores, *vs.* whitish lores. Gray body color and ventral markings are similar. Uppertail coverts are gray/black, *vs.* white; tail similarly marked. Underwing is black with white spots, *vs.* mostly whitish with fine gray barring.

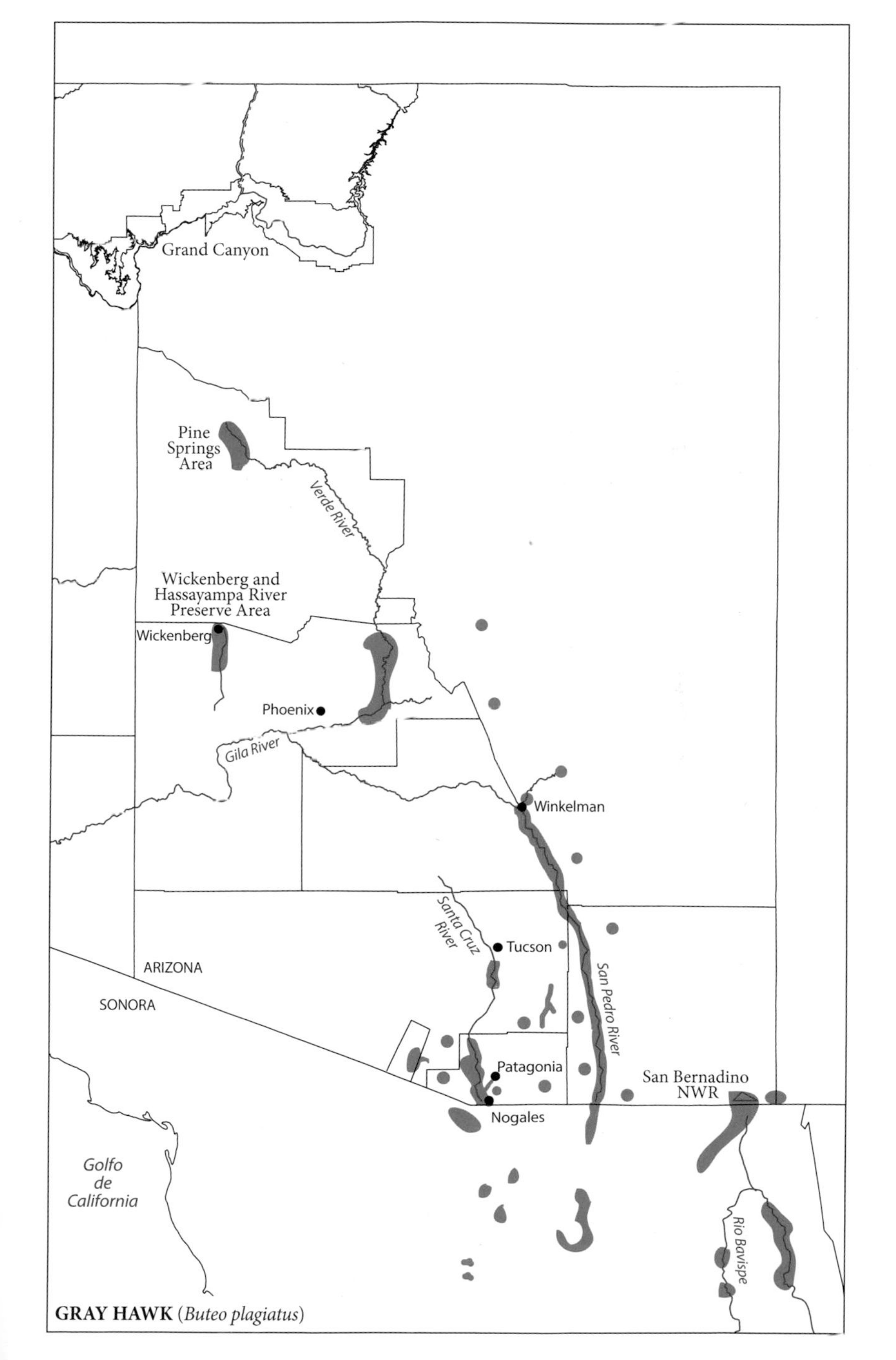

GRAY HAWK (*Buteo plagiatus*)

Figure 1. San Pedro River, Cochise Co., Ariz. (Jun.)
Nest territory with large Fremont Cottonwood adjacent to thornbush habitat.

Figure 2. Bentsen–Rio Grande Valley St. Pk., Hidalgo Co., Tex. (Apr.)
Year-round habitat in mesquite woodland.

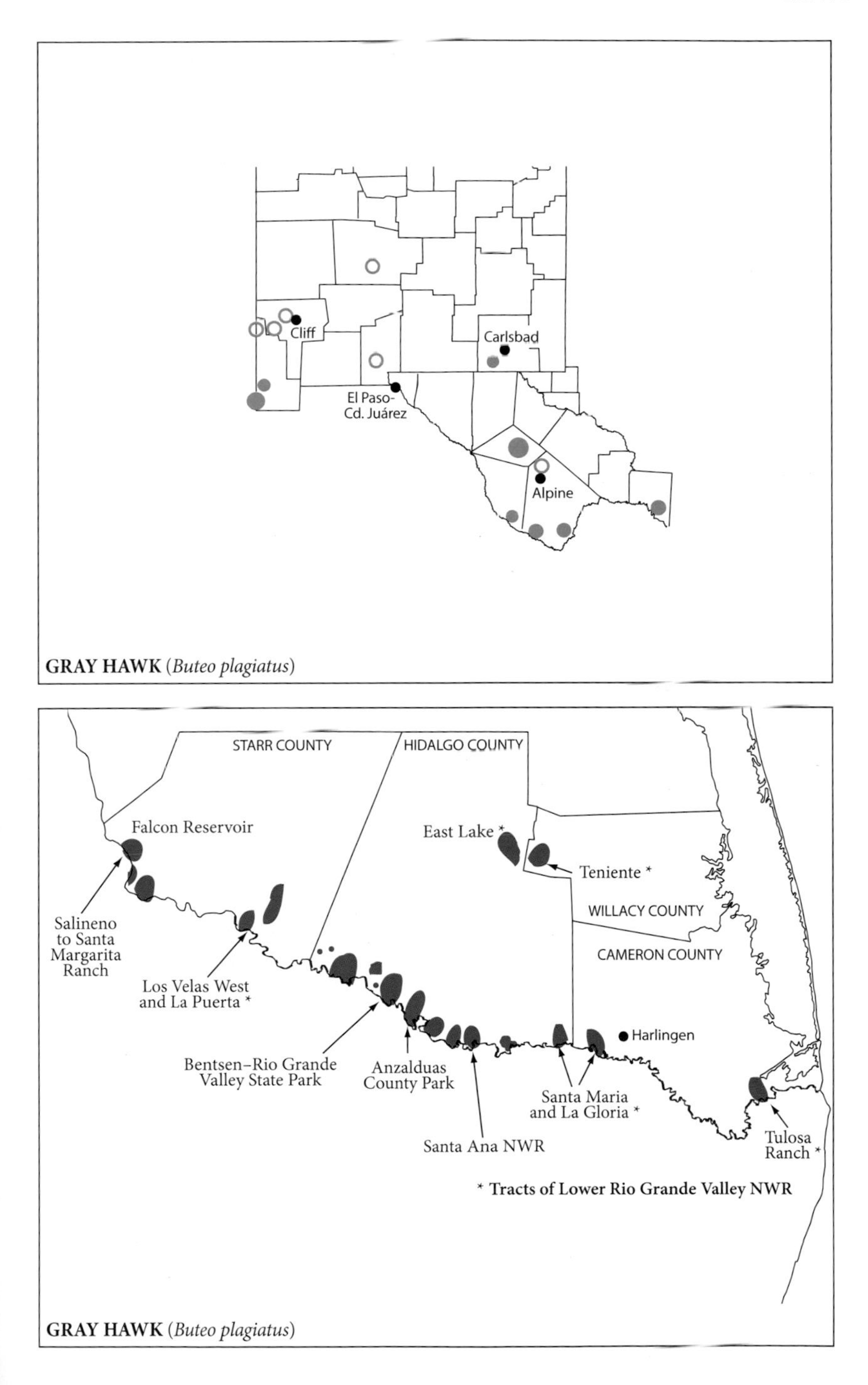

GRAY HAWK (*Buteo plagiatus*)

GRAY HAWK (*Buteo plagiatus*)

Plate 82. SHORT-TAILED HAWK (*Buteo brachyurus*)

Ages: Juvenile and adult. Juvenile plumage held for 1st year. In flight, wings are long, bowed on trailing edge, and have fairly pointed tips. *When perched, wingtips equal tail tip. White forehead and outer lores form mask; inner lores are usually dark but can be pale.* Eyes are dark brown. *Juvenile.*—Underside of wings has moderately wide, gray band on trailing edge. *On light morph, basal portions of primaries and secondaries are white; distal portions are gray. On dark morph, white on basal area of primaries; secondaries are all-gray.* Both color morphs typically have full tail banding; subterminal band is equal in width or a bit wider than inner bands. *Adult.*—Underside of wing has wide, black band on trailing edge; *gray secondaries and inner ½ of primaries; white outer ½ of primaries.* Tail has 2–4 complete, fairly wide, black inner bands and wider, black subterminal band. *Note:* All ages in West, but especially light morph juveniles, typically have *much* more defined tail banding than Fla. birds. **Subspecies:** *B. b. fuliginosus* from s. Ariz. through Panama; also in Fla. *B. b. brachyurus* is in S. America. **Color morphs:** Light and dark morphs (no cline). Juvenile dark morph has variable ventral patterns. **Size:** L: 15–17" (38–43cm), W: 32–41" (81–104cm). **Habits:** Tame. Perches for short stints in tree canopies when feeding and in inclement weather; perches less commonly on exposed branches, very rarely on utility poles (suburban Tucson, Ariz.). **Food:** Strictly aerial hunter of sparrow-size to jay-size birds perched on outer branches and bushes; less commonly chipmunks, lizards. **Flight:** Highly aerial. Wings held flat, with wingtips bent upward when soaring/kiting. Hunts by kiting on hillsides; vertical dives often interspersed with stints of kiting or "parachuting" vertically downward. Prey may be eaten in flight. Does not hover. **Voice:** Drawn out, high-pitched *keee* at nest sites; emits *kree, kree, kree* at other times.

82a. Juvenile light morph: *White mask. Inner lores are dark. Dark eye line is bordered by pale supercilium and cheek; dark malar mark. Note:* Throat may have thin, dark center streak (*only* in West).

82b. Juvenile dark morph: *White mask.* Inner lores, and all of rest of head is dark brown.

82c. Juvenile, 4-banded tail (dorsal): Bands may be V-shaped. *Note:* Light morph shown.

82d. Juvenile, 5-banded tail (dorsal): Bands may be straight across each feather. *Note:* Dark morph shown.

82e. Juvenile light morph: *Paler-headed than 82a, with distinct thin, dark eye line and dark malar mark.* Streaked flanks and often streaked on mid-belly. Upperparts are solid brown.

82f. Juvenile dark morph, mottled type: *Dark bib on breast; mottled belly.*

82g. Juvenile light morph tail (ventral): 2 or 3 dark inner bands show; subterminal band is wider. Outer thin dark bands may not align with inner bands.

82h. Juvenile dark morph tail (ventral): 2 or 3 dark inner bands show; subterminal band is wider.

82i. Juvenile light morph: *Gray on distal ½ of secondaries.* Wingtips are pointed.

82j. Flight silhouette: *Soaring/kiting. Wings are held on flat plane, with wingtips bent upward.*

82k. Juvenile dark morph, mottled type: Dark head and bib, with white-speckled coverts and belly. *Secondaries are gray; white on inner primaries.*

82l. Juvenile dark morph, all-dark type: Mainly dark body and coverts; may have white speckling on axillaries. *Secondaries are gray; white on much of primaries. Note:* Uncommon type.

82m. Adult light morph: *White mask. There is a rufous patch on side of neck.*

82n. Adult light morph: *Rufous side-of-neck patch. Tail has 4 or 5 dark bands.*

82o. Adult light morph, aberrant type: Tawny underparts, with some breast and flank streaking. Mid-throat has thin streak. *Note:* Adult that wintered in Tucson, Ariz., 2008–11 (images from Jan. 2009).

82p. Adult dark morph: *Dark brown body, including undertail coverts.*

82q. Adult light morph: *White/tawny body contrasts with gray secondaries and inner ½ of primaries.*

82r. Adult light morph tail (ventral): *Fully banded, with wider, dark subterminal band. Note:* Full tail banding on light morph seen in West; rarely in Fla.

82s. Adult dark morph: *White mask*; rest of head is dark brown.

82t. Adult light morph: *Dark brown upperparts, including uppertail coverts; darker flight feathers. Tail has 4 or 5 dark bands.*

82u. Adult dark morph: *Dark upperparts; flight feathers may be darker. Tail has 4 or 5 dark bands.*

82v. Adult dark morph: *Dark body, including undertail coverts. Secondaries and inner ½ of primaries are gray. Multiple tail bands and wider black subterminal band.*

a
b
c
d
e
f
g
h
i
j
k
l
m
n
o
p
q
r
s
t
u
v

SHORT-TAILED HAWK (*Buteo brachyurus*)

HABITAT: Arizona. *Summer.*—High montane "sky islands" with forests of tall coniferous trees. *Winter.*—Suburban areas with groves of tall trees of any type. Also have been found in this habitat in se. N.Mex. **West Texas.** *Summer.*—Found in coniferous woodlands at montane elevations. **Central Texas.** *Summer.*—Hilly, arid locales with large tracts of mature, tall, deciduous trees (mainly oaks). **East/South Texas.** *All seasons.*—Humid, low-elevation, flat terrain with semi-open tracts of moderate-size and/or tall trees; nested in a tall palm. **STATUS: Arizona.** 1st sight record in the state was in Aug. 1985; 1st photo documentation in summer of 1999. Recently fledged youngsters, indicative of breeding, have been seen almost annually since summer of 2001 in Chiricahua Mts., Cochise Co. A light morph pair successfully fledged 2 youngsters there in summer of 2007—1st documented nesting for Ariz., only state outside of Fla. where this species nests; nested at same location in 2010 (Snyder et al. 2010). Probable nesting occurs in Huachuca Mts., Cochise Co. Recent summer records for Catalina Mts., Pima Co. Records in the state occur from late Mar. to late Oct. An aberrantly marked adult light morph inhabited suburban Tucson during winters of 2008–11, the state's 1st winter record. **Texas.** Annual visitor to the state. 1st record was in Jul. 1989; seen annually since 1994, mainly in Nueces Co. and areas bordering lower Rio Grande Feb.–Oct.; no records Nov.–Jan. Largest numbers are seen in Apr. It is irregular in summer in Big Bend Nat'l. Pk., Brewster Co.; 1 record for Jeff Davis Co. It occurs irregularly to Bandera Co. and has been noted once each in nearby Bexar and Uvalde Cos., and once each farther north in Comal and Hays Cos. Most northerly record is from Smith Co. in Oct. 2009. **New Mexico.** 1 sight record in late May 2005 in Animas Mts., Hidalgo Co. (Williams et al. 2007). *Color morph status.*—Mainly light morphs but a few dark morphs are seen in Tex. Ariz. birds have been light morph, except 1 early record of a dark morph. N.Mex. record is a light morph adult. **NESTING:** Based on Snyder et al. (2010), female tends to all nest duties, including feeding young and nest defense, during whole nesting cycle; male supplies food only. Clutch is of 2 eggs. **Arizona.** Late Mar.–Aug. The 2007 Ariz. nest was placed on totally exposed treetop of live, dominant-size Arizona Pine (*Pinus arizonica*). **Texas.** Only state breeding attempt was Mar. 31–May 27, 2004, in Cameron Co., when nest built in *Sabal* palm by light morph mated to light morph Swainson's Hawk was destroyed by a storm. **MOVEMENTS:** Not well documented, especially in Tex., where birds can be seen spring–fall. *Spring.*—Majority of Tex. records are of adults in Apr., but seen as early as Feb. (1 record) and Mar. (3 records) in se. counties. A few records of adults and 1-year-olds also exist later in spring, and a few in summer in s. Tex., which may be northward-dispersing birds out of Mexico. Ariz. adults seem to move into the state during Mar.; 1-year-olds probably arrive later, in Apr.–May. *Fall.*—Data from hawk watch at Hazel Bazemore Co. Pk., Nueces Co., Tex., show they are fairly regular late Aug.–late Oct., with most records in Sep.–Oct. Ariz. population probably moves during same period. **COMPARISON:** *Short-tailed Hawk light morph juvenile.* **Red-shouldered Hawk (juvenile).**—All underparts are streaked or barred, *vs.* some flank, belly streaking. When perched, wingtips are far shorter than tail tip, *vs.* equal to tail tip. In flight, tawny or white window/panel shows on base of primaries, *vs.* lack of pale window/panel. Commonly perches along roadways on posts, utility wires, *vs.* used utility poles only in Tucson, Ariz. **Broad-winged Hawk (juvenile).**—Lacks thick dark eye line, *vs.* thick dark eye line. Wingtips are shorter than tail tip, *vs.* equal to tail tip. Holds wings flat *vs.* wings flat with upturned tips, and regularly kites. All-white flight feathers ventrally, *vs.* gray secondaries. Commonly perches along roadways on branches and wires, *vs.* used poles only in Tucson. **Red-tailed Hawk (juvenile).**—Wingtips are shorter than tail tip. Tan panel on dorsal primaries and greater primary coverts and whitish window on ventral side, *vs.* all-dark dorsal wing and lack of ventral window. Kites but also hovers, *vs.* only kites. Perches on open branches, poles, posts, *vs.* used poles only in Tucson. *Short-tailed Hawk light morph adult.* **Swainson's Hawk (adult, lighter morphs).**—White lores and white forehead are similar. Partial bib on some birds is similar to rufous neck patch on adult Short-tailed. Underwing has uniformly dark gray flight feathers, *vs.* gray secondaries and primaries white basally. Holds wings in dihedral in flight, frequently hovers, *vs.* wings flat with upturned tips, and never hovers. Wingtip-to-tail-tip ratio is identical when perched. Commonly perches on ground or open perches, *vs.* never perches on ground. For Swainson's juvenile, especially lightly marked birds, wing and flight data as noted for adult. *Short-tailed dark morph.* **Swainson's Hawk (darker morphs, all ages).**—Dark forehead, *vs.* white forehead. Adults have rufous underwing coverts, *vs.* all-dark underwing coverts. Undertail coverts are pale, *vs.* all-dark coverts. **Zone-tailed Hawk (juvenile).**—Similar wingtip-to-tail-tip ration when perched. Wings in dihedral and tipsy side-to-side flight, *vs.* flat wings with upturned tips, stable flight. Both species have white foreheads. **"Western" and "Harlan's" Red-tailed Hawks (juveniles).**—Eyes are pale, *vs.* dark brown. Wingtips are distinctly shorter than tail tip, *vs.* equal to tail tip. Kites but also hovers, *vs.* kites. Wingtips rounded, *vs.* pointed. Regularly flaps wings, *vs.* rarely flaps wings.

Figure 1. Carr Canyon, Huachuca Mts., Cochise Co., Ariz. (Jun.)
High elevation "sky island" montane summer (and possible breeding) habitat. Photo by Ned Harris

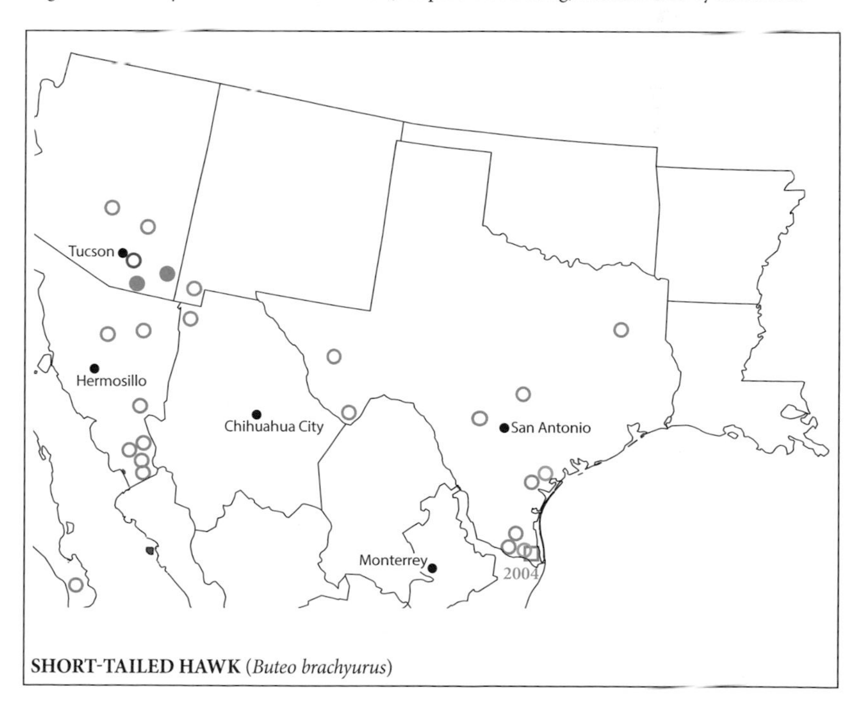

Plate 83. ZONE-TAILED HAWK (*Buteo albonotatus*)

Ages: Juvenile and adult. Juvenile plumage is retained for 1st year. In flight, wings are long and moderate width, with squared-off tips. *When perched, wingtips equal or extend past tail tip.* Upper mandible has small, pale blue spot on lower base. *Forehead and lores are white and form broad mask.* Eyes are dark brown. Cere is yellow. Legs are short, and feet are far short of tip of undertail coverts in flight. *Juvenile.—* Body is brownish black, which fades and wears to more brownish tone by spring. Plumage is variably speckled with white on nape, hindneck, and underparts; some lack speckling. *Underside of flight feathers is white and fully barred; narrow dark band on trailing edge. Tail has 4 or 5 V-shaped, black bands on brown upperside; bands on basal region are thicker. Adult.—Body is grayish black. Upperwing is more brownish on females.* Underside of flight feathers is gray and fully barred, with wide black band on trailing edge. Tail is black, with somewhat sexually dimorphic gray bands on dorsal side and white bands on ventral side. *Dorsal side of tail:* Both sexes have 1 moderate-width gray band and 1 thin, gray inner band; females may have 2 gray inner bands. When fanned or molting, much white shows on inner webs. *Ventral side of tail:* Both sexes show 1 moderate-width white band when closed. Both sexes have 2 thin, white inner bands, but on male innermost band lacks white on outer web. Female can have up to 4 full white bands. **Subspecies:** Monotypic. **Color morphs:** None. **Size:** L: 19–21" (48–53cm), W: 48–55" (123–140cm). Females average larger. **Habits:** Tame. Aggressive at nest sites. Uses concealed and open natural objects for perches; rarely uses utility poles or wires. **Food:** Low-altitude aerial hunter of terrestrial lizards, smaller birds, and rodents. **Flight:** Very aerial raptor. Soars and glides with wings held in high dihedral in tipping and rocking fashion, much as Turkey Vulture (and often seen where there are vultures). **Voice:** Vocal only at nest site, emitting nasal drawn-out *keeyah.*

83a. Juvenile: Small blue spot on lower base of upper mandible. *White forehead and lores form mask.* Eye is dark brown. Some have white speckles on nape and breast. Head can have brownish cast.

83b. Juvenile tail (dorsal): *V-shaped black bands are thicker on basal part.*

83c. Juvenile tail (dorsal): *V-shaped black bands have offset pattern; thicker on basal part.*

83d. Juvenile (fresh plumage): *Brownish black with white speckling. Wingtips equal tail tip.*

83e. Juvenile (worn plumage): *Dark brown with minimal white speckling. Wingtips equal tail tip.*

83f. Juvenile (fresh plumage): *Uniformly brownish black, including upperwings. Tail bands are wider in basal area.*

83g. Juvenile (fresh plumage): *White flight feathers are fully barred with thin bars; gray smudge on base of outer primaries. 2 or 3 thin, dark inner tail bands; subterminal band is much wider.*

83h. Adult: Small blue spot on lower base of upper mandible. *White forehead and lores form mask.*

83i. Adult male: Grayish-black body, including upperwings. *Tail has 1 wide, 1 thin gray band (both sexes).*

83j. Adult tail, fanned (dorsal): White inner webs are exposed when tail is widely fanned and/or in molt.

83k. Adult female: Uniformly grayish-black body; female often has more brownish on wing than male. *Wingtips extend beyond tail tip. Either sex can have pattern of 1 wide band and 1 thin band on dorsal side of tail.*

83l. Flight silhouette: High dihedral when soaring, gliding; rocks back and forth. White mask is often visible at a distance.

83m. Adult tail, closed (ventral): In both sexes, may show 1 moderate-width white band when closed.

83n. Adult female tail, closed (dorsal): *3 gray bands; innermost band often only on middle feathers.*

83o. Adult tail, fanned (ventral): *Both sexes can show 1 wide white band and 2 partial, thin, white inner bands; innermost white band does not extend onto outer web of outermost feather on most males.*

83p. Adult female tail feather (ventral): Innermost thin, white band extends onto outer web on female tail.

83q. Adult female tail, fanned (ventral): *4 bands, with 3 complete, thin, white inner bands, extending onto outer web of outermost feather set.*

83r. Adult male: *Flight feathers are gray and fully barred; wide black band on trailing edge. 2 white tail bands on males only.* Undertail coverts extend past feet.

a
b
c
d
e
f
g
h
i
j
k
l
m
n
o
p
q
r

ZONE-TAILED HAWK (*Buteo albonotatus*)

HABITAT: *Summer.*—Semi-open and wooded arid regions with areas along streams and rivers supporting single trees of groves, from flat lowlands to rugged topographic areas of hills and canyons up to moderate montane elevations, to 7,000' (2100m) in Ariz. and N.Mex. Foraging birds use open as well as forested habitat, up to 8,500' (2600m). *Winter.*—Arid, semi-open and wooded lowland areas with flat or hilly terrain are favored in the few areas where this hawk winters in U.S. **STATUS:** Very uncommon. Sparsely distributed even in core areas of Ariz., N.Mex., and Tex. Bulk of population is in Ariz., which has 150–200 pairs; upwards of 50 pairs each are in sw. N.Mex. and in Trans-Pecos and Edwards Plateau regions of Tex. A few pairs inhabit s. Calif., s. Nev., and sw. Utah. Range is expanding northward in most areas. **NESTING:** Begins in late Mar. and ends by late Aug. Large but flimsy nest is composed of thin sticks and is placed 40–100' (12–30m) high in primary or outer secondary fork in upper part of shaded canopy of tall, live tree. It may be isolated tree, in small group of trees, or in large gallery forest stand. Edwards Plateau population in Tex. often builds nests on cliff ledges. Nests may be reused and become fairly large and sturdy. (Nest is similar to that of Common Black Hawk, which often nests in close proximity). Clutch is 2 eggs, incubated by female. **MOVEMENTS:** Short- to moderate-distance migrant; Calif. birds may be resident. Most winter in Mexico; a few often winter in s. Tex. Migrates singly but may be among flocks of Broad-winged Hawks, Swainson's Hawks, or Turkey Vultures. *Spring.*—Breeding adults move north late Feb.–early Apr., but adult-plumaged females are seen at a cen. N.Mex. hawk watch throughout May. Most breeders are on territory by mid-Mar. Returning 1-year-olds head north in May–Jun. *Fall.*—Late Aug.–early Nov. for all ages; most movement is noted in late Sep.–Oct. *Extralimital dispersal.*—Birds disperse northward to cen. Calif. in late Aug.–early Sep., overwintering in locations north of known breeding locales. They are irregular north into e. Utah and all of Colo., especially along foothills of Rocky Mts. and riparian areas of se. Colo. There is 1 record for Keith Co., Nebr. **COMPARISON.** ***Zone-tailed juvenile.*** **Swainson's Hawk (adult dark morph).**—Tail pattern is similar. Underwing coverts are mainly rufous, *vs.* black. Underside of flight feathers is gray, *vs.* white. Undertail coverts are white/tawny, *vs.* all-black. **"Western" and "Harlan's" Red-tailed Hawks (juveniles, darker morphs).**—Tail patterns can be similar on dorsal and ventral sides. Wingtips are much shorter than tail tip, *vs.* equal or longer than tail. Wings are held on fairly flat plane and flight has stable manner, *vs.* wings in dihedral and tipsy manner. ***Zone-tailed adult.*** **Turkey Vulture (both ages, in flight).**—Similar dihedral wing position and tipsy mannerism, but head is small (all-red on adults, all-gray on juveniles), *vs.* large head with yellow cere and white forehead. Legs and feet are white, *vs.* yellow. Underside of flight feathers is uniformly gray, *vs.* barred. When closed, tail is wedge-shaped and plain gray, *vs.* banded and square-tipped. **Common Black Hawk (adult).**—Forehead is black, *vs.* white. When perched, wingtips are a bit shorter than tail tip, *vs.* equal to or longer than tail tip. Flying birds exhibit very broad wings held on flat plane, *vs.* moderate-width wings held in dihedral. **Swainson's Hawk (adult dark morph).**—Dark forehead and dark inner lores, *vs.* white forehead and all-white lores. Wingtip-to-tail-tip ratio identical when perched. Similar dihedral position and tipsy flight mannerism, but underwing coverts are usually rufous, *vs.* all-black. Undertail coverts are white/barred, *vs.* all-black. **"Harlan's" Red-tailed Hawk (adult).**—Harlan's with whitish- or grayish-banded tails can be similar, but have numerous dark and light bands, *vs.* 1–4 bands at most. Harlan's with broad terminal band and white inner tail can be similar, especially when tail is closed, as inner tail will appear as white band. Forehead and white lores may be similar. Wingtips are shorter than tail tip, *vs.* equal to or longer than tail. Wings are broad and held on flat plane, *vs.* held in dihedral. **Rough-legged Hawk (adult dark morph).**—Overlap Oct.–Apr. Species are similar, with their white foreheads, long wings, and white-banded tail pattern. Legs are fully feathered, *vs.* bare lower legs. Regularly hovers in flight, *vs.* no hovering. Flight is stable, *vs.* tipsy mannerism.

Figure 1. Aravaipa Canyon, Graham Co., Ariz. (Apr.)
Classic breeding habitat in rugged canyon with tall sycamores for nest sites.

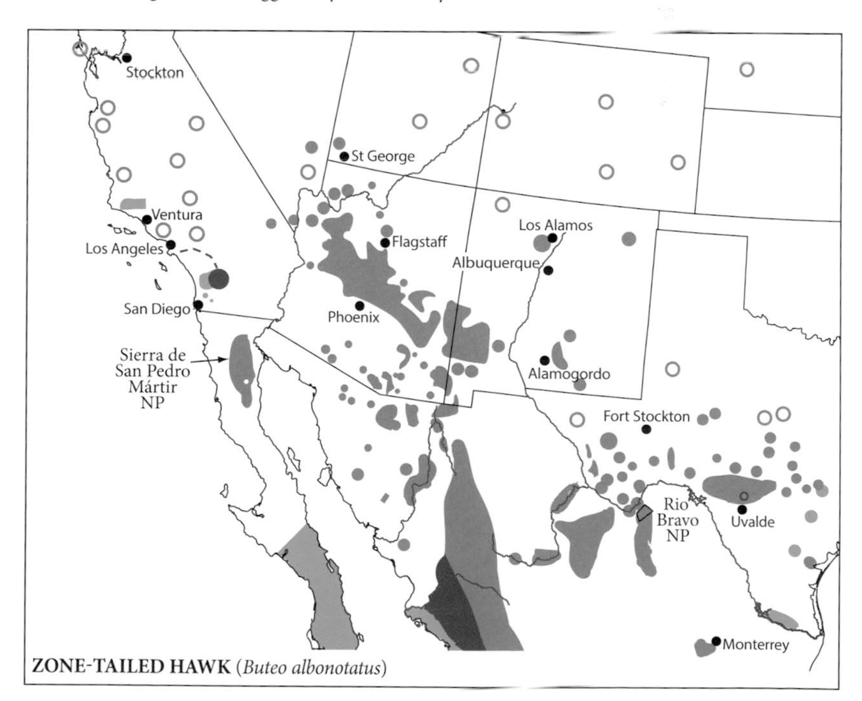

Plate 84. CRESTED CARACARA (*Caracara cheriway*)

Ages: Juvenile, 1-year-old, and adult. *Bill is pale blue. Large expanse of bare facial skin on lore, cere, and cheek regions varies in color with age and mood. Dark crown; shaggy crest can be raised or lowered, depending on mood and temperature. Large, white patch on base of outer 6 primaries shows on dorsal and ventral sides of wing. Legs are very long. Long, white, thinly banded tail has wide, dark terminal band. Juvenile.*—Plumage held for 1st year. *Facial skin varies from very pale blue or purple to dark pink. Crown is dark brown.* Eyes are dark brown or medium brown. Tawny head and neck; *lower neck and back are streaked with brown.* Body plumage is brown. *Legs and feet are pale gray. 1-year-old.*—Plumage is attained in 2nd year. *Facial skin as on juvenile in early to middle period of age class (early/mid-stage) but may turn pale yellow or pinkish orange in latter part (late stage). Crown is dark brown.* Eyes are medium brown or tan. *Tawny head and neck. Back, belly, and flanks are dark brown. Some birds have solid dark-brown belly and flanks; others have thin, pale barring on belly and forward part of flanks.* Legs and feet vary from pale yellow to medium yellow. *Adult.*—Plumage is acquired as 2-year-old. Eyes vary from *medium brown to pale yellow or pale gray. Facial skin is mainly orange but may be bright yellow with orange around eyes or vivid reddish orange. Head and upper neck are white; lower neck, breast, and back are neatly barred with black*; breast and lower neck can be tawny. Body plumage is black. Legs and feet are bright yellow. **Subspecies:** Monotypic. Occurs in Okla. and Tex., south to Amazon Basin; in East, isolated in Fla. **Color morphs:** None. **Size:** L: 21–24" (53–61cm), W: 46–52" (117–132cm). Females average larger. **Habits:** Fairly tame. This is a very terrestrial species. Perches on exposed objects but seeks shaded areas during hot midday. All ages perform head-throwback display. Feet are quite dexterous when searching for food on ground. When feeding, caracaras are highly aggressive toward other scavenging birds of prey. **Food:** Mainly carrion but also small live prey dislodged from fires or plowing. **Flight:** Powered flight consists of long periods of shallow, steady wingbeats. *Glides with wings bowed downward.* May soar to great altitudes but appears unsteady, with periodic flapping stints. Does not hover or kite. **Voice:** Mainly silent. Staccato, guttural rattle emitted during head-throwback display. Also emits *cack* notes.

84a. Juvenile: *Flushed, pale blue facial skin; often pinkish below eyes. Brown crown is fully raised when cool or when fluffing feathers (rousing). Sides of head and neck are tawny; lower neck is streaked.*

84b. Juvenile: *Dark pink facial skin. Brown crown is moderately raised on nape. Sides of head and neck are tawny; lower neck is streaked. Note:* 1-year-old can have similar facial color, as on *84h.*

84c. 1-year-old (late stage): *Facial skin may turn pale yellow with orange or pink under eyes. Brown crown is partly raised on nape.* Sides of head are white or tawny; lower neck is tawny and spotted/barred.

84d. Adult: Eye is medium brown. *Facial skin is bright yellow. Black crown raised on nape (in aggression).*

84e. Adult: Eye is pale yellow/pale gray. *Facial skin is orange. Black crown is lowered.*

84f. Juvenile: *Tawny lower neck and breast are streaked.* Plumage of back, belly, and dorsal side of wings is brown. *Very long, gray legs.*

84g. Juvenile, head-throwback display: Head is quickly thrown onto the back.

84h. 1-year-old (early/mid-stage): *Tawny head and neck, with irregular spotting/barring from lower neck to breast and back.* Plumage of back, belly, and dorsal side of wings is brown. *Very long legs are pale yellow. Note: Dark pink facial skin, as on 84b.*

84i. 1-year-old (late stage): *Facial skin is pale yellow, as on 84c. Head and upper neck are tawny, with irregular spotting/barring from lower neck to breast and back.* Plumage of back, belly, and dorsal side of wings is brown. *Very long legs are medium yellow. Note:* Some have thin, pale barring on belly.

84j. Adult: *Vivid reddish-orange facial skin. Sides of head and upper neck are white; lower neck, breast, and upper back are tawny and neatly barred. Rest of body plumage is black. Very long, bright yellow legs.*

84k. Flight silhouette: *Gliding. Wings are bowed downward.*

84l. Juvenile: *Large, white primary patch. Lower neck and breast are streaked.* Body is brown.

84m. Adult: *Large, white primary patch. Lower neck and breast are neatly barred.* Body is black.

84n. Adult: *Black wings have large white patch on base of outer 6 primaries. Note:* Younger ages similar but have brown wings.

a
b
c
d
e
f
g
h
i
j
k
l
m
n

CRESTED CARACARA (*Caracara cheriway*)

HABITAT: Arizona. Arid, seasonally hot lowlands with semi-open palo verde–Saguaro thorn-scrub. **Texas.** Varies from seasonally hot and arid in interior semi-open mesquite thornbush areas in s. region to seasonally hot and humid coastal areas and interior n. regions of open and semi-open grazing lands, savannas, and moderate agricultural areas—the lattermost especially seasonally, during burning, harvesting, or plowing. Less commonly found in suburban areas (San Antonio, Tex.; J. Economidy pers. comm.). **STATUS: Arizona.** Uncommon to common but very local; only 20–25 pairs, found only in Pima Co. on Tohono O'odam Nation reservation. **Texas.** Common, stable, and widespread in core range in s. part of state, but absent in extensive agricultural regions of Rio Grande valley and some coastal counties. North of this region, inhabits scattered, isolated habitat patches. Range is expanding northward to se. Okla. This is a very adaptable species and acclimates to marginal habitat. Disperses regularly to purple dashed line. **Louisiana.** Nests in Cameron and Calcasieu Parishes. **NESTING: Arizona.** Late Mar.–Aug. **Texas.** They may not nest until 3 years old. Begins Jan.–Mar.; ends May–Jul. Large stick nest is built by both sexes (only falcon species to build its own nest) in top section of tall bush or tree; in Ariz. often in fork of Saguaro. Nest is placed 8–50' (2.4–15m) high. Nest is not lined with greenery, only prey debris; nests are often reused and can become quite large. Clutch is of 1–4 eggs, usually 2 or 3. Youngsters stay with parents for up to 7 months, and family groups are a common sight. **MOVEMENTS:** Mainly sedentary, but young birds, especially, are prone to short- to long-distance dispersal. Ariz. birds regularly disperse to Pinal and Cochise Cos. There is regular dispersal from main breeding areas in Tex. northward and westward in the state and throughout Okla. *Extralimital dispersal records.*—Calif., Colo., Idaho, Kans., Minn., Nev., Oreg., Alta. **COMPARISON: Black Vulture.**—Similar large white patches on primaries but has black neck and short black tail, *vs.* long white neck and long white tail with black band on tip. **Bald Eagle (immatures).**—Many have similar pale heads with darker crowns and pale necks, but bills are dark or yellowish, *vs.* blue. Yellow legs are short, *vs.* very long. In flight, white patches are on inner primaries and axillaries, *vs.* large white patch on outer primaries and black axillaries.

Figure 1. Pima Co., Ariz. (Apr.)
Year-round arid palo verde–Saguaro thorn-scrub habitat.

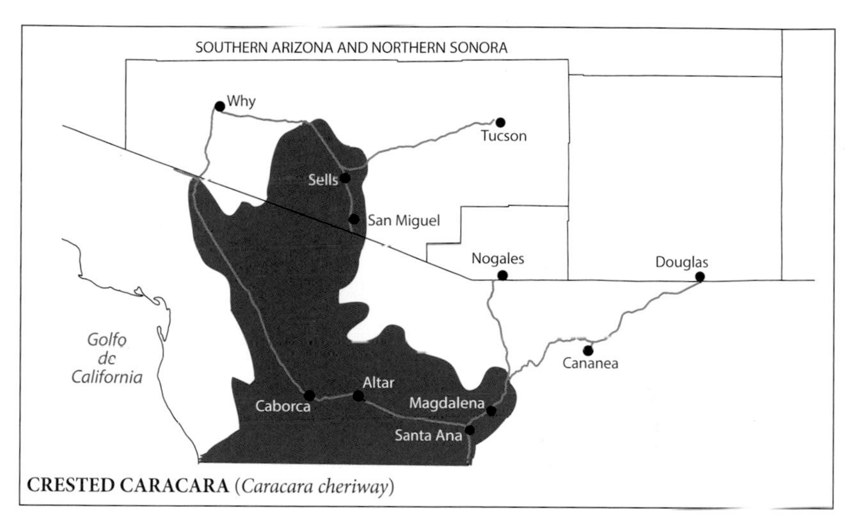

CRESTED CARACARA (*Caracara cheriway*)

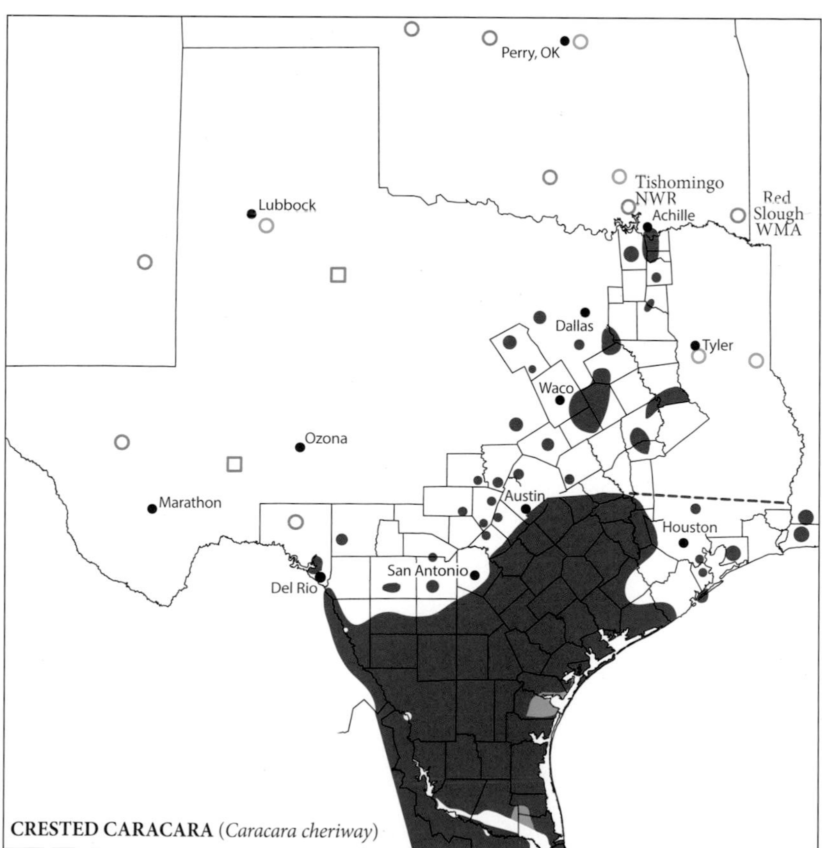

CRESTED CARACARA (*Caracara cheriway*)

Plate 85. APLOMADO FALCON (*Falco femoralis*)

Ages: Juvenile and adult. Juvenile plumage is retained for several months or 1 full year. May engage in partial 1st-year molt in winter–early spring, then enter 1st prebasic molt at 1 year of age. Adult plumage is attained as 1-year-old. As in all falcons, molt begins on head and progresses down body, with wings and tail molting last. Pale, fleshy orbital skin encircles dark brown eye. Thin, black malar mark connects to neat, dark, thin eye stripe. *Crisply defined, tawny or white, long, moderate-width supercilium extends in V shape onto nape. In fresh plumage, lower part of V-shaped supercilium stripe is rufous.* Head has broad, tawny or white forehead patch and solid dark crown. *Narrow band across belly and black flanks form dark bellyband.* Wings are long, with pointed tips in flight, and tail is very long. *When perched, wings are much shorter than tail tip. Dark underwing is uniformly patterned with thin white bars on dark gray flight feathers; there is a broad tawny band on front edge of wing. Moderately wide, white bar adorns trailing edge of secondaries.* Legs and feet are yellow. *Juvenile.*—Cere and orbital skin are pale blue when recently fledged but turn pale yellow thereafter. Pale plumage areas on head and breast are tawny. Upperparts are dark brown. Breast is moderately or heavily marked with thick, dark streaks, typically more heavily marked on females. Black bellyband has thin tawny streaks. *Rufous-tawny leg feathers, lower belly, and undertail coverts fade to tawny by end of age cycle. Dorsal tail pattern is highly variable, with 6–7 very thin, gray or white, full or partial bands. Adult.*—Cere and orbital skin are yellow, brighter on male. Breast is unmarked on male and usually lightly streaked in center on female; 1-year-old male (not depicted) has sparse, thin streaking on breast (Wheeler 2003b, Fig. 553; Clark and Schmitt 2017). Upperparts are gray, but are more bluish on male. Thin, white-fringed edges on most belly and flank feathers. *Rufous leg feathers, lower belly, and undertail coverts. Tail has 5–6 thin, white bands.* **Subspecies:** *F. f. septentrionalis* in U.S. and Mexico; 2 other subspecies in Cen. and S. America. **Color morphs:** None. **Size:** L: 14–18" (36–46cm), W: 31–40" (79–102cm). Females are distinctly larger. **Habits:** Fairly tame. Perches mainly on exposed objects. Mated pairs remain together year-round. Prey may be pursued on foot. **Food:** Avian prey from sparrow-size to Mourning Dove–size birds and large flying insects (e.g., cicadas). Gulf coast resident diet may include amphibians and crustaceans; may pirate prey (often non-avian) from raptors and herons. Pairs often hunt cooperatively. **Flight:** Soars and glides with wings held on flat plane. Hunting flights are fast and direct. May soar to high altitudes. **Voice:** High-pitched, toylike, squeaky, rapid chattering: *cack-cack-cack. Note*: Examples from Cameron Co., Tex. (AMNH; field photos).

85a. Juvenile (recently fledged): Cere and orbital skin are pale blue. *Rear part of long supercilium stripe is rufous.* Pale plumage areas are tawny; dark areas are dark brown.

85b. Juvenile (fall–winter): Pale yellow cere and orbital skin. *Supercilium stripe is long.* Crown is dark brown.

85c. Adult female (fresh plumage): Crown is gray. *Rear of long supercilium stripe is rufous.* Pale areas of head and breast are pale tawny in fresh plumage, white when worn. Center of breast may be lightly streaked.

85d. Adult: White, long, V-shaped supercilium stripes. Pale plumage areas are white.

85e. Juvenile (recently fledged): *Rich tawny legs and undertail coverts. Breast is moderately streaked. Black bellyband. Secondaries have broad, white rear edge. Tail has 7 very thin, white bands. Note:* Lightly streaked breast is typical of males.

85f. Juvenile: *Tawny leg feathers and undertail coverts. Breast is heavily streaked. Black bellyband. Secondaries have broad white rear edge. Tail variation has partial bands on outer tail feathers; 2 deck feathers are solid dark.*

85g. Female in partial 1st-year molt/1st prebasic molt (May, Tex.): Head, hindneck, and upper back molted into adult plumage. Breast has remnant juvenile thick streaks and new adult female thin streaks. *Rest of body is juvenile feathering, including faded, pale tawny legs and undertail coverts. Tail variation (juvenile) has 6 very thin bands.*

85h. Adult female: Sparsely streaked mid-breast. Upperparts are gray. *Black bellyband. Rufous legs and undertail coverts. Black tail has 6 thin, white bands. Note:* Some females lack breast streaking.

85i. Adult male: Unmarked white breast. *Upperparts are blue-gray. Black bellyband.*

85j. Juvenile: *White bar on trailing edge of secondaries. Black bellyband. Tail is long, with very thin, pale bands.*

85k. Juvenile: *Brown upperparts with white bar on trailing edge of secondaries. Tail has 7 very thin bands.*

85l. Adult male: *Blue-gray upperparts; white bar on trailing edge of secondaries. Tail has 5 thin bands.*

85m. Adult: *White bar on trailing edge of secondaries. Black bellyband. Lower belly, undertail coverts are rufous.*

a
b
c
d
h
i
e
f
g
k
j
l
m

APLOMADO FALCON (*Falco femoralis*)

HABITAT: South Texas. *Year-round.*—Present-era released stock (see "Status," below) inhabits humid, sea-level terrain in open grassy prairie expanses on large barrier islands and coastal mainland. Widely scattered, mainly single, or small clusters of, tree-size yuccas, shrubs, small trees, or utility poles are used for perching, roosting, and nest sites. *Note:* All areas also have numerous, widely scattered human-made nest-box structures placed atop 7.5' (2.3m) -tall predator-proof poles (Hunt et al. 2013). **New Mexico and West Texas.** *Possible year-round.*—Semi-arid, high-elevation, mainly flat terrain in wide-open shrub-free grassy plains or broad valleys situated between hills and mountains of Chihuahuan Desert. Tree-size Soaptree and Torrey Yuccas (*Yucca elata* and *Y. treculeana*)—favored nest structures—are sparsely studded either throughout or on edges of grasslands. Utility poles, windmills, and Honey Mesquite may also be sparsely distributed within or on edges of grasslands. In disturbed habitat of woody-shrub biome, territories may encompass large, open, shallow depressions and/or basins of regionally endemic Tobosa Grass (*Pleuraphis mutica*) known as Tobosa swales; yucca often grows in grass-shrub transition zone adjacent to open swales (R. Meyer pers. comm.). Historic and recent *natural* breeding occurs in s. N.Mex. at around 4,100' (1250m); historic and temporary release-era breeding occurred in w. Tex. at 4,400' (1350m). **Chihuahua, Mexico.** *Year-round.*—Inhabits semi-arid, high-elevation, expansive grassland plains, valleys, and Tobosa swales of Chihuahuan Desert, as noted above, except topography is interspersed with more hills and mountains. Habitat is becoming fragmented in many prime breeding areas due to influx of irrigated agriculture, or it is becoming overgrazed and encroached with extensive shrub growth. Nest-site elevation varies from 3,900' (1200m) to 5,200' (1600m) on flat terrain but up to 6,200' (1900m) on low-slope hillsides (Macías-Duarte et al. 2004). *Note:* The falcon can effectively hunt its avian prey *only* in open habitat with little or no shrub and/or tree growth. **STATUS: U.S.** Federally listed as endangered since 1986. Historically, 2 widely separated U.S. populations were contiguous with Mexican populations: (1) s. Texas and Gulf Coast of Mexico, and (2) Chihuahuan Desert of se. Ariz., s. N. Mex., and w. Tex. and the desert states of n.-cen. Mexico. Last known *natural* resident population in U.S. was on King Ranch in Kenedy Co., Tex., in early 1950s (Hector 1987). There are only 10 documented breeding records of *natural* (non-introduced) pairs in U.S. between 1941 and 2017: 1 in s. Tex. and 9 in s. N.Mex. (8 records have occurred in N.Mex. since 2001). Besides irregular sight records of mainly single birds in s. and sw. Tex. and regular sight records and previously irregular to now regular breeding in s. N.Mex., the falcon never reestablished a viable *natural* breeding population in any of its former range. *Captive-release (hacking) program:* To jumpstart defunct U.S. breeding population, The Peregrine Fund (TPF), USFWS, and volunteering private landowners were involved in a program to release captive-bred falcons into the wild. The program was active 1993–2013, with 1,900 young Aplomado Falcons—of Veracruz, Mexico lineage—being released. Safe Harbor Agreement (SHA) permits were granted for large acreage landowners in Texas. The Nonessential Experimental Population (NEP) provision of section 10(j) of the U.S. Endangered Species Act was used for N. Mex. Success was quickly attained in prey-rich subtropical coastal s. Tex., resulting in a stable and viable breeding population. The program, however, failed to establish a population in the less-prey-rich temperate Chihuahuan Desert of w. Tex. and s. N.Mex. *Note:* All released birds carry a USFWS aluminum band and a color-marked, numbered leg band. **Mexico.** In order to comprehend plight of U.S. population, an understanding of the 2 geographically separate Mexican populations is necessary. In 2002, Mexico listed the species as "subject to special protection," a downgrade in protection status from its 1996 listing as an endangered species (Young et al. 2004). *Gulf coast population:* Historically, the range extended from s. Tex. south through Chiapas and Yucatán; now n. Veracruz south to Chiapas and Yucatán (W. S. Clark pers. comm.). Hector (1987) said there was a habitat void in coastal prairie south of Brownsville, Tex., of about 125 mi. (200km); now there is a void of at least 300 mi. (480km) between the newly restored s. Tex. population and the northernmost *natural* Veracruz population. *Chihuahuan Desert population:* Historically, the falcon was resident in all contiguous Chihuahuan Desert of Chihuahua and Coahuila, and perhaps small parts of Durango and Sonora (Young et al. 2004). Now it is found in a swath only about 100 mi. (160km) wide within the desert's borders across n. and ne. Chihuahua (Young et al. 2004). *Note:* Grassland habitat still exists in s. Chihuahua, Coahuila, and Durango, but falcons were not detected in recent winter passerine surveys by Rocky Mountain Bird Observatory (RMBO 2011). The falcon was undoubtedly present in reasonably healthy numbers in n. Chihuahua, because falcons from this contiguous desert population fed dispersal into historic w. U.S. range long after U.S.-breeding population in this region was gone. Present-era population was discovered in 1992 in 2 major centers in n. Chihuahua: around El Sueco, south of Villa

Ahumada (Valles Centrales region), and around Tinaja Verde, near Coyame (Montoya 1995, Montoya et al. 1997). Other isolated breeding pairs and single birds were located in surveys in late 1990s in n. and ne. Chihuahua, only about 32 mi. (51km) from w. Tex. border and within 8 mi. (13km) of sw. N.Mex. border (Young et al. 2004). Until the late 20th century, the Aplomado Falcon haven of vast yucca-studded grasslands of n. Chihuahua escaped the overgrazing and agricultural demise that had occurred much earlier in the species' U.S. range. Change, however, has come to n. Chihuahua, and it does not bode well for the survival of the remnant Chihuahuan Desert population. By 2006, 501,600 ac. (203,000ha) had been converted to irrigated agriculture in prime falcon habitat in Valles Centrales in n.-cen. Chihuahua; by 2011, 672,200 ac. (272,000ha) had been lost to irrigated agriculture (Pool et al. 2014). The 35 falcon territories that previously (1996–2002) existed around El Sueco and Tinaja Verde (Macías-Duarte et al. 2004) were down to 11 territories in 2013, with very low breeding success—affected, too, by prolonged drought (Rodríguez and Calderón 2013). On the s. end of the falcon's current range, 50 mi. (80km) southwest of Ojinaga, Chihuahua, over 230,000 ac. (93,000ha) have been converted to irrigated agriculture since 1994 (Von Oldershausen 2014). *Note:* Habitat loss also greatly affects the large number of grassland-dependent passerines that winter on desert grasslands (Macías-Duarte et al. 2004; RMBO 2011). **U.S. Status History.**—The falcon's presence can be divided among 4 locales: s. N.Mex., s. Tex., w. Tex. (Trans-Pecos and adjacent area), Ariz. Occupation periods are divided into 3 eras: HISTORIC ERA (1852–1952).—Date of 1st museum-taken bird/sight record to last *natural* U.S. breeding record. RECENT ERA (1953–1992).—Period of sight records but no breeding records. PRESENT ERA (1993–2017).—Period of captive-bred releases/breeding (all data through 2012 based on Hunt et al. 2013); and period of more regular *natural* sight records/renewed breeding. *Note*: All weather related data in following text is based on NOAA (2016). **Southern New Mexico:** This area underwent severe cattle and sheep overgrazing that began in the 1870s but grew worse from the 1880s to the early 1900s, following the Indian Wars period. Curtin et al. (2002) wrote, "perennial grasses were reduced to bare ground." Drought was common; severe dry periods were noted in 1885, 1891–94, 1909–12, 1917–22, Dust Bowl era of 1933–37, and even worse dry spell of 1950–56 (Merlan 2010). Fragmented, large tracts of regenerated habitat and potential breeding locales currently exist in historic Doña Ana, Grant, Hidalgo, and Luna Cos. (S. Williams III pers. comm.). This is the *only* state that still has *natural* nesting, which, for at least 1 pair, has lately become an annual occurrence in Luna Co. HISTORIC ERA.—Data mainly from S. Williams III (unpubl. data) and Hector (1987). 1st U.S. museum bird was taken from Luna Co. in Aug. 1852; sight record, also from Luna Co., in Mar. 1853. There were 17 additional sight/museum records to 1939 in Doña Ana, Grant, Hidalgo, Luna, Otero, Sierra, and Socorro Cos. Several breeding pairs were noted in 1908 and 1909 in Doña Ana Co.. 4 pairs, although not documented as breeding, were seen in 1917 and 1918. There was 1 sight record for the 1930s (1939, in Hidalgo Co.). There were no records during the 1940s and only 3 sight records in the 1950s (all in 1951, in Luna Co.). The last U.S. nesting record—for the next 50 years—was in Luna Co., in 1952 (2nd year of severe 1950s drought). The falcons were probably fairly common but local prior to 1920; thereafter they were uncommon to very uncommon and quite local. RECENT ERA.—Data based on S. Williams III (unpubl. data) and Hector (1987). There were 2 records each for the 1960s and 1970s; a pair was in Lea Co. in 1962, although no breeding was known. Sightings picked up in several s. counties in 1980s, with annual sight/photographic records of single birds through 1992, except in 1989, when none occurred. A pair was also seen in Eddy Co. in 1988, again with no known nesting. It was thought that repopulation was beginning to occur in the state (S. Williams III pers. comm.). (These records align with likely dispersal of agriculturally displaced birds, which was beginning to occur in Chihuahua, Mexico.) All birds were seen in historic breeding counties. They were very uncommon to even rare during this period. DDT probably had moderate effects in this area. PRESENT ERA.—Annual sightings of *natural* stock, often of more than 1 bird, continued from 1993 to mid-2000s; single falcons were seen in 1997 and 2000 at extreme n. part of former range in Bernalillo Co. (S. Williams III unpubl. data). This notable dispersal period is also likely result of agriculture-dislodged *natural* Mexican stock. A *natural* pair that was discovered in Luna Co. in Oct. 2000 (may have nested earlier that year) had 2 unsuccessful breeding attempts in 2001 (in normal precipitation period); bred successfully, fledging 3, in 2nd attempt in 2002 (at beginning of severe but short drought); and bred unsuccessfully in 2004 (failure likely due to disturbance; Meyer and Williams 2005; S. Williams III, R. Meyer pers. comm.). Falcons became regular but still uncommon during this era. Captive-release program occurred in 2006–12, with 337 falcons released by TPF on government/private lands (using NEP option) in Grant, Luna, Sierra, and Socorro Cos. All released falcons either dispersed or perished (Hunt et al. 2013). An experimental feeding program assisted successful breeding pairs in 2007 and 2011 on Armendaris Ranch in Sierra Co. (Sweikert and Phillips 2015). Unsuccessful nesting occurred on Armendaris Ranch in 2008, 2009, and 2010 (B. Mutch, P. Juergens/TPF pers. comm.). There were short periodic droughts in 2006–10, severe drought in 2011–14, and severe winter in 2009–10. *Note:* The release program was terminated in 2012, with no surviving population or breeding success. The falcons seemingly could not survive with persistent drought and minimal prey base. Winter passerine census in 2011 showed this area to have one of the lowest densities and numbers of birds of all Chihuahuan Desert locations (RMBO 2011). *Update:* An unmarked *natural* pair nested in the same area in Luna Co. unsuccessfully in 2013, and successfully in 2014 (fledged 1), 2015 (fledged 1), and 2016 (fledged 2, but only 1 survived past fledging period). (Data from B. Mutch, P. Juergens/TPF pers. comm.; R. Meyer pers. comm.) A pair in this same area fledged 3 young in late May 2017 (B. Mutch pers. comm.). The 2013 failure was drought-related (R. Meyer pers. comm.); 2013 and 2014 were drought years, and 2015 had above normal precipitation. *Final note:* New Mexico has historically been Aplomado Falcon's favorite w. U.S. haunt. Adequate regenerated habitat exists in previously noted border counties; however, Luna Co. has always been the choice location for the falcon since its 1st documentation in the state in 1852. Luna Co. is currently the favored destination in the state for agriculture-displaced falcons out of n. Chihuahua, Mexico. Time will tell whether Chihuahuan-dispersed falcons will gain a foothold in this county and other historically occupied border counties. Breeding-season prey availability has seemingly been sufficient during most recent breeding attempts, and winter passerine surveys by RMBO (2011) showed rather high numbers and densities in these counties—comparable to falcon-inhabited areas in Chihuahua, Mexico. The prospects are good for a small, reestablished resident *natural* population in this state. *Note*: An unbanded juvenile was photographed late Oct. 2017 in Luna Co. (P. Juergens, B. Mutch/ TPF pers. comm.). **South Texas:** Historic era.—Fairly common to common in some counties, especially Cameron Co. Between 1890 and 1914, F. B. Armstrong collected 84 sets of eggs (typical clutch is 2 or 3 eggs; his sets were abnormally composed of 4 eggs) and 27 falcons, mainly from Cameron Co., for museums; R. D. Camp was also a major collector of eggs and falcons in Cameron Co., through the 1920s; others collected for museums, too, in various counties, and the last falcon was taken in Kleberg Co. in 1949 (Hector 1987). These numbers show how common the birds once were in s. Tex., but this extended period of collection no doubt had some negative effect on this regional population. Massive habitat alteration and loss occurred with agricultural activities and oil exploration. Agriculture exploded in San Patricio Co.—in midst of prime falcon habitat—in 1910–30 and kept rapidly expanding (Guthrie 2016). Falcon numbers steadily dwindled throughout the 1940s, and the last documented *natural* nesting in Tex. occurred in Brooks Co. in 1941. Up to 9 birds were present on 12,000 ac. (4,850ha) of still-undisturbed prairie on King Ranch in Kenedy Co. in the early 1950s (Hector 1987). Recent era.—Data from Hector (1987). Highly irregular sight records occurred throughout era. The falcons were very uncommon to rare, with no breeding records. After the onslaught of habitat loss and intense collecting of the previous era, DDT dealt the final blow to this once-thriving population of falcons. The deadly pesticide was undoubtedly used on nearby agriculture fields as well as accumulated by the influx of n. migrant passerines (the falcon's prey) from e. U.S., where it was extensively used. There were 4 records for 1954–57, 1 for the 1960s, 3 for the 1970s (a nonbreeding pair was in Starr Co. in 1971), and 1 for the 1980s. Most records were from coastal areas from Cameron Co. north to Aransas Co. These birds may have been the last survivors or, more likely, Gulf coast Mexican birds dispersing northward. Present era.—TPF released 839 falcons at 2 locations in 1993–2004: near Rockport (mainly on isolated Matagorda Island, Calhoun Co.; and San José Island, Aransas Co.), as a "north" location and near Brownsville, in historic Cameron Co. (mainland Laguna Atascosa N.W.R. and adjacent private land areas, using SHA permits), for a "south" location. 1st mainland releases failed in what was deemed classic historic prairie habitat (based also on habitat of current *natural* populations in Veracruz, Mexico) that was bordered or interspersed with small to large wooded tracts (Hunt et al. 2013). It was quickly discovered that this classic falcon habitat also (in Tex.) supported Great Horned Owls (see "Reasons for the Aplomado Falcon's Demise and Inability to Reestablish Population," further below). Subsequent releases were still done in classic falcon habitat but in more open areas with minimal tall woody growth on barrier islands and outer coastal mainland areas that were owl-free. 1st breeding of released stock occurred in 1995. The population quickly grew to a high of 37 pairs in 2002 but has since leveled off and has consistently stayed at around 30 pairs, fairly evenly split between both areas. After a long hiatus, more releases occurred in 2012, when 35 falcons were released at a new "north" location on n. Mustang Island, Nueces Co. In 2013, 30 more falcons were released on Mustang Island, and 22 were released at a new "south" location on the Laguna Atascosa N.W.R. unit on South Padre Island (TPF 2014). *Note:* The release program was terminated in 2013. All told, 926 falcons were released. The program was successful, although with a lower-than-anticipated but stable population (original goal was 60 pairs). High mortality of adult males in the "south" region, resulting in many territorial females that lack mates or are mated to 1-year-old males, is a concern; also, urban sprawl in Brownsville region may affect prime historic habitat in Cameron Co. (Hunt et al. 2103). The population will be monitored and bar-sided nest boxes will be continually refurbished, and new ones erected when needed. This region has an abundant year-round prey base, and this stable food source greatly enabled the release program's success. *Update:* An adult pair was on territory on Mustang Island winter of 2015–16; it nested in a White-tailed Hawk nest on a low bush, but failed, in May 2016; a pair was also on territory on North Padre Island in spring of 2016 (B. Mutch/TPF pers. comm.). No pairs have successfully nested on Mustang or Padre Island since historic times. (The only island nesting has been on Matagorda Island, in "north" region.) *Final note:* A viable *natural* population is unlikely to have reestablished itself, since the closest population in Veracruz is quite distant. The intervention of TPF let this species thrive in s. Tex. once again. The current survival of the released "north" population on the barrier islands is attributed, in part, to the falcons' quick adaptation to bar-sided nest boxes—which virtually all pairs now use and which *substantially* reduce mortality compared to predator-accessible natural nest structures (Hunt et al. 2013). **West Texas:** HISTORIC ERA.—This yucca-studded grassland of the n. Chihuahuan Desert underwent intense habitat alteration (though little agricultural conversion) that ended any breeding potential of the falcon early on. 1st large ranching operation began in the Trans-Pecos in 1857, and massive overstocking and overgrazing, mostly of cattle but also of more habitat-destructive goats and sheep, occurred in 1880s–1890s; severe overgrazing—with fewer cattle but more drought-tolerant sheep and goats—peaked in the 1940s and numbers of these grazers remained high until the 1960s (Richardson 2003). Periodic droughts have always affected this region (earliest drought on record in 1870); subsequent droughts align with dates noted previously in "Southern New Mexico." Data that follow are based on Hector (1987) unless noted. 1st Trans-Pecos record was in 1860 in Pecos Co.; a juvenile male was taken for a museum in Brewster Co. in 1890. The *only* actual breeding records for w. Tex. were from 1900 in Jeff Davis Co. and from 1904 east of this region in Midland Co.—farthest east for the species in this area—where 3 nests were found. (Nests were recorded during a long period of normal precipitation.) An undated breeding record exists for Ector Co. (Oberholser 1974), also east of the Trans-Pecos and, like Midland Co., in the Permian Basin in an area that underwent large-scale habitat destruction due to oil and gas exploration and massive expansion of Honey Mesquite onto grasslands. J. K. Strecker, who located the Midland Co. nests, found the falcon to be "frequently seen" in Brewster, Culberson, Jeff Davis, and Midland Cos. in 1904. RECENT ERA.—Data based on Hector (1987). There were only highly irregular sight records, and no breeding records, during this era in Brewster (most records), Culberson, El Paso, Jeff Davis, and Presidio Cos. Single birds were seen in 1935, 1937, and 1938; no records in the 1940s; 3 or so birds during the 1950s, with a pair, although no documented breeding, in Presidio Co. in 1950; none in the 1960s; 3 singles in the 1970s; none in the 1980s. 1 bird stayed for a few months in winter of 1991–92 in Jeff Davis Co. (Peterson and Zimmer 1998). DDT probably had a moderate effect here. PRESENT ERA.—1 bird was in Culberson Co. in 1996 (Peterson and Zimmer 1998). Breeding birds were found in late 1990s, 32 mi. (51km) from Tex. border—in similar habitat—in Chihuahua, Mexico (Young et al. 2004). TPF released 637 falcons in 2002–11 in n. Brewster, Jeff Davis, sw. Culberson, s. Hudspeth, and ne. Presidio Cos. on private properties (utilizing SHA permits) that had habitat similar to that of nearby falcon-inhabited areas of Chihuahua. 1st breeding of released stock occurred in 2007 near the town of Valentine, in historically used Jeff Davis Co. (It took 3 years longer than it took for 1st pair to nest in s. Tex. releases, likely because of inadequate prey base.) Subsequent breeding also took place near Valentine, with a peak of 10 pairs in 2009; the number of breeding pairs dropped quickly thereafter, to 2 pairs in 2010, 1 pair in 2011, and none in 2012 (Hunt et al. 2013). The 2007–9 period had fairly normal precipitation, with short drought spells the latter 2 years; severe drought occurred in 2011–14. The winter of 2009–10, affected by El Niño, was harsh and likely began depleting the prey base; that, coupled with severe drought, did not permit breeding. All falcons either dispersed or perished (Hunt et al. 2013; B. Mutch pers. com.). Winter passerine census in 2009 in the Trans-Pecos by RMBO (2011) showed moderate numbers, but a dramatic reduction occurred in 2010 and 2011, in line with the severe winter/drought data. *Note:* The release program was terminated in 2011, with no population established and no further breeding beyond 2011. As long as there is a viable population in Chihuahua, Mexico, there will be occasional sight records in this region; however, it is uncertain whether breeding will occur again. *Note*: An unbanded adult was photographed in w. Presidio Co. in late Aug. 2017 (Cain et al. 2017). **Arizona:** HISTORIC ERA.—Data based on Hector (1987). There were sightings from Cochise and Pima Cos., but breeding occurred only in Cochise Co., in 1887, with 5 nesting pairs. RECENT ERA.—There are very few, highly irregular sight records. PRESENT ERA.—No records. **NESTING:** Begins Jan.–Mar. but as late as May on 2nd attempts. Breeding pairs typically are adults at least 2 years of age; can be a mixed pair of 1-year-old and adult, or sometimes a pair of 1-year-olds. Like other falcons, this species does not build a nest; it uses abandoned nests built by other large birds. In s. Tex., mainly uses nests of Chihuahuan Raven (*Corvus cryptoleucus*), White-tailed Hawk, Crested Caracara, and White-tailed Kite. In Luna Co., N.Mex. (and Chihuahua, Mexico), primarily uses old nests of the raven and Swainson's Hawk, but also Red-tailed Hawk and, uncommonly, Harris's Hawk and White-tailed Kite. Nests are typically in tall bushes, small trees, and tree-size plants such as yucca. Soaptree and Torrey Yuccas—which deter ground predators with their long, spiny leaves—are most commonly used sites, especially in N.Mex.; nests in Honey Mesquite are also frequently used but suffer higher mortality due to easy predator access (Hunt et al. 2013). Also uses old raven nests on utility poles (N.Mex.) and inactive windmills (Chihuahua). Low bushes have been used as nest sites but with high mortality; ground nests on dune grass have also been noted in Matagorda Island, Tex. (Hunt et al. 2013). Most nests are 4–17' (1.2–5.2m) high, but higher on utility poles. Most pairs on island areas and some mainland locales in s. Tex. use human-made bar-sided nest boxes, with solid wooden tops and bottoms (bars are spaced widely enough for the falcon to access but too close for larger avian predators to enter and prey on eggs or nestlings). The boxes are placed about 7.5' (2.3m) high on poles with metal flashing on the bottom area to prevent access by mammalian predators. Clutch is of 2 or 3 eggs, incubated by both sexes. Juveniles leave natal areas by early fall (late Sep. in 2002 Luna Co. nesting; Meyer and Williams 2005). *Note:* Raptors establish nest territories and build nests where prey is most abundant. Being a falcon and unable to build a nest, this species relies on happenstance to find a suitable abandoned nest that *also* is in midst of an adequate year-round prey base; this is often a difficult scenario to attain. **MOVEMENTS:** Nonmigratory. Adult pairs remain on territory year-round. Adults in quest of new territories—especially agriculture-displaced Chihuahua, Mexico, birds—and juveniles are very prone to dispersal in any direction. Chihuahua birds have dispersed to s. N.Mex. and s. Tex. for eons. A youngster banded in Chihuahua in May 1999 was found 180 mi. (290km) away, in Otero Co., N.Mex., in mid-Sep. of that year. U.S. stock historically also shifted south into Chihuahua: A 2012-released juvenile traveled south into nw. Chihuahua, to near Nuevo Casas Grandes, 140 mi. (225km) south of Deming, N.Mex. (apparent release site); the bird's telemetry signal became inoperative on Dec. 26, 2012 (Hunt et al. 2013). Telemetry-tracked birds in s. Tex. move north or south of release origins, or east from barrier islands to mainland (TPF 2014). They are irregularly seen from late Aug. through fall at the hawk watch at Hazel Bazemore Co. Pk. in Nueces Co., Tex. **Range Maps:** *South Texas.*—Released stock has been documented successfully breeding on islands only on Matagorda and San José Islands. Released birds also nest on coastal mainland east and north of Brownsville in Cameron, Willacy, and s. Kenedy Cos. Limited surveys by TPF have not yet detected breeding on near-coastal areas of Kleberg and Kenedy Cos., but possibility exists (Hunt et al. 2013). Areas north of Matagorda Island mark irregular recent sight records. *Texas, New Mexico, and Mexico.*—The 2001, 2002, 2004 and 2013–17 Luna Co., N.Mex., nesting sites are in vicinity of the purple dot in N.Mex. Range in n. Chihuahua is more extensive than in U.S., but birds are found sparingly, from Coyame to Casas Grandes and Janos, and north to U.S. border. Highest nesting density has been around El Sueco and Coyame. As of the late 1990s, birds have been in s. Coahuila (R. Padilla Bora pers. comm.). There are currently no nesting birds in w. Tex. **COMPARISON: Mississippi Kite (all ages).**—In flight, both species share broad white bar on trailing edge of wing. They also share brown (juvenile) or gray (adult) dorsal color. Kite lacks any head features, *vs.* single dark malar stripe. Tail is square or notched, *vs.* rounded. **American Kestrel.**—2 black stripes on sides of head, *vs.* 1 black stripe. Upperparts are rufous, *vs.* brown (juvenile) or gray (adult). Pale underwing lacks white trailing edge, *vs.* blackish with broad white edge. **Merlin.**—Single face stripe is ill-defined, *vs.* distinct black stripe. Underparts are streaked, *vs.* black bellyband. Dark underwing *vs.* dark with white rear edge. **Peregrine Falcon.**—Some juveniles have similar head pattern, with dark facial stripe and pale supercilium line. Underparts are streaked (juvenile) or barred on belly (adult), *vs.* black bellyband. When perched, wingtips equal tail tip, *vs.* much shorter than tail tip. **Prairie Falcon.**—Head pattern is similar. Juveniles have black flanks but streaked belly, *vs.* black bellyband. Wingtips are barely shorter than tail tip, *vs.* much shorter than tail tip.

Figure 1. Chihuahuan Desert, N.Mex. (Sep.)
Classic year-round grassland habitat in Chihuahuan Desert. Nest site on top of tall Soaptree Yucca. Photo by Raymond Meyer

Figure 2. Cameron Co., Tex. (May)
Bar-sided nest box. This design allows falcons to fit between the bars, but larger predators cannot. Most pairs in s. Tex. use these safe predator-proof nest boxes. Photo by Brian Mutch/The Peregrine Fund.

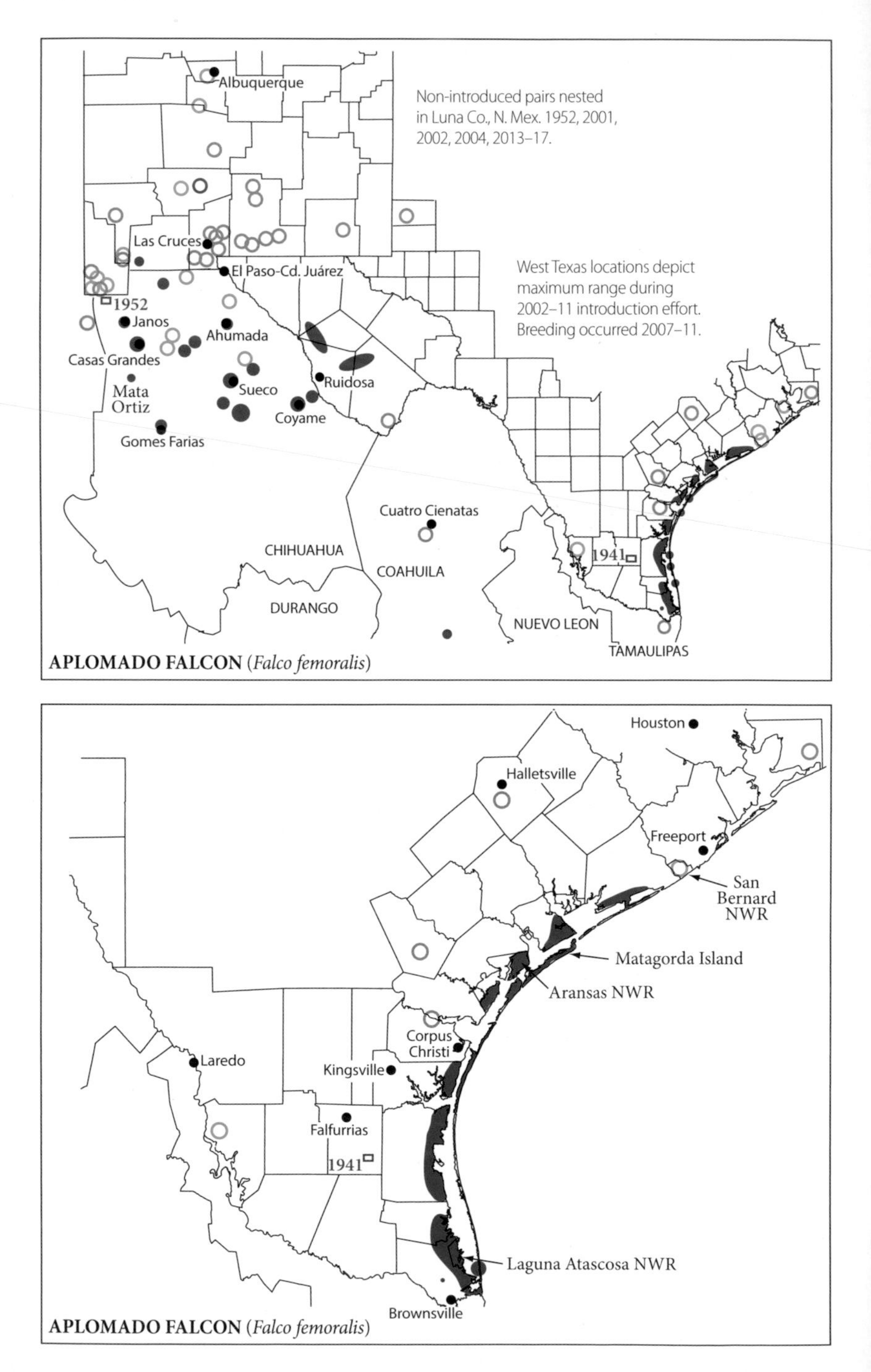

APLOMADO FALCON (*Falco femoralis*)

APLOMADO FALCON (*Falco femoralis*)

Reasons for the Aplomado Falcon's Demise and Difficulties in Reestablishing Populations

1. Habitat degradation.—Alteration and/or obliteration of habitat occurred in all of the Aplomado Falcon's range. *1a. Overgrazing:* The Chihuahuan Desert of s. N.Mex. and w. Tex. was severely overgrazed, which allowed for rapid massive intrusion of woody shrub and tree growth and non-native grasses onto open grasslands. Native drought-tolerant perennial grasses, such as Blue Grama (*Bouteloua gracilis*), provide better soil protection, habitat, and seed production for seed-eating avian prey than invasive, non-native grasses; if destroyed or disturbed too much, Blue Grama can take up to 50 years to regenerate (BluePlanetBiomes.org. 2000). *1b. Fire suppression:* Range fires promote natural grass regeneration on grasslands and prevent encroachment of habitat-destructive woody shrubs and trees onto the grasslands. For protection of life, livestock, and property, fires have been subdued for over a century. *1c. Agriculture:* South Tex., especially, underwent massive habitat destruction during intense agricultural conversion. However, when its falcon population disappeared, s. Tex. still retained ample grassland/prairie habitat on barrier islands, large interior and coastal private ranches, and refuges. These remnant habitat locales are currently being utilized by reintroduced falcons. The U.S. portion of the Chihuahuan Desert grasslands was seemingly altered too much to support a stable population, although regenerated habitat in historic regions is now regularly utilized by small numbers of visiting and/or nesting falcons. *1d. Industrialization:* Large-scale oil and gas explorations and their attendant facilities have encroached on the cen. Gulf coast of s. Tex. and the Permian Basin of se. N.Mex. and w. Tex., creating a negative impact on former grassland habitat.

2. Food supply.—As resident birds, breeding pairs of Aplomado Falcons need a *vast*, year-round food base of avian and/or large-size flying insect prey. This is available consistently in humid subtropical s. Tex. but inconsistently in the arid, temperate Chihuahuan Desert. When food is in ample supply falcons breed, when it is not, they may not breed—or if they do, success rate is often compromised. Macías-Duarte et al. (2004) found that abundance of resident avian prey species was more important to falcon breeding success than the influx of wintering avian species in Chihuahua, Mexico. *2a. Drought:* Especially in the Chihuahuan Desert, drought occurs frequently and creates temporary negative effects on the falcon population. Short droughts have minor impact, but long droughts can have a major negative impact. Drought reduces seed production in grasses (Blue Grama goes dormant), which is vital for adequate food for resident and especially wintering avian prey species. Drought also reduces the insect population required by some summer prey species. *Note: 2b. Severe winter weather (Chihuahuan Desert):* Regular subfreezing temperatures and light snow may affect the desert; heavy snow occurs on occasion. Any severe winter weather disperses prey species and reduces the falcon's prey base. *2c. Reduced avian prey abundance:* With the drastic habitat alteration that occurred in the U.S. portion of the Chihuahuan Desert, resident avian species are likely not found in the numbers and densities that occurred in historic times, and many areas do not seem to be able to support regular contemporary falcon occupancy—only temporary visiting stints. Also, a large percentage of grassland-dependent passerine (prey) species that breed on the Great Plains winter in the Chihuahuan Desert, and their numbers have been decreasing annually with loss of habitat on breeding grounds (Macías-Duarte et al. 2004; RMBO 2011). The Chestnut-collared Longspur (*Calcarius ornatus*) is consistently the most abundant prey species wintering in the falcon's Chihuahuan Desert range, and its numbers—though still reasonably large—have been reduced by 84% since the mid-1960s (RMBO 2011).

3. Peripheral range.—Historic U.S. distribution was at the very n. periphery of the falcon's range, and such populations are less resilient than those in core areas. South Tex. was once a thriving region for this falcon but had become extralimital range for the species by the 1960s; w. Tex. had become extralimital by 1910, s. N.Mex. by the late 1920s.

4. Museum collecting.—Aplomado Falcons were subjected to massive professional and amateur egg and skin (specimen) collecting for museums from the late 1800s to the 1950s, particularly in s. Tex. (Falcons typically replace egg sets with new clutches, unless pairs are disturbed too much—then nests are abandoned.) Being on periphery of range, as noted above, supply of replacement birds is greatly limited.

5. DDT.—Although use of organochlorine pesticide started in the late 1940s, when the Aplomado Falcon population was already in sharp decline, the pesticide greatly affected this bird-eating species, as it did Peregrine Falcon, Merlin, and Cooper's and Sharp-shinned Hawks. Effects of DDT ingestion quickly prevented successful nesting, especially of the remnant s. Tex. population that had endured habitat loss. The U.S. ceased DDT use in 1972; Mexico did so in 2000. *Note:* The resident populations of 2 Aplomado Falcon study areas in Chihuahua, Mexico, show minimal amounts of organochlorine pesticide traces—far below the amount that would affect breeding success (Mora et al. 2011).

6. Great Horned Owl.—A mortal enemy of the falcon, this very adaptable owl (*Bubo virginianus*) has perhaps expanded its habitat usage since historic times. It prevented falcon population reestablishment, by precluding survival of captive-released falcons, in what was historic prime falcon habitat in portions of mainland s. Tex. The owl is the reason falcon reintroduction has taken place on remote Tex. barrier islands. The islands have superb prey base for the falcons but lack tall woody vegetation that the owls prefer.

Cameron Co., Texas (May 2000)
An adult male Aplomado Falcon of captive-release origin guarding his nest territory.

Bibliography

Alaska Nature Tours. 2005. Checklist for the Birds of Haines and the Alaska Chilkat Bald Eagle Preserve. http://www.alaskanaturetours.net/BirdChecklist2008.pdf [Oct. 26, 2013].

American Ornithologists' Union. 1997. Forty-first Supplement to the American Ornithologists' Union Check-list of North American Birds. *Auk* 114 (3): 242–52.

American Ornithologists' Union. 2007. Forty-eighth Supplement to the American Ornithologists' Union Check-list of North American Birds. *Auk* 124 (3): 1109–15.

American Ornithologists' Union. 2012. Fifty-third Supplement to the American Ornithologists' Union Check-list of North American Birds. *Auk* 129 (3): 573–88.

American Ornithologists' Union. 2015. Fifty-sixth Supplement to the American Ornithologists' Union Check-list of North American Birds. *Auk* 132 (3): 748–64.

American Ornithologists' Union. 2016. Fifty-seventh Supplement to the American Ornithologists' Union Check-list of North American Birds. *Auk* 133 (3): 544–60.

Anchorage Audubon Society. 1993. Birding Anchorage: A Checklist (1993). http://www.anchorageaudubon.org/?page_id=2 [Oct. 26, 2013].

Arizona Game and Fish Department. 2016. Condors and Lead. http://www.azgfd.gov/w_c/california_condor_lead.shtml [July 25, 2016].

Arnold, K.A. 2001. *The Texas Breeding Bird Atlas.* Ferruginous Hawk. Texas A&M University System, College Station and Corpus Christi, TX. https://txbba.tamu.edu [Nov. 19, 2017].

Asselin, N. C., M. S. Scott, J. Larkin, and C. Atuso. 2013. Golden Eagles (*Aquila chrysaetos*) Breeding in Wapusk National Park, Manitoba. *Canadian Field-Naturalist* 127: 180–84.

Atlas of the Breeding Birds of Ontario. 2006. Bird Studies Canada, Environment Canada's Canadian Wildlife Service, Ontario Nature, Ontario Field Ornithologists, and Ontario Ministry of Natural Resources. http://birdsontario.org/atlas/index.jsp; Red-tailed Hawk, Golden Eagle: http://www.birdsontario.org/atlas/maps.jsp?lang=en [Jun. 17, 2015].

Bailey, B. H. 1916. Krider's Hawk (*Buteo borealis krideri*) in Alaska. *Auk* 33 (3): 321.

Bechard, M. J., and T. R. Swem. 2002. Rough-legged Hawk (*Buteo lagopus*). In *Birds of North America*, no. 641. A. Poole and F. Gill, eds. Acad. Nat. Sci., Philadelphia, PA/Am. Ornithol. Union, Washington, DC.

Bloom, P. H., and W. S. Clark. 2001. Molt and Sequence of Plumages of Golden Eagles and a Technique for In-Hand Ageing. *North Am. Bird Bander* 26 (3): 97–116.

Bloom, P., M. D. McCrary, J. M. Scott, J. M. Papp, K. J. Sernka, et al. 2015. Northward Summer Migration of Red-tailed Hawks Fledged from Southern Latitudes. *J. Raptor Res.* 49 (1): 1–17. http://www.bioone.org/doi/full/10.3356/jrr-14-54.1 [Sep. 3, 2015].

BluePlanetBiomes.org. 2000. Blue Grama Grass. http://www.blueplanetbiomes.org/blue_grama-grass.htm [Mar. 27, 2016].

Boal, C. W., M. D. Giovanni, and B. N. Beall. 2006. Successful Nesting by a Bald Eagle Pair in Prairie Grasslands of the Texas Panhandle. *Western North Am. Naturalist* 66 (2): 246–50.

Breeding Bird Atlas. 2009. Current Atlases: Colorado Breeding Bird Atlas II (2007–2011), Minnesota Breeding Bird Atlas (2009–2013), Ohio Breeding Bird Atlas II (2006–2010), 2nd Pennsylvania Breeding Bird Atlas (2004–2008), West Virginia Breeding Bird Atlas II (2009–2014). Cornell Lab of Ornithology http://bird.atlasing.org/ [Jun. 6, 2010].

Cain, A., D. LeBlanc, D. Dittmann, L. Wygoda, S. W. Cardiff, and T. Randle. 2017. Checklist S38984860, Sat. Aug. 26, 2017 8:10 AM. ebird.org. ebird.org/ebird/view/checklist/S38984860 [Oct. 17, 2017].

Calef, G. W., and D. C. Heard. 1979. Reproductive Success of Peregrine Falcons and other Raptors at Wager Bay and Melville Peninsula Northwest Territories. *Auk* 96: 662–74.

California Bird Records Committee. 2017. Common Black Hawk. www.californiabirds.org/queryDatabase.asp?partial=on&species=Common+black+hawk [Aug. 20, 2017].

California Dept. of Fish and Wildlife. 2016. Nonlead Ammunition in California. https://www.wildlife.ca.gov/hunting/nonlead-ammunition [Jul. 25, 2016].

Carrière, S., and S. Matthews. 2013. Peregrine Falcon Surveys along the Mackenzie River, Northwest Territories, Canada. File Report No. 140. Environment and Natural Resources Government of the Northwest Territories. http://www.arlis.org/docs/vol1//L/864936994.pdf [Apr. 2, 2014].

Chandler, R. M., P. Pyle, M. E. Flannery, D. J. Long, and S. N. G. Howell. 2010. Flight Feather Molt of Turkey Vultures. *Wilson Bulletin* 122 (2): 354–60.

Clark, W. S. 2007. Taxonomic status and distribution of Mangrove Black Hawk *Buteogallus (anthracinus) subtilis*. *Bull. Brit. Ornithol. Club* 127 (2): 110–17.

Clark, W. S. 2009. Extreme Variation in the Tails of Adult Harlan's Hawks. *Birding* 41 (1): 30–36.

Clark, W. S. 2016. Ageing Immatures of Accipitrid Raptors Using Remige Molt Sequences. Unpubl. PowerPoint. 43 pp.

Clark, W. S., and P. H. Bloom. 2005. Basic II and Basic III Plumages of Rough-legged Hawks. *J. Field Ornithol.* 76 (1): 83–89.

Clark, W. S., and P. Pyle. 2015. Commentary: A Recommendation for Standardized Age-Class Plumage Terminology for Raptors. *J. Raptor Res.* 49 (4): 513–17.

Clark, W. S, and J. Schmitt. 2017. *Raptors of Mexico and Central America.* Princeton University Press, Princeton, NJ.

Clark, W. S., and B. K. Wheeler. 1987. *A Field Guide to Hawks of North America.* Houghton Mifflin, Boston, MA.

Clark, W. S., and B. K. Wheeler. 2001. *A Field Guide to Hawks of North America.* 2nd edition. Houghton Mifflin, Boston, MA.

Clum, J. J., and T. J. Cade. 1994. Gyrfalcon (*Falco rusticolus*). In *Birds of North America*, no. 114. A. Poole and F. Gill, eds. Acad. Nat. Sci., Philadelphia, PA/Am. Ornithol. Union, Washington, DC.

Collins, P. C. and R. D. Reynolds. 2005. Ferruginous Hawk (*Buteo* regalis), A Technical Conservation Assessment. Prepared for the USDA Forest Service, Rocky Mountain Region, Species Conservation Project, Sept. 2, 2005: http://www.fs.fed.us/r2/projects/scp/assessments/ferruginoushawk.pdf [Nov. 18, 2017].Colorado Bird Records Committee. 2017. Common Black Hawk. coloradobirdrecords.org/Reports/SpeciesDetail.aspx?id=100 [Aug. 19, 2017].

COSEWIC. 2008. COSEWIC assessment and update status report on the Ferruginous Hawk, *Buteo regalis*, in Canada. Committee on the Status of Endangered Wildlife in Canada. Ottawa. vii, 24 pp. www.sararegistry/gc.ca/virtual_sara/files/cosewic/sr_ferruginous_hawk_0808_e.pdf [Nov. 15, 2017].

Court, B. 2008. Species Profile: Watching and Studying Peregrine Falcons at Tankin Inlet, Nunavut, Canada. *BirdWatching*. https://www.birdwatchingdaily.com/featured-stories/peregrine-falcon/ [Aug. 10, 2017].

Curtin, C. G., N. F. Sayre, and B. D. Lane. 2002. Transformations of the Chihuahuan Borderlands: Grazing, Fragmentation, and Biodiversity Conservation in Desert Grasslands. *Enviro. Sci. and Policy.* 218: 1–14.

Doyle, F. 2008. Breeding Success of the Goshawk (*A. g. laingi*) on Haida Gwaii/Queen Charlotte Islands 2008. Prepared For: Gwaii Forest Society, Haida Gwaii, B.C. VOT 1SO and Western Forest Products, Inc. Campbell River, B. C. V9W 8C9, March 2009. Pdf. [Accessed Oct. 13, 2017].

Doyle, F. 2016. Haida Gwaii's last remaining Northern Goshawks threatened, biologist says. CBC News (Canadian Broadcasting Corp.) www.cbc.ca/amp/1.3829904 [Accessed Oct. 15, 2017].

Dickerman, R. W., and F. C. Parkes. 1987. Subspecies of the Red-tailed Hawk in the Northeast. *Kingbird* 37: 57–64.

Dunn, E. H., A. D. Brewer, A. W. Diamond, E. J. Woodsworth, and B. T. Collins. 2009. Red-tailed Hawk (*Buteo jamaicensis*) 337.0. *Canada Atlas of Bird Banding*, vol. 3: *Raptors and Waterbirds 1921–95*. Special Publication, Canadian Wildlife Service. http://www.ec.gc.ca/aobc-cabb/index.aspx?lang=En&nav=bird_oiseaux&aou=337.

eBird. 2017. California Condor (map). eBird.org. https://ebird.org/ebird/map/calcon?bmo=1&emo=12&byr=2011&eyr=2015 [accessed Nov. 16, 2017].

Edelstam, C. 1984. Patterns of Molt in Large Birds of Prey. *Ann. Zool. Fennici.* 21: 271–76.

Ericson, P. G. P., C. L. Anderson, T. Britton, A. Elzanowski, U. S. Johansson, M. Kallersjo, et al. 2006. Diversification of Neoaves: Integration of Molecular Sequence Data and Fossils. *Biology Letters* 2: 543–47.

Erikson, R. A., R. A. Hamilton, and S. N. G. Howell. 2001. New Information on Migrant Birds in Northern and Central Portions of the Baja California Peninsula. In *Birds of the Baja California Peninsula: Status, Distribution, and Taxonomy.* Monographs in Field Ornithology no. 3. American Birding Assoc., Colorado Springs, CO.

Ferguson-Lees, J., and D. A. Christie. 2001. *Raptors of the World.* Christopher Helm/A & C Black, London.

Franke, A. 2013. Sun., Jun. 2, 2013; Sat., Jun. 23, 2012; Tue., Apr. 17, 2012. *ArcticRaptors.ca Blog: Nunavut's Raptor Study.* http://blog.arcticraptors.ca/?m=1 [Aug. 11, 2017].

Gaede, P. A., D. Kisner, and H. Ranson. 2010. Northern Goshawk: First Nesting Record for Santa Barbara County and Current Breeding Status in Southern California. *Western Birds* 42 (4).

Golumbia, T. E. 2000. Introduced Species Management in Haida Gwaii (Queen Charlotte Islands). Proceedings of Conference on the Biology and Management of Species and Habitats at Risk, Kamloops, B.C., 15–19 Feb. 1999. L. M. Darling, editor. B.C. Ministry of Environment, Lands and Parks, Victoria, B.C. and University College of the Cariboo, Kamloops, B.C. 490 pp., Vol. 1.

Good, R. E., M. Nielson, H. H. Sawyer, and L. I. McDonald. 2004. Population Level Survey of Golden Eagle (*Aquila chrysaetos*) in the Western United States. U.S. Fish and Wildlife Service, Arlington, VA.

Great Basin Bird Observatory. 2017. Nevada Species Accounts: Ferruginous Hawk *Buteo regalis.* https://www.gbbo.org/bird-conservation-plan/ [PDF]. Nov. 18, 2017.

Guthrie, K. 2016. San Patricio County. Handbook of Texas Online, Texas State Historical Association. http://www.tshaonline.org/handbook/online/articles/hcs04 [Mar. 19, 2016].

Hackett, S. J., R. T. Kimball, S. Reddy, R. C. K. Bowie, E. L. Braun, M. J. Braun, et al. 2008. A Phylogenomic Study of Birds Reveals Their Evolutionary History. *Science* 320: 1763–67.

Hayden, T. 2012. First Gray Hawk Sighting in California? *Santa Barbara Independent*, Nov. 27, 2012. http://www.independent.com/news/2012/nov/27/first-gray-hawk-sighting-california/ [Apr. 10, 2014].

Hector, D. P. 1987. The Decline of the Aplomado Falcon in the United States. *Am. Birds* 41 (3): 281–89.

Hofmann, D. A. 2016. *Buteogallus anthracinus* Common Black-Hawk (Santa Rosa, CA, Dec. 18, 2016). Flickr. https://www.flickr.com/photos/23326361@N04/31373180890 [Aug. 19, 2017].

Houston, S. C. 1967. Recoveries of Red-tailed Hawks Banded in Saskatchewan. *Blue Jay* 25: 109–11.

Houston, C. S., D. R. Barber, M. J. Bechard, J. Mandel, K. Bildstein, et al. 2006. New Information on Turkey Vulture Movements. Scientific paper presented at Symposium SS-08 Raptor Migration: Ecology and Conservation in the New World, IV North American Ornithological Conference, World Trade Center, Veracruz, Mexico, Oct. 3–7, 2006.

Houston, C. S., P. D. McLoughlin, J. T. Mandel, M. J. Bechard, M. J. Stoffel, et al. 2011. Breeding Home Ranges of Migratory Turkey Vultures near Their Northern Limit. *Wilson J. Ornithol.* 123 (3): 472–78.

Hug, L. 2013. Common Black-Hawk Nesting in Northern California. Abstracts: Scientific Presentations. 38th Annual Conference of the Western Field Ornithologists. A Joint Conference with Washington Ornithological Society, Olympia, WA, Aug. 22–25, 2013.

Hull, J. M., et al. 2010. Population Structure and Plumage Polymorphism: The Intraspecific Evolutionary Relationships of a Polymorphic Raptor, *Buteo jamaicensis harlani*. *BMC Evolutionary Biology.* http://www.biomedcentral.com/1471-2148/10/224 [Mar, 13, 2013].

Humphrey, P. H., and K. C. Parkes. 1959. An Approach to the Study of Molts and Plumages. *Auk* 76: 1–31.

Hunt, W. G., J. L. Brown, T. J. Cade, J. Coffman, M. Curti, E. Gott, et al. 2013. Restoring Aplomado Falcons to the United States. *J. Raptor Res.* 47 (4): 335–51.

Idaho Fish and Game. 2005. *Idaho Conservation Data Center*: Ferruginous Hawk (*Buteo regalis*). [PDF]. [Nov. 16, 2017].

Jarvis, E. D., S. Mirarah, A. J. Aberer, F. Li, P. Houde, C. Li, et al. 2014. Whole-Genome Analyses Resolve Early Branches in the Tree of Life of Modern Birds. *Science* 346: 1320–31.

Johnsgard, P. A. 1990. *Hawks, Eagles, and Falcons of North America*. Smithson. Inst. Press, Washington, DC.

Johnson, J. A., R. Thorstrom, and D. P. Mindell. 2007. Systematics and Conservation of the Hook-billed Kite Including the Island Taxa from Cuba and Grenada. *Animal Conservation* 10: 349–59.

Jones, R. 2005. Status and Distribution of Black Vulture in Arizona, with Notes on Bird Finding. *Arizona Birds Online* 1 (2). http://www.azfo.org/journal/blvu.html [Sep. 30, 2008].

Jorgensen, J. 2007. 2007 Bald Eagle Nesting Survey (Nebraska). Nebraska Game and Park Commission, Lincoln, NE.

Liguori, J. 2005. *Hawks from Every Angle.* Princeton University Press, Princeton, NJ.

Liguori, J. 2011. *Hawks at a Distance.* Princeton University Press, Princeton, NJ.

Liguori, J., and B. L. Sullivan. 2010. A Study of Krider's Red-tailed Hawk. *Birding* 64 (2): 38–45.

Liguori, J., and B. L. Sullivan. 2014. Northern Red-tailed Hawk (*Buteo jamaicensis abieticola*) Revisited. *North Am. Birds* 67 (3): 374–83.

Little, J. B. 2016. Condors Reach New Milestone of Survival, Thanks to Tree-Climbing Biologists. *National Geographic*, Jul. 18, 2016. http://news.nationalgeographic.com/2016/07/condor-climbers.html [Jul. 19, 2016].

Macías-Duarte, A., A. B. Montoya, W. G. Hunt, A. Lafón-Terrazas, and R. Tafanell. 2004. Reproduction, Prey, and Habitat, of the Aplomado Falcon (*Falco femoralis*) in Desert Grasslands of Chihuahua, Mexico. *Auk* 121 (4): 1081–93.

Manitoba Breeding Bird Atlas. 2014. Golden Eagle. Bird Studies Canada. http://birdatlas.mb.ca/mbdata/maps.jsp?lang=en [Mar. 6, 2014].

Mehus, S., and M. Martell. 2010. A Wintering Population of Golden Eagles in Southwestern Wisconsin and Southeastern Minnesota. *Passenger Pigeon* 72 (2): 135–41.

Merlan, T. 2010. *Historic Homesteads and Ranches in New Mexico: A Historic Context.* Historic Homestead Workshop, Sep., 25, 2010. Historic Preservation Division, Office of Cultural Affairs, State of New Mexico, Santa Fe, NM.

Meyer, R. A., and S. Williams III. 2005. Recent Nesting and Current Status of Aplomado Falcon (*Falco femoralis*) in New Mexico. *North Am. Birds* 59 (2): 352–59.

Mindell, D. P. 1985. Plumage Variation and Winter Range of Harlan's Hawk (*Buteo jamaicensis harlani*). *Am. Birds* 39: 127–33.

Mlodinow, S. G. 2011. First Records of the Short-tailed Hawk and Gray Hawk for the Baja California Peninsula. *Western Birds* 42: 183–87.

Montoya, A. D. 1995. Habitat Characteristics, Prey Selection, and Home Range of the Aplomado Falcon in Chihuahua, Mexico. M.S. thesis, New Mexico State Univ., Las Cruces, NM.

Montoya, A. D., P. J. Zwank, and M. Cardena. 1997. Breeding Biology of Aplomado Falcons in Desert Grasslands of Chihuahua, Mexico. *J. Field Ornithol.* 68: 135–43.

Mora, M. A., et al. 2011. PBDEs, PCBs, and DDE in Eggs and Their Impacts on Aplomado Falcons (*Falco femoralis*) from Chihuahua and Veracruz, Mexico. *Environmental Pollution* 159 (12): 3433–38. https://doi.org/10.1016/j.envpol.2011.08.025 [Mar. 6, 2016].

National Oceanic and Atmospheric Administration (NOAA). 2016. National Climate Data Center: Palmer Drought Severity Index (precipitation maps, Jan. 2001–Dec. 2015). https://www.ncdc.noaa.gov/temp-and-precip/drought/historical-palmers/psi/200101-201512 [Mar. 29, 2016].

New Mexico Avian Conservation Partners. 2017. New Mexico Avian Conservation Partners. Population Size (1998 data): Ferruginous Hawk (*Buteo regalis*).www.avianconservationpartners-nm.org [PDF]. [Nov. 18, 2017].

New Mexico Department of Game and Fish. 2008. Threatened and Endangered Species of New Mexico, 2008 Biennial Review, Dec. 4, 2008, p. 87. http://www.wildlife.state.nm.us/download/conservation/threatened-endangered-species/biennial-reviews/2008-Biennial-Review-Executive_Summary_and_Full_Text.pdf [Mar. 29, 2010].

Nicholson, J. 2014. Rare Bird of Prey Not Seen Here in More Than a Century. *Winnipeg Free Press*, May 28, 2104. http://www.winnipegfreepress.com/local/out-of-the-blue-260878551.html [Jul. 5, 2014].

North American Bird Conservation Initiative, U.S. Committee. 2009. State of the Birds, United States of America, 2009. U.S. Dept. of Interior: Washington, DC. 36 pp. http://www.stateofthebirds.org/2009/overview [Mar. 31, 2010].

Oberholser H. C. 1974. *The Birdlife of Texas*, vol. 1. Univ. of Texas Press, Austin.

Osborn, S. A. H. 2007. *Condors in Canyon Country: The Return of the California Condor to the Grand Canyon Region.* Grand Canyon Association, Grand Canyon Village, AZ.

Palmer, R. S. (ed.). 1988. Diurnal Raptors. In *Handbook of North American Birds*, vols. 4 and 5. Yale Univ. Press, New Haven, CT.

Peregrine Fund, The (TPF). 2014. Aplomado Falcon Safe Harbor Agreement. Idaho Land Conservation Assistance Network. www.stateconservation.org/idaho//local-resources/Aplomado-Falcon-Safe-Harbor-Agreement/9598 [Mar. 9, 2016].

Peterson, J., and Zimmer, B. R. 1998. *Birds of the Trans-Pecos.* Univ. of Texas Press, Austin.

Pike, J. E., K. L. Garrett, and A. J. Searcy. 2014. The 38th Annual Report of the California Bird Records Committee: 2012 Records. *Western Birds* 45: 246–75.

Pool, D. B., A. O. Panjabi, A. Macías-Duarte, and D. M. Solhjem. 2014. Rapid Expansion of Croplands in Chihuahua, Mexico, Threatens Declining North American Grassland Bird Species. *Biological Conservation* 107: 274–81. https://doi.org/10.1016/j.biocon.2013.12.019 [Mar. 11, 2016].

Poole, K. G. 2011. Update on the Northwest Territories/Nunavut Raptor Database, January 2011. Prepared for Dept. of Environment, Government of Nunavut and Dept. of Environment and Natural Resources, Government of the Northwest Territories. Aurora Wildlife Research, Nelson, BC.

Porsild, A. E. 1943. Birds of the Mackenzie River Delta. *Can. Field Nat.* 57 (1): 19–35.

Powell, A. 2017. Swallow-tailed Kite Migration: A Ten Thousand Mile Odyssey. Swallow-Tailed Kites. Org. http://www.swallow-tailedkites.org/2017/?m=1 [April 25, 2017].

Preble, E. A. 1908. *North American Fauna No. 27: A Biological Investigation of the Athabaska-Mackenzie Region.* U.S. Dept. of Agric., Washington, DC.

Preston, C. R., and R. D. Beane. 1993. Red-tailed Hawk (*Buteo jamaicensis*). In *Birds of North America*, no. 52. A. Poole and F. Gill, eds. Acad. Nat. Sci., Philadelphia, PA/Am. Ornithol. Union, Washington, DC.

Pyle, P. 2005a. First-cycle Molts in North American Falconiformes. *J. Raptor Res.* 39 (4): 378–85.

Pyle, P. 2005b. Remigial Molt Patterns in North American Falconiformes as Related to Age, Sex, Breeding Status, and Life-History Strategies. *Condor* 107: 823–34.

Pyle, P. 2013. Evolutionary Implications of Synapomorphic Wing-Molt Sequences among Falcons (Falconiformes) and Parrots (Psttaciformes). *Condor* 115 (3): 593–602.

Pyle, P., A. Engilis Jr., and T. G. Moore. 2004. A Specimen of the Nominate Subspecies of the Red-shouldered Hawk from California. *Western Birds* 35: 100–104.

Québec Breeding Birds Atlas. 2010–14. Regroupment QuébecOiseaux, Canadian Wildlife Service of Environment Canada, and Bird Studies Canada. 2nd atlas project. http://www.atlas-oiseaux.qc.ca/atlas_en.jsp [Jun. 17, 2015]. [Data obtained from the website of the atlas for Red-tailed Hawk, Golden Eagle.]

Raptor Research Center. 2017. Peregrine Falcon Migration Routes. Boise State University. https://raptorresearchcenter.boisestate.edu/peregrine-falcon--migration-routes/ [Aug. 10, 2017].

Richardson, C. 2003. *Trans-Pecos Vegetation: A Historical Perspective.* Texas Parks and Wildlife, Trans-Pecos Wildlife Management Series, lflt. no. 7, Jun. 2003.

Ridgway, R. 1890. Harlan's Hawk a Race of Red-tail, and Not a Distinct Species. *Auk* 7 (2): 205–6.

Rocky Mountain Bird Observatory [RMBO]. 2011. Wintering Grassland Bird Densities in Chihuahuan Desert Grassland Priority Areas, 2007–2011. Tech. Report #1-NEOTROP-MXPLAT-10-2. Rocky Mountain Bird Observatory, Brighton, CO.

Rodríguez, R., and P. Calderón. 2013. Saving the Aplomado Falcon in the Grasslands of Chihuahua, Mexico. Bird Conservancy of the Rockies, Nov. 7, 2013. www.birdconservancy.org/saving-the-aplomado-falcon-in-the-grasslands-of-chihuahua-mexico/ [Feb. 29, 2016].

St. John, K. 2014. Follow an Arctic Peregrine on Migration [Island Girl of Baffin Island]. Outside My Window (blog), Sep. 29, 2014. http://www.birdsoutsidemywindow.org/2014/09/29/follow-an-arctic-peregrine-on-migration/ [Aug. 10, 2017].

Salter, R. E., M. A. Gollop, S. R. Johnson, W. R. Koski, and C. E. Tull. 1980. Distribution and Abundance of Birds on the Arctic Coastal Plain of Northern Yukon and Adjacent Northwest Territories, 1971–1976. *Canadian Field-Naturalist* 94: 210–38.

Sealy, S. 2012. Voucher Specimens of Red Squirrels Introduced to Haida Gwaii (Queen Charlotte Islands), British Columbia. *Wildlife Afield* 9 (1): 59–65.

Seegar, W. S., M. A. Yates, G. E. Doney, J. P. Jenny, T. C. M. Seegar, C. Perkins, and M. Giovanni. 2015. Migrating Tundra Peregrine Falcons Accumulate Polycyclic Aromatic Hydrocarbons along Gulf of Mexico Following *Deepwater Horizon* Oil Spill. *Ecotoxicology*, vol. 24 (5): 1102–11.

Smallwood, J. A., C. Natale, K. Steenhof, M. Meetz, C. D. Marti, et al. 1999. Clinal Variation in the Juvenal Plumage of American Kestrels. *J. Field Ornithol.* 70: 425–35.

Snyder, N. F. R., H. A. Snyder, N. Moore-Craig, A. D. Flesch, R. A. Wagner, and R. A. Rowlett. 2010. Short-tailed Hawks Nesting in the Sky Islands of the Southwest. *Western Birds* 41 (4): 202–30.

Starin, D. 2016. Condors and Carcasses. *Natural History* [n.d.] www.naturalhistorymag.com/perspectives/082655/condors-and-carcasses [Mar. 16, 2016].

Steenhof, K., M. R. Fuller, M. N. Kochert, and K. K. Bates. 2005. Long-range Movements and Breeding Dispersal of Prairie Falcons from Southwest Idaho. *Condor* 107: 481–96.

Sullivan, B. L., and J. Liguori. 2009. Active Flight Feather Molt in Migrating North American Raptors. *Birding* July: 34–45

Sullivan, B. L., and J. Liguori. 2010. A Territorial Harlan's Hawk (*Buteo jamaicensis harlani*) in North Dakota with Notes on Summer Records of This Subspecies from the Northern Great Plains. *North Am. Birds* 64 (3): 368–72.

Sweikert, L., and M. Phillips. 2015. The Effect of Supplemental Feeding on the Known Survival of Reintroduced Aplomado Falcons: Implications for Recovery. *J. Raptor Res.* 49 (4): 389–99.

Swick, N. 2014. #ABArare – Mississippi Kite – Manitoba. ababloq, American Birding Association, Aug. 3, 2014. http://blog.aba.org/2014/08/abarare-mississippi-kite-manitoba.html [Aug. 9, 2014].

Todd, W. E. C. 1950. A Northern Race of Red-tailed Hawk. *Annals of the Carnegie Museum* (Pittsburgh) 31: 289–96.

U.S. Department of the Interior. 2015. California Condor Recovery Program: Annual Reporting for Calendar Year 2015 as of Dec. 31, 2015. Fish and Wildlife Service, Hopper Mountain NWR Complex, Ventura, CA.

U.S Fish and Wildlife Service. 1992. Notice on Finding Petition to List the Ferruginous Hawk. *Fed. Reg.* 57: 37507-37513 [Aug. 19, 1992].

U.S. Fish and Wildlife Service. 2007. Bald Eagle Soars off Endangered Species List (6/28/2007). *ScienceDaily*, June 28, 2007. https://www.sciencedaily.com/releases/2007/06/070628101017.htm [June. 29, 2007].

U.S. Fish and Wildlife Service. 2008. U.S. Fish and Wildlife Service Lists the Desert Bald Eagle as Threatened under the Endangered Species Act. USFWS Division of Public Affairs, Mar. 18, 2008. https://www.fws.gov/news/ShowNews.cfm?newsId=C32D63BF-F9C9-3364-86287F0244484C9B [Mar. 21, 2008].

U.S. Fish and Wildlife Service. 2012. Endangered and Threatened Wildlife and Plants: Listing the British Columbia Distinct Population Segment of the Queen Charlotte Goshawk under the Washington Dept. of Fish and Wildlife. 2012. Annual Report: Ferruginous Hawk (*Buteo regalis*). Washington Dept. of Fish and Wildlife. www.wdfw.gov/conservation/endangered/species/ferruginous_hawk.pdf [Nov. 17, 2017].

Endangered Species Act; Final Rule. Federal Register, vol. 77, no. 148 (Aug. 1, 2012), 45870.

U.S. Fish and Wildlife Service. 2016. Eagle Permits; Revisions to Regulations for Eagle Incidental Take and Take of Eagle Nests; Proposed Rule. Federal Register, vol. 81, no. 88 (May 6, 2016), 27934.

Utah Division of Wildlife Resources. 2016. California Condors: Rescued from the Brink of Extinction. Feb. 8, 2016. wildlife.utah.gov/California-condors-rescued-from-the-brink-of-extinction.html [Jul. 25, 2016].

Varland, D. E., T. L. Fleming, and J. B. Buchanan. 2008. Tundra Peregrine Falcon (*Falco peregrinus tundrius*) Occurrence in Washington. *Washington Birds* 10: 48–57.

Von Oldershausen, S. 2014. Mexico's Mennonites: An Oasis Grows in Chihuahua. Big Bend Now, Nov. 20, 2014. bigbendnow.com/2014/11/mexicos-mennonites-an-oasis-grows-in-chihuahua/ [Apr. 9, 2016].

Washington Dept. of Fish and Wildlife. 2012. Annual Report: Ferruginous Hawk (*Buteo regalis*). www.wdfw.gov/conservation/endangered/species/ferruginous_hawk.pdf [Nov.16, 2017].

Webster, D. J. 2006. Birds of Sitka, Alaska. Sitka Nature. http://sitkanature.org/sitka-birds/webster/ [Oct. 26, 2013].

Wheeler, B. K. 2003a. *Raptors of Eastern North America.* Princeton University Press. Princeton, NJ.

Wheeler, B. K. 2003b. *Raptors of Western North America.* Princeton University Press. Princeton, NJ.

Wheeler, B. K., and W. S. Clark. 1995. *A Photographic Guide to North American Raptors.* Academic Press. San Diego, CA.

Wheeler, B. K., and W. S. Clark. 2003. *A Photographic Guide to North American Raptors.* Princeton University Press, Princeton, NJ.

Williams, S. O., III. 2000. History and current status of Bald Eagles nesting in New Mexico. *New Mexico Ornithol. Soc. Bull.* 28: 43–44.

Williams, S. O., III, J. P. Delong, and W. H. Howe. 2007. Northward Range Expansion by the Short-tailed Hawk, with First Records for New Mexico and Chihuahua. *Western Birds* 38 (1): 2–10.

Williams, S. O., III, and D. J. Krueper. 2008. The Changing Status of the Gray Hawk in New Mexico and Adjacent Areas. *Western Birds* 39: 202–8.

Wyoming Game and Fish Dept. 2011. Wyoming Species Account, Ferruginous Hawk *Buteo regalis*, pp. 129-131. Wyoming Game and Fish Dept. PDF. [Nov. 16, 2017].

Young, K. E., B. C. Thompson, A. L. Terrazas, A. B. Montoya, and R. Valdez. 2004. Aplomado Falcon Abundance and Distribution in the Northern Chihuahuan Desert of Mexico. *J. Raptor Res.* 38 (2): 107–17.

Red-tailed Hawk Online Images

Bardon, K. 2013. Pembina Valley Raptors 2013 (4–27 Apr.): Red-tailed Hawk light morph [online images]. PBase. http://www.pbase.com/karlbardon/redtail_light [Dec. 31, 2016].

Barlow, R. 2015. Hawk Images: Raymond Barlow Nature and Wildlife Photographer [online images]. PBase. http://www.pbase.com/raymondjbarlow/hawks [Nov. 11, 2014].

Bauschardt, K. 2012. Adult in overhead flight, Edmonton, AB, Jul. 2, 2012. [online image]. Flickr. https://www.flickr.com/photos/kurt-b/7505618508/in/album-72157629283533048/ [Apr. 2, 2015].

Borlé, M. 2016a. American Crow and Harlan's Red-tailed Hawk, Edmonton, AB, Jun. [14] 2016 [online image]. FlickRiver. http://flickriver.com/photos/130790757@N04/27186717183/ [Jan. 1, 2017].

Borlé, M. 2016b. Harlan's Red-tailed Hawk, Morinville, AB, Jun. [23] 2016 [online image]. Flickr. https://www.flickr.com/photos/130790757@N04/27275029403/ [Jan.1, 2017].

Borlé, M. 2016c. Juvenile, Red-tailed Hawk, Edmonton, AB, Sep. [1] 2016 [online image]. This young hawk is 1 of 3 offspring of a dark "Harlan's" male and light "Eastern" female that nested west of Edmonton. The other 2 young ones are dark birds. Flickr. https://flickr.com/photos/130790757@N04/29304366592 [Jan. 7, 2017].

Borlé, M. 2016d. Northern Red-tailed Hawk (*Buteo jamaicensis abieticola*), north of Edmonton, AB, Apr. 2016 [online image]. Flickr. https://www.flickr.com/photos/130790757@N04/26448577936 [Dec. 16, 2016].

Borlé, M. 2016e. Red-tailed Hawk, St. Albert, AB, Jul. 2016 [online image]. Flickr. https://www.flickr.com/photos/130790757@N04/28299875782 [Dec. 29, 2016].

Borlé, M. 2016f. Red-tailed Hawk with prey. Lloydminster, AB, Aug. 2016 [online image]. Flickr. https://www.flickr.com/photos/130790757@N04/28527562954 [Dec. 16, 2016].

Brokelman, A. 2010. Red-tailed Hawk Nest 2009–2017. Female Red-tailed Hawk brings food to the nest, Jun. 17, 2010 [online image]. Ann Brokelman Photography. redtailnest.blogspot.com/2010/06/female-red-tailed-hawk-brings-food-to.html?m=1 [Nov. 4, 2014].

Carrolan, T. 2009. Red-tailed Hawk in Onondaga Co., NY, in Feb. [online image, title page]. Hawk, Art, Science, Hawks Aloft: Redtails R Us. http://www.hawksaloft.com/RedtailsRus.html [Sep. 4, 2015].

Conlin, D. 2015. Wild St. Albert, Red-tailed Hawk [online images]. Conlin Photography. http://www.wildstalbert.com/keyword/Red tailed hawk/; also: http://www.wildstalbert.com/keyword/alberta/i-nTRMHbz/A [Aug. 15, 2015].

Cornell Lab. 2012. Big Red lays Egg #3! YouTube video from Cornell Lab of Ornithology Web Cam, posted by NestWatch07, Mar. 22, 2012. https://m.youtube.com/watch?v=Rw3RnrW4BrA [Dec. 16, 2016].

Cosby, D. 2014a. 2014 Red-tailed Hawk pair, Mar. 23, 2014 [online image]. DianaCosby.com. http://www.dianacosby.com/images/RedTailedHawks.jpg [Aug. 26, 2015].

Cosby. D. 2014b. Red-tailed Hawk, Nov. 14, 2014 [online image]. DianaCosby.com. http://dianacosby.com/images/RedTailedHawk.jpg [Aug. 8, 2015].

Cosby, D. 2015a. 2015 Red-tailed Hawk pair, Jun. 28, 2015 [online image]. DianaCosby.com. http://www.dianacosby.com/images/RedTailedHawkpair.jpg [Aug. 26, 2015].

Cosby, D. 2015b. Adult male Red-tailed Hawk (early Sep.) [online image]. DianaCosby.com. http://www.dianacosby.com/images/RedTailedHawk4.jpg [Oct. 15, 2015].

Cosby, D. 2016. Adult Red-tailed Hawk, May 13, 2016 [online image]. DianaCosby.com. http://www.dianacosby.com/images/RedTailedHawk5132016.jpg [May 13, 2016].

Deal, J. 2011. Scarlett, 1989 to 2011 [online image]. Wildlife Center of Virginia. https://www.wildlifecenter.org/news_events/news/scarlette-1989-2011 [Apr. 9, 2015].

Dice, W. 2012. Red-tail Hawk nest, Lincoln Park Commons Pond, Kettering, OH (Chicks), YouTube video posted on May 8, 2012 [online images/video]. https://www.youtube.com/watch?v=dctKI7JYXaY [Mar. 5, 2015].

Edwards. B. 2012. A parent of the 26-day-old Red-tailed Hawk flies over as Al DeGroot bands the chick in its nest near Redwater on Jun. 23, 2012 [online image]. *Edmonton Journal.* http://www.edmontonjournal.com/parent+tailed+hawk+flies+over+DeGroot+bands+chick+near+Redwater+June+2012/6831153/story.html [Dec. 20, 2016].

Ennis, J. 2010. Red-tailed Hawk, Brunswick Co. NC, Sep. 25, 2010 [online images]. Carolina Bird Club. https://www.carolinabirdclub.org/gallery/Ennis/rtha_2.html [Mar. 24, 2015].

Erickson, L. 2012. Red-tailed Hawk, Mar. 23, 2012, Cornell Lab of Ornithology Web Cam [online image]. Flickr. https://m.flickr.com/#/photos/lauraerickson/6862259344/ [Jun. 3, 2015].

Eriksson, P. 2011. Red-tailed Hawk (*Buteo jamaicensis*). In-flight photo of a Red-tailed Hawk in AB, Canada, Jun. 7, 2011 [online image]. Patrick Eriksson Photography. http://www.patrickeriksson.com/index.php?album=birds/birds-of-prey/red-tailed-hawk&image=red-tailed-hawk-buteo-jamaicensis-ab-canada-05.jpg [Nov. 12, 2016].

Goggin, L. 2015. Christo with a piece of bark, Mar. 21, 2015. Flickr. https://www.flickr.com/photos/goggla/16278591324 [Sep. 5, 2017].

Hand, A. J. 2003a. Red-tailed Hawk (pair), Jan. 22, 2003 [online image]. Friends of Sherwood Island State Park, Westport, CT. http://friendsofsherwoodisland.org/birds/red-tailed-hawk-2/. Can also be accessed in "Birds of Prey": http://friendsofsherwoodisland.org/birds-of-the-park/photographs-by-a-j-hand/ [Jan. 8, 2017].

Hand, A. J. 2003b. Red-tailed Hawk taking off, Aug. 2, 2003 [online image]. Friends of Sherwood Island State Park, Westport, CT. http://friendsofsherwoodisland.org/birds/red-tailed-hawk-3/. Can also be accessed in "Birds of Prey": http://friendsofsherwoodisland.org/birds-of-the-park/photographs-by-a-j-hand/ [Jan. 8, 2017].

Herriot, T. 2010. Red-tailed Hawk on the move [online image]. Trevor Herriot's Grass Notes (blog), Sep. 30, 2010. http://trevorherriot.blogspot.com/2010/09/red-tailed-hawks-on-move.html?m=1 [Dec. 16, 2016].

Iliff, M. 2012. Red-tailed Hawk, Palmer, AK, May 12, 2012 [online images]. Checklist S10829032, eBird. http://ebird.org/ebird/view/checklist?subID=S10829032 [Jan. 17, 2015].

Jantunen, J. 2014. Yukon Red-tailed Hawks [online images]. Flickr. https://m.flickr.com/#/photos/90072707@N03/sets/72157632915829248/ [Jun. 15, 2015].

Karim, L. 2005. Charlotte (Junior Mom) on a Linden inside the Heckscher Playground, Sep. 1, 2005 [online image]. Palemale.com. http://www.palemale.com/se120.html [May 10, 2015].

Karim, L. 2006. Pale Male Jr., May 6, 2006 [online image]. Palemale.com. http://www.palemale.com/may62006.html [May 8, 2015].

Karim, L. 2007. Pale Male with a fresh green branch for the nest, May 4, 2007 [online image]. Palemale.com. http://www.palemale.com/may42007.html [Sep. 4, 2015].

Karim, L. 2011. Daily Images from 2011: Oct. 8, Oct. 21, Nov. 12, Nov. 19, Nov. 30, Dec. 20, Dec. 21, Dec. 27, Dec. 29 [online images]. Palemale.com. http://www.palemale.com/2011.html [Sep. 4, 2015].

Kube, R. 2012. Red-tailed Hawk Pair DSC_1216, Horsecreek Rd, nw of Cochrane, AB, May 3, 2012 [online image]. Flickr. https://www.flickr.com/photos/30638967@N03/7004897782/in/pool-300mm [Jun. 15, 2015].

Lichter, G. 2010. Prairie Sentinel #2: Red-tailed Hawk, Cochrane, AB [online image; #42 of 50]. Flickr. http://flickrhivemind.net/Tags/hawk,hawkredtailed/interesting. Can also be accessed: http://farm2.static.flickr.com/1255/5154191918_d2586bbaeb_b.jpg [Jun.15, 2015].

Lincoln Park News Blogspot. 2012. Red-tailed Hawks of Kettering, OH [online images]. Lincoln Park Commons Pond News, Kettering, OH. http://lincolnparkpondnews.blogspot.com/2012/08/the-red-tail-hawks-of-kettering-oio.html [Feb. 3, 2015].

Add to bibliography: Louisiana Ornithological Society. 2017. White-tailed Hawk: Number of accepted White-tailed Hawk records for Louisiana=21 as of August 2017. www.losbird.org/lbrc/wtha.html [Nov. 2, 2017].

Lynch, W. 2016. Red-tailed Hawk (*Buteo jamicanesis*) roosting in a snowy aspen forest during its autumn migration, AB, Canada [online image]. Getty Images. http://www.gettyimages.com/detail/photo/red-tailed-hawk-roosting-in-a-snowy-aspen-high-res-stock-photography/177802894 [Dec. 15, 2016].

Maire, A. 2013a. Red-tailed Hawk, Becancour, QC, Feb. 5, 2013 [online image]. Flickr. https://www.flickr.com/photos/moustiques/8447687793/in/album-72157617458859787/ [Apr. 16, 2016].

Maire. A. 2013b. Red-tailed Hawk, Yamachiche, Mauricie, QC, Dec. 18, 2013 [online image]. Flickr. https://www.flickr.com/photos/moustiques/11438847743/in/album-72157617458859787/ [Apr. 16, 2016].

Maire, A. 2016. Albums: Raptors and Owls [online images]. Flickr. https://www.flickr.com/photos/moustiques/albums/72157617458859787/ [Apr. 16, 2016].

Martin, C. 2012. Red-tailed Hawk, Bragg Creek, AB, May 2012 [online image]. Christopher Martin Photography. http://chrismartinphotography.files.wordpress.com/2012/05/red-tailed-hawk-c2a9-2012-christopher-martin-5632.jpg?w=652 [Apr. 2, 2015].

Martin, C. 2014. Turner Valley, AB, May 2014 [online image]. Christopher Martin Photography http://chrismartinphotography.files.wordpress.com/2014/04/fence-launch-c2a9-christopher-martin-693192.jpg [Jul. 1, 2015].

Maybank, B. 2012. Red-tailed Hawk-P1020364–11 Nov. 2012, Scots Bay, Kings Co., NS. Flickr. https://www.flickr.com/photos/blakemaybank/12120192353/in/album-72157640108458563/ [Apr. 16, 2016].

Michaels, E. 2015a. Pale Male, NYC [online images]. Ellen Michaels Photography. http://www.ellenmichaelsphotos.com/v/hawks_pale_+male/ [Jan. 13, 2015].

Michaels, E. 2015b. Red-tailed Hawk (in flight) [online image]. Ellen Michaels Photography. http://www.ellenmichaelsphotos.com/v/hawks_pale_+male/red-tailed_hawk_9761.jpg.html [Jan. 13, 2015].

Miller, C. 2012. History of Red-tail Hawk Family, Part 3, May 12, 2012; with Gray Squirrel [online image]. Pluff Mud Perspective (blog). http://pluffmudperspectives.blogspot.com/2012/06/history-of-red-tailed-hawk-family-part.html?m=1 [Mar. 7, 2015].

Nikographer. 2006. Red-tailed Hawk, Brighton Dam, MD, May 24, 2006 [online image]. Flickr. https://m.flickr.com/#/photos/-jon-/152741969/ [Dec. 4, 2014].

Podrasky,M. 2011. Red-tailed Hawk profile, New Hudson, MI, May 7, 2011 [online image]. Flickr. https://m.flickr.com/#/photos/30474454@N08/8244779882 [Mar. 25, 2015].

Polucci, L. 2012. A squirrel, a hawk, and a train [online image]. Decatur Metro.com. http://www.decaturmetro.com/2012/07/09/a-squirrel-a-hawk-and-a-train/ [Apr. 16, 2015].

Pope, R. 2012. Fledgling [adult] Red-tailed Hawk, Delaware Water Gap NRA, Jun. 6, 2012 [online image]. Flickr. https://m.flickr.com/#photos/richpope/7348962918 [Mar. 5, 2105].

Russell, J. 2010. Alaska Expedition 2010 - Day 5–6, May 17, 2010. "Western" Red-tailed Hawk (*B. j. calurus*) darker morph (dark intermediate morph), west of Edmonton [online image]. Married to a Birder (blog) http://marriedtoabirder.blogspot.com/2010/05/alaska-expedition-2010-day-5-6.html [Jan.21, 2017].

S., D[ennis]. 2014. Red-tailed Hawk above Zane Grey Museum, Lackawaxen, PA [online image]. Photograph by Dennis S. posted to Trip Advisor.com. http://media-cdn.tripadvisor.com/media/photo-s/06/81/ec/72/red-tailed-hawk-above.jpgmed [Jun. 2, 2015].

Saunders, N. 2009. Red-tailed Hawk [online image]. Saskatchewan Bird & Nature (blog). www.saskbirder.com/2009/06/more-shorebirds-n-stuff.html?m=1 [Jun. 24, 2015].

Schmunk, R. 2010. Highbridge Park Red-tail Dad "George," NYC, Mar. 21, 2010 [online image]. Flickr. https://m.flickr.com/photos/rbs10025/4452276151/ [Mar. 15, 2015].

Schmunk, R. 2014a. Cathedral Hawk–3700, Norman, he of the obsidian eyes, NYC, Jun. 30, 2014 [online image]. Flickr. https://m.flickr.com/#/photos/rbs10025/14547350465/ [Apr. 23, 2015].

Schmunk, R. 2014b. JHW Hawk (8412), J. Hood Wright Park, NYC, May 3, 2014 [online image]. Flickr. https://www.flickr.com/photos/rbs10025/14097446301/in/album-72157639709252704/ [Mar. 15, 2015].

Schmunk, R. 2014c. JHW Hawk nest (8981), feeding time, J. Hood Wright Park, NYC, May 10, 2014 [online image]. Flickr. https://www.flickr.com/photos/rbs10025/14156532544 [Aug. 22, 2015].

Schmunk, R. 2014d. Red-tailed Hawk (6901) "George," Highbridge Park/Swindler Cove, NYC, Apr. 19, 2014 [online image] Flickr. https://www.flickr.com/photos/rbs10025/13942288993 [Apr. 16, 2015].

Schneider, G. 2008a. Red-tailed Hawk (*Buteo jamaicensis*) flying back to the nest with prey, May 15, 2008 [online image]. Gary Schneider Photography. http://www.gschneiderphoto.com/gallery3/birds/raptors/redtailedhawk/red-tailed_hawk-3417 [Jun. 6, 2015].

Schneider, G. 2008b. Red-tailed Hawk (*Buteo jamaicensis*) flying back to the nest with field vole for its young, May 15, 2008 [online image]. Gary Schneider Photography. http://www.gschneiderphoto.com/gallery3/var/albums/birds/raptors/redtailedhawk/red-tailed_hawk-screaming-3438.jpg?m=1419979480ww [Jun. 7, 2015].

Sim, M. 2011. Bird Profile: Red-tailed Hawk, Jul. 26, 2011 [online image]. Birds Calgary. http://www.birdscalgary.com/2011/07/26/bird-profile-red-tailed-hawk/ [Jul. 7, 2015].

Stoner, D. 2010. Raptors of November [online image]. Delaware Nature Society. http://blog.delawarenaturesociety.org/2010/11/05/raptors-of-november/ [Apr. 10, 2015].

Wainman, B. 2011a. "Soft Landing." Red-tail Hawk, female of pair, near Pambrun, SK (approx. 25 mi/40km southeast of Swift Current, SK), Canada, Jun. 2011 [online image]. Fine Art America. http://fineartamerica.com/featured/soft-landing-blair-wainman.html [Dec. 9, 2016].

Wainman, B. 2011b. "Up and Away." Red-tail Hawk, male of pair, in flight, vigilant over the nest (same location as "Soft Landing" image), Jun. 29, 2011. Fine Art America. http://fineartamerica.com/featured/up-and-away-blair-wainman.html [Dec. 9, 2016].

Washington Square Park Blog. Oct. 10, 2012. Bobby and Rosie, NYC [online image]. http://www.washingtonsquareparkblog.com/2012/10/10/INSTEAD-OF-WASHINGTON-SQUARE-PARKS-MAINTENANCE-ISSUES-BEING-ADDRESSED-INEFFECTIVE-AND-HAWK-KILLING-RODENTICIDE-PLACED-IN-PARK-BY-CITY/ [Jul. 9, 2017].

Wildlife Center of Virginia. 2011. Red-tailed Hawk, Scarlette [online image]. Wildlife Center Program on January 29 at Virginia Beach Wild Birds Unlimited. https://www.wildlifecenter.org/news_events/news/wildlife-center-program-january-29-virginia-beach-wild-birds-unlimited [Apr. 9, 2015].

Yolton, D. B. 2008a. Highbridge Park, NYC, Red-tailed Hawks, early images of George [online images]. Urban Hawks. http://urbanhawks.blogs.com/photos/uncategorized/2008/04/24/042408hb02.jpg; and Highbridge Hatched? (April thread showing barred axillaries, barred legs of George): http://urbanhawks.blogs.com/urban_hawks/2008/04/index.html [Nov. 6, 2016].

Yolton, D. B. 2008b. Charlotte, NYC, Nov. 16, 2008 [online images]. Urban Hawks. http://urbanhawks.blogs.com/.a/6a00d83451c30169e2010535fb006e970c-pi; and Young vs. Established (last 7 images of thread for November 2008): http://urbanhawks.blogs.com/urban_hawks/2008/11/index.html [Jun. 7, 2015].

Yolton, D. B. 2014. Tompkins Square, NYC, couple [online images]. Urban Hawks. http://urbanhawks.blogs.com/urban_hawks/2014/10/index.html; http://urbanhawks.blogs.com/.a/6a00d83451c30169e201bb07a18223970d-pi [Oct. 28, 2014].

Yolton, D. B. 2015. Adult day in Washington Square Park, NYC, Oct. 20, 2015 [online images/video]. Urban Hawks. http://urbanhawks.blogs.com/urban_hawks/2015/10/adult-day-in-washington-square-park.html [Apr. 15, 2016].

Yolton, D. B. 2016. Washington Square Park, NYC, Apr. 6, 2016 [online images]. Urban Hawks. http://urbanhawks.blogs.com/urban_hawks/2016/04/washington-square-park.html [Apr. 15, 2016].

Zipp, J. 2015. Red-tailed Hawks (*Buteo jamaicensis*), New Haven, CT [online images]. Jim Zipp Nature Photography. http://www.jimzippphotography.com/cpg/thumbnails.php?album=60 [Jun. 11, 2015].

Abbreviations

States and Provinces

United States

State	Abbreviation	
Alabama	Ala.	AL
Alaska	Alaska	AK
Arizona	Ariz.	AZ
Arkansas	Ark.	AR
California	Calif.	CA
Colorado	Colo.	CO
Connecticut	Conn.	CT
Delaware	Del.	DE
Florida	Fla.	FL
Georgia	Ga.	GA
Idaho	Idaho	ID
Illinois	Ill.	IL
Indiana	Ind.	IN
Iowa	Iowa	IA
Kansas	Kans.	KS
Kentucky	Ky.	KY
Louisiana	La.	LA
Maine	Maine	ME
Maryland	Md.	MD
Massachusetts	Mass.	MA
Michigan	Mich.	MI
Minnesota	Minn.	MN
Mississippi	Miss.	MS
Missouri	Mo.	MO
Montana	Mont.	MT
Nebraska	Nebr.	NE
Nevada	Nev.	NV
New Hampshire	N.H.	NH
New Jersey	N.J.	NJ
New Mexico	N.M.	NM
New York	N.Y.	NY
North Carolina	N.C.	NC
North Dakota	N.D.	ND
Ohio	Ohio	OH
Oklahoma	Okla.	OK
Oregon	Ore.	OR
Pennsylvania	Pa.	PA
Rhode Island	R.I.	RI
South Carolina	S.C.	SC
South Dakota	S.D.	SD
Tennessee	Tenn.	TN
Texas	Tex.	TX
Utah	Utah	UT
Vermont	Vt.	VT
Virginia	Va.	VA
Washington	Wash.	WA
Wisconsin	Wisc.	WI
West Virginia	W.Va.	WV
Wyoming	Wyo.	WY

Canada

Province/Territory	Abbreviation
Alberta	Alta.
British Columbia	B.C.
Labrador	Lab.
Manitoba	Man.
New Brunswick	N.B.
Newfoundland	Nfld.
Northwest Territories	N.W.T.
Nova Scotia	N.S.
Nunavut	Nunavut
Ontario	Ont.
Prince Edward Island	P.E.I.
Québec	Que.
Saskatchewan	Sask.
Yukon	Y.T.

Museums

AMNH American Museum of Natural History, New York, N.Y.
BMNH Bell Museum of Natural History, Minneapolis, Minn.
CMN Canadian Museum of Nature, Ottawa,
CMNH Carnegie Museum of Natural History, Pittsburgh, Pa.
LSM Lousiana State Museum, Patterson, La.
RBCM Royal British Columbia Museum, Victoria, B.C.
UAM University of Alaska Museum, Fairbanks, Alaska
USNM U.S. National Museum of Natural History, Washington, D.C.

Measurements

ac. acre
cm centimeter
ha hectare
km kilometer
m meter
mi mile
mm millimeter

Index